D0349453

SCIENCE AND TECHNOLOGY IN HISTORY

SCIENCE AND TECHNOLOGY IN HISTORY

An Approach to Industrial Development

Ian Inkster

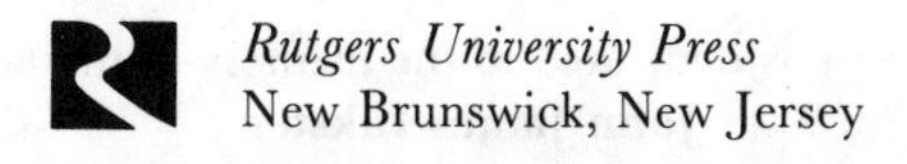

Rutgers University Press
New Brunswick, New Jersey

First published in the United States of America by Rutgers University Press, 1991.

First published in the United Kingdom by Macmillan Education, Ltd., 1991.

Library of Congress Cataloging-in-Publication Data
Inkster, Ian.
Science and technology in history : an approach to industrial development / Ian Inkster.
p. cm.
Includes bibliographical references and index.
ISBN 0-8135-1680-3 : $50.00
1. Technology—History. 2. Science—History. 3. Technology transfer—History. 4. Industrialization—History. I. Title.
T18.I56 1991
609—dc20 90-24557
CIP

Printed in Hong Kong

This book is dedicated to the memory of my father
John James Inkster

Contents

List of Tables ix

List of Figures xiii

Preface xiv

1 INTRODUCTION: SCIENCE, TECHNOLOGY AND ECONOMIC DEVELOPMENT 1
- 1.1 The Importance of Technological Change 1
- 1.2 The Historical Evidence 2
- 1.3 The Sources of Invention and Innovation 8
- 1.4 The Importance of Invention and Innovation 11
- 1.5 The Adoption and Diffusion of Technologies 13
- 1.6 The Transfer of Technologies 20
- 1.7 Science, Technology and The Social Matrix 27
- 1.8 Science, Technology and Underdevelopment 29

2 MENTAL CAPITAL – TRANSFERS OF KNOWLEDGE IN EIGHTEENTH-CENTURY EUROPE 32
- 2.1 The Scientific Enterprise Phase (1) – Communities and Academies 32
- 2.2 The Scientific Enterprise Phase (2) – Science in Culture, 1780–1800 39
- 2.3 Eighteenth-century Idea-Merchants: the Transfer of Science 46

2.4 Change Agents: the Transfer of Technology 49
2.5 Conclusions: the Conditions of Success 55

3 SCIENCE AND TECHNOLOGY IN THE BRITISH INDUSTRIAL REVOLUTION 60
3.1 Introduction: Definitions 60
3.2 Invention, Technological Change and Industrialisation 66
3.3 Connections: Science and Technology 69
3.4 Activists: the Sources of Scientific and Technical Information 72
3.5 Audiences: the Diffusion of Scientific and Technical Information 78
3.6 Applications: the Social Location of Technological Change 80
3.7 Conclusions: Information, Institutions and Industrial Revolution 86

4 THE SCIENTIFIC ENTERPRISE: INSTITUTIONS AND THE DIFFUSION OF KNOWLEDGE IN THE NINETEENTH CENTURY 89
4.1 Introduction: Science, Technology and Institutions 89
4.2 The Organisation of Science – National Trends 95
4.3 Human Capital: the Diffusion and Application of Knowledge 101
4.4 Private Enterprise 110
4.5 The State 116
4.6 Beyond the Pale: the Organisation of Science in Japan 123
4.7 Conclusions: Institutions and Industrialisation 128

5 TECHNOLOGY, ECONOMIC BACKWARDNESS AND INDUSTRIALISATION – GENERAL SCHEMA 131
5.1 Context: the Demonstration of Forwardness 131
5.2 The Location of Production 134
5.3 Catching Up 137
5.4 Modelling 'Response' 139

5.5 Impassive Missionaries? Relative Backwardness and the Mechanisms of Technology Transfer 142

6 INDUSTRIALISATION: WINNERS AND LOSERS 167
6.1 Introduction 167
6.2 The Phaseology of Late Industrialisation 168
6.3 Aspects of Phase Transition 170
6.4 Technological Progress 174
6.5 The Response of the State 175
6.6 Development Projects: Railroads 176
6.7 Winners and Losers 181

7 TECHNOLOGY, ECONOMIC BACKWARDNESS AND INDUSTRIALISATION – THE CASE OF JAPAN 184
7.1 Japan, Technology and Late Industrialisation 184
7.2 Transfer Mechanisms 187
7.3 Boundary Conditions 193
7.4 The Appropriation of Technology: Case Studies 197
7.5 Conclusions 202

8 SCIENCE, TECHNOLOGY AND IMPERIALISM: (1) INDIA 205
8.1 Imperialism 205
8.2 Room for Debate 209
8.3 Transfer, Diffusion and the British 218
8.4 Conclusions 225

9 SCIENCE, TECHNOLOGY AND IMPERIALISM: (2) CHINA AND BEYOND 227
9.1 The Foreign Presence 227
9.2 Room for Debate Once More 228
9.3 Transfer Mechanisms 236
9.4 'Enlarged Intercourse' – India, China and Japan 242

10 CENTRE AND PERIPHERY: SCIENCE AND TECHNOLOGY IN AMERICA AND AUSTRALIA 248
10.1 Science, Technology and Recent Settlement 248

10.2 From Colonies to Industrial Economy – the American Case 250
10.3 The 'American System': Science, Technology and Economic Growth 254
10.4 Local Imperatives and the Natural History Enterprise: Australia 264
10.5 Summary and Conclusions 268

11 TWENTIETH-CENTURY AFTERMATHS: SCIENCE, TECHNOLOGY AND ECONOMIC DEVELOPMENT 271
11.1 Development and Underdevelopment 271
11.2 Science, Technology and Institutions 274
11.3 Technology Transfer: Japan and India 278
11.4 Alternatives: Appropriate Technology and the Chinese Model 284
11.5 The Distribution of Science and Technology Resources 288
11.6 The Institutional Context (1): Japan and the USA 290
11.7 The Institutional Context (2): the Newly Industrialising Countries 294
11.8 Exits: On Pursuing the Past 297

Appendix 1: Inventions and Innovations – Britain c. 1700–1800 304

Appendix 2: Key to Patents Subject Typology 307

Notes 308

Name Index 375

Subject Index 381

List of Tables

1.1 Inventions and innovations in the USSR (1960s) 13
1.2 World trade and industrial production, annual average percentage growth, 1720–1977 22
2.1 Scientists, technologists and higher education in England, percentages 36
2.2 The chemical enterprise of the eighteenth century 40
2.3 Patented technique in Britain, 1689–1829 41
3.1 University education of 498 applied scientists and engineers 73
3.2 The Scottish production of British medical graduates, 1701–1850 75
3.3 Evening instruction and the artisans, 1850–51 (percentages in brackets) 81
3.4 The social location of British patenting in the 1790s 84
3.5 The geographical location of British patenting in the 1790s 85
3.6 Approximate subject typology of British patenting in the 1790s 86
4.1 Dichotomies in the organisation of science 95
4.2 Natural science in the German university system (student numbers) 97
4.3 Staff and students in chemistry at Berlin and Hanover *Technische Hochschulen* (1885–1900) 97

4.4 184 patentees in five English cities in 1860 104
4.5 Britain in 1841 – occupations, skills and skilling 104
4.6 Institutes of the British Engineers, 1818–1912 108
4.7 Professional technical associations in the scientific enterprise of Manchester, 1845–1900 (no. of institutions) 109
4.8 Employment in the German chemical industry, 1882–1907 115
4.9 Scientific enterprise and strategic patenting in the six largest organic chemical firms in Germany, 1886–1900 115
4.10 Funding of ten leading institutions in England and Germany in 1899 118
4.11 Staff–student ratios in five major chemistry departments in Germany, c. 1899 119
4.12 Science and technology at Stuttgart Polytechnic 120
4.13 Patterns of national scientific enterprise prior to 1914 129
5.1 Location of 1174 major production centres worldwide, c. 1887 135
5.2 Major production in 235 urban centres in France, Germany and Russia, c. 1887 (percentages) 136
5.3 Volume of industrial production, Index, UK in 1900 = 100 137
5.4 Per capita levels of industrialisation, Index, UK in 1900 = 100 137
5.5 Production of steel in the four main producers, 1865–1912 (Index, UK in 1865 = 100, volume) 138
5.6 Patents granted in 15 nations, 1842–61 163
5.7 Patent applications in Britain, US and Germany, 1877–94 163
5.8 The German patent system, 1877–94; categories of patents granted 164
5.9 Patent applications in Britain originating from 50 major cities, 1867–9 165
5.10 Foreign activity in the Australian patent system, 1886–1918 166
10.1 American urban patenting in Britain, 1867–9 (11 cities; 824 patents) 263

10.2 Patenting in New South Wales and the Australian Commonwealth, 1886–1904 (selected years; 6112 patents; percentages) 267
10.3 Patent registrations by foreign residents in Australia, five-year periods, 1904–18, percentages 268
11.1 Technology imports to Japan as a percentage of technology exports from Japan (1971–81), in ¥ billion measures 281

List of Figures

1.1 The sources of increase in output 7
1.2 Assembly line and parallel groups 11
1.3 Techniques of production and technical innovation 15
1.4 Diffusion graphs – (a) relocation diffusion and (b) expansion diffusion 17
2.1 Transfer and diffusion of technologies 57
2.2 Feedback mechanisms in technology transfer 58
7.1 Patterns of industrialisation 203

Preface

THIS book is about the changing relationships between science, technology and economic development from the eighteenth century to the present time. Such an embracing subject demands that the text swings between social history and economic history, between cultures and markets, between Europe and Asia and between information (embodying ideas) and artefacts (including machines). In performing this exercise a very large and complex body of secondary literature is drawn upon, ranging from material on popular science in Hanoverian England to modernisation strategies in post-Maoist China. I very much hope that the undergraduate and postgraduate students for whom the book is intended will bear with me and eventually find some enlightenment and even a touch of entertainment. Quite simply, the task of this book is to uncover the dynamics of industrial change.

An understanding of this book does not require any advanced training in modern science or engineering, nor a profound knowledge of the mysteries and models of modern neo-classical economics. But it does require some real concentration. I think that students should use this text as a book which can be broken into at various points. Important historical themes, such as those of industrial revolution, technology transfer or the institutional setting of knowledge diffusion are broached in chapters whose major topics may vary between eighteenth-century Europe or late nineteenth-century Japan. A coherent story is told, a chronology is followed, and in many ways the text is quite conventional. But no student should feel that he or she must read 'through' this book. I am not a Tom Sharpe on one hand or an Eric Hobsbawm on the other. This is a book to be used several times rather than loved once.

The general approach of this book is stridently comparative and interactive. Here are to be found no 'single causes' of any large event or process nor any mathematical models. But there are many (my reviewers will say, too many) attempts at explanation and several disclosures of at times quite complicated causal sequences. Although we cover an awful lot of ideas and much general history in the first two chapters of this book, the major historical problems of the text are not formally or finally defined at the outset and subsequently discovered in a series of titillating historical episodes. Historians (and I am still hopeful enough to add, and social scientists) do not apply theory to history. Rather, if they are somewhat patient and reasonably insightful, they may use history to develop theory. In particular, the task of economic history – one of the central dimensions of the book – is not merely to fructify economic theory, to test models against what is purported to be a past reality. For all too often the finger exercises of such fructification pose a threat to the integrity of the data as we know it. Even worse, too much attention to existing theory may inhibit the essential task, the redefinition of and continual search for more data, more evidence. If model or schema building of any sort is to be valid, it must allow in its essence for the possibility of variance, of intractability, of that material which is by nature fractious, and for the feasibility of explaining such variations within boundaries discovered through serious empirical examination. A belief in the veracity of historical detail, rooted in time and place, has governed the sometimes eclectic nature of this book. I have attempted to present fairly the material and the viewpoints of a great many representative historians and social scientists by the selection from a range of possible themes of those which to my mind best illustrate the essential dynamics in modern history.

As a flick through the pages will show, there are several diagrams and quite a number of tables. At points the discussion is summarised or evaluated in terms of a number of enumerated, accumulative factors or elements. Nearly two decades of teaching modern history to undergraduate and postgraduate students convinces me that such breaks in the formal text, although primarily designed to harness data or argument in a useful manner, also help to sharpen the attention of the reader. I hope that these features of the book will make it useful not only as a reading or text for essays and examinations, but as a source of tutorial discussion and general debate.

All textbooks, however high they aim or eccentric they become, owe debts to several people, and the present book is no exception. Intellectually, the commanding works of Joseph Schumpeter, Alexander Gerschenkron, Albert Hirschman, Nathan Rosenberg and R. R. Nelson have been decisive in influencing the manner in which I come to this whole subject area. Sidney Pollard, now of Bielefeld, but in the early 1970s my research supervisor at Sheffield University, has exerted a quiet, now distant but transforming presence. To him more than any other scholar I owe my sense of time pressing. My debt to other academics is apparent in the footnotes, and I trust that none will feel seriously or harmfully misrepresented.

My thanks also go to the staff of several institutions, including the librarians and archivists of the Diet Library and the Institute of Developing Economies, both in Tokyo, the Indian International Centre, New Delhi, the Bodleian Library, Oxford, and the Mitchell and State Public Libraries of Sydney. No photocopier or fiche can replace their personal interest. My publishers have provided stalwart service and my secretary Charleen Borlase has worked much beyond the call of duty. At the end, conventionally, I must thank my wife and three children, who continue to put up with a lot. In this case, the convention which places the most at the end is well observed. The volume is dedicated to the memory of my father.

IAN INKSTER

1. Introduction: Science, Technology and Economic Development

Concepts are plans for action; they are programmes for research . . . And they will be known by their fruits.

Alexander Gerschenkron, 1968

The notion of a production function – the spectrum of all known techniques of production – is by itself a metaphysical concept.

Mark Blaug, 1961

1.1 THE IMPORTANCE OF TECHNOLOGICAL CHANGE

WHEN *X grows* it expands in size. When *X develops* it grows and changes. Most writers refer to *economic growth* as being the sustained rise of national income or of national income per head. Historical data suggests that the growth in output of such major nations as the USA, Britain or Japan has been at rates far beyond those that could be accounted for by an increase in the supply of such conventional *inputs* as labour, land, and capital. Over the period with which this book deals, scarce economic resources have been used with increasing efficiency.

Here we refer to *economic development* as embracing economic growth, an expansion of productive potential associated with measurable structural changes in the economy (e.g. from agriculture to commerce, from manufacturing to services), and adjustment in social institutions (see 1.5 to 1.8 below) of a type which permits the maintenance of social control mechanisms and national sovereignty, and an improvement in the real standard of living of the mass of the population. If all of this occurs in one phase of a nation's history, an implication is that economic development has reflected substantial improvements in the utilisation of resources (i.e. productivity), often associated with the *invention* and utilisation of new technologies, with the widespread *diffusion* of new or existing technologies, or with the *transfer* of superior technologies to the nation from other places.

A technology may be very simple, involving little expenditure of tangible assets. A technology may be organisational rather than mechanical, procedural rather than chemical. In 1830 in the port of Liverpool, a system of signals was invented by which each local merchant could direct his incoming vessels long before docking. In so far as this reduced overall costs of transport or handling, the new process represented technological progress based on an inexpensive alteration in procedure-organisation, rather than upon the introduction of new equipment or machinery. On the other hand, the reduction of overall costs associated with the silk manufacturing of John and Thomas Lombe in early eighteenth-century England required a large investment. This was needed in order to place a clutch of existing machine *techniques* into a new organisational framework. A combination of larger-scale production, continuous flows between machines and improved production control reduced the price of the finished product. Although little was introduced in the way of new machine technique, a superior technology had been created.[1]

The role of technological change in economic development may be exaggerated (see 1.2 below), but it may also be underestimated. Technological progress not only increases the efficiency of a firm or nation, it produces entirely new products and services. Whether these satisfy old demands in new ways or create new needs entirely, the production of an ever-changing bundle of goods affects the quality of life in ways which are difficult to quantify. In terms of consumer satisfaction, electric lighting, the telephone, radio and penicillin had no exact precursors. Silicon chip technology has meant that it now takes inside one minute to perform the number of logical operations which it would have taken an individual's whole life to complete not so very long ago. Longevity, reduction of disease and pain, improvements in shelter, diet, leisure, information and communication and increased control over human reproduction itself are all products of new technologies and *their* products – antibiotics, tranquilisers, anaesthetics, contraceptives, canning, refrigeration, sewing and washing machines, televisions. This product-widening feature of technological change is of course directly related to economic development in the long run. If old industries do eventually face diminishing returns (as argued in some places by Simon Kuznets and Joseph Schumpeter), and if industries are to an extent product-bound, then product rather than process innovation becomes a principal element allowing a shift of resources into new industries, those which will grow rapidly and efficiently.

The *endowments* of a nation or region are given; its *resources* or assets are not. Technological change can convert endowments into resources. New exploratory, extraction and handling techniques may allow the exploitation of badly-located or low-grade materials previously ignored – indeed, the concept of 'low grade' must be technological and subject to change rather than natural and subject to stasis. New manufacturing technologies may convert hitherto useless deposits into assets (petroleum, uranium) or may allow a saving on the usage of recognised resources. New recovery techniques can convert wastes and tailings into resources. Technology creates resources of land and space. The resources of the American mid-West were merely natural endowments until railroadisation opened up the region to the centres of demand in the east. Prior to the turnpikes of the eighteenth century, Britain was a large place; after the railway it grew small. During the 1860s, Finland's endowments of timber were refashioned as resources when steam shipping converted from the paddle wheel to the propellor, which reduced the price of Finnish timber in Britain by some 25 per cent and marked an economic boom in the exporter nation. Technological change has released vast tracts of land as inputs into production. From a variety of transport improvements to the cotton gin, barbed wire and grain harvesters, technical changes have lifted the land constraint on development. So too have improved geological and mapping techniques.

Technical change has been insidious and unplanned, and nowhere is this better shown than in the history of agriculture since the mid-nineteenth century. Although much of the new chemical and machinery input into the sector was deliberately produced for agriculture, the enormous growth of production and specialisation worldwide during the late nineteenth century was more the result of the combined effects of innovations in railroads, the iron steamship, refrigeration and barbed wire. Spinoffs abounded. In the actual *application* of even the new agricultural chemistry, it was the gas-works who yielded cheap ammonium sulphate, the Gilchrist–Thomas furnaces which from the 1880s provided basic slag and the Haber–Bosch process of the 1900s which permitted the inexpensive fixation of atmospheric nitrates. American agriculture benefited from the diffusion of the steel plough, the cultivator, the reaper and barbed wire fencing, which in combination allowed new varieties of natural grain products to be exploited. With high land-to-labour ratios, American improvements centred on mechanisation, which increased the acreage cultivated by a single farmer. In much of Asia, wet-paddy rice growing benefited essentially from inputs of a biological or biochemical type. To fertilisers were added breeding and rearing techniques and inputs of herbicides and insecticides.[2]

Of course, technical change may be disastrous (see 1.6 and 1.7). Twentieth-century technologies have allowed the substitution of new natural or artificial resources for old ones (oil for coal, synthetic fibres for cotton, wool and silk, synthetic for natural rubber, plastics for leather and metals). Sectors or nations committed to the supply of increasingly redundant materials become vulnerable to falling demand and prices, from which they may not easily escape, (see 1.7 and 1.8 below). The ability to adjust to new circumstances may well depend on a combination of the character of the endowment-resource base, the responsiveness of key government and private institutions and the availability of appropriate technologies.

1.2 THE HISTORICAL EVIDENCE

In 1956 Moses Abramovitz used quantitative methods to show that only a small fragment of United States economic growth in the years 1870–1950 could be accounted for by an increased *quantity* of labour and capital inputs, and that growth was more a function of improvements in the efficiency of resource usage. Furthermore,

> Since we know little about the causes of productivity increase, the indicated importance of this element may be taken to be some sort

of measure of our ignorance about the causes of economic growth in the United States and some sort of indication of where we need to concentrate our attention. . . . To identify the causes which explain not only the rates at which our opportunities to raise efficiency increase but also the pace at which we take advantage of those opportunities will, no doubt, remain the central problem in both the history and theory of our economic growth.

Fifteen months later Robert Solow derived a method which isolated shifts in the aggregate production function (i.e. technological progress) from other factors in the growth of the US economy for the years 1909–49, and concluded that 'Gross output per man-hour doubled over the interval, with 87.5% of the increase attributable to technical change and the remaining 12.5% to increased use of capital'. At the sectoral level, through case studies of shipping and railroads Gilfillan and Fishlow depicted the enormously high productivity stemming from a series of minor, almost indistinguishable improvements over a long period.[3] Discontinuities in *production* may come from the cumulative effect of many actions, rather than from the dramatic impact of a major invention. In the twentieth century, the growth of artificial fibres and petroleum-refining industries have been due more to cost reductions attached to a series of *improvements* in 'major' innovations than to the initial cost reductions of the original breakthroughs.[4]

From the early work of the 1950s and 1960s, economists had seemingly provided some measure, however conditional, for the variety of claims which had been made for the role of technological progress in economic development, claims stemming from the work of historians, engineers and sociologists. But a basic problem of this macro-aggregate approach to technological change is that it normally depends on a production function for the economy which postulates that the return to a unit of each factor of production (labour, capital) will approximate its marginal product, indicated in data on rents, wage rates and profits. This requires an assumption of competitive market conditions, at all times questionable, and during periods of fundamental structural change probably untenable. As Solow and Temin have suggested, we might treat the results of such formulae as parables, that is, as of rough validity.[5] Furthermore, the exact contribution of technological progress is difficult to specify, as the production function normally fails to capture *quality* improvements in labour and investment goods, the contribution of inputs into agriculture (above), the degree to which conventional inputs are fully employed, and assumes constant returns to scale, this leaving the 'residual' of growth as a composit *containing* technological progress and a reflection of all computational errors. In a series of works E. F. Denison has utilised a more complicated accounting process which includes as 'inputs' (directly measured) not only labour, capital and land, but also quality changes, hours of employment, expertise and education, changes in incentives and in the age and sex composition of the workforce, and allows for shifts in resource utilisation and positive economies of scale. Despite such whittling-away at the 'residual', Denison's work still leaves a large part in the growth process for technological progress, even when this is narrowly defined as the productive impacts of the 'advance of knowledge'.[6] An important point to acknowledge is that at any time, only a small proportion of a nation's *total* capital stock will be especially amenable to technological change, for the bulk of it is held in the form of residential and farm structures, consumer durables, government and institutional structures and inventories. Perhaps less than 10 per cent of total capital in Britain or the USA in the nineteenth century was made up of machinery, machine tools and equipment, the areas normally associated with technological change.[7] In nations such as China or India, where vast amounts of capital took the form of hydraulic structures, the proportion left over as machinery was almost certainly even smaller. Whilst this may tell us something of the comparative economic histories of Asia and Europe in the eighteenth and nineteenth centuries (see Chapters 8 and 9 below) it may also indicate that the increased efficiency of a relatively small proportion of any nation's capital was of paramount importance in the overall development process.

Figure 1.1 is drawn directly from the very fine survey by the economic historian, J. D. Gould.[8] For our purposes inputs which fall under sub-headings C1 and C2 may be defined as elements in technological change at large. Management innovations include incentive pay schemes, time and motion, improvements in layout and work flow, (see below, 1.3 and 1.7). Technological change as Rosenberg's 'increments to the stock of useful knowledge' are caught directly as G1 and G3, which may both include advances or applications arising from social as well as natural science.[9] We might hazard that many policymakers believe that through time the ratio

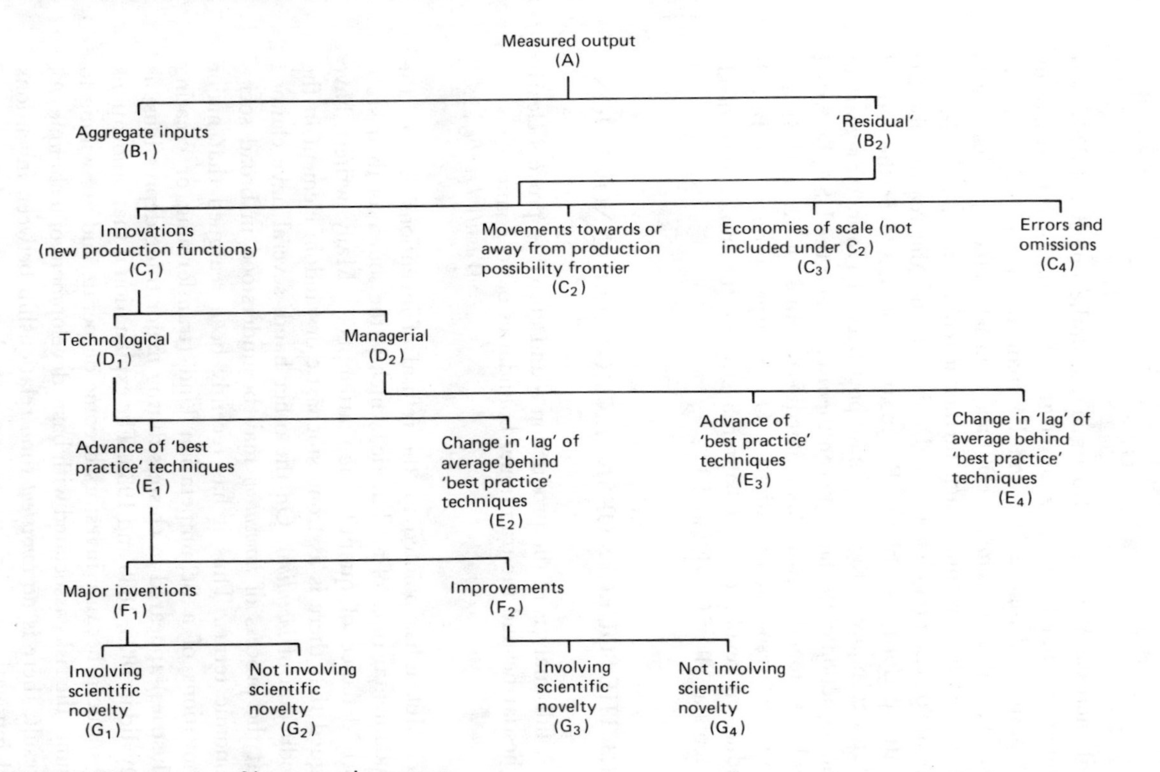

FIGURE 1.1 The sources of increases in output

$$\frac{G1 \times G3}{G2 \times G4}$$

will increase. We would also argue that in the long swing F2 is more important than F1, that economic growth of a nation is more a result of positive change at E2 and E4 than at E1 and E3, and that managerial improvements (D2) are themselves technologies as well as frameworks within which further machine and chemical-based technological progress arises. Denison found the reasons for the better performance of the European economies over the United States economy in the immediate post-Second World War years in the reduction of the lag between themselves and the USA, i.e. at E2 and E4, together with associated economies of scale (C3) and improved resource allocations.[10] Our interpretation of the British industrial revolution in Chapter 3 stresses the importance of elements at all of G1, E2 and E3.

1.3 THE SOURCES OF INVENTION AND INNOVATION

> A full analysis of the production – and growth – process clearly lies far beyond the traditional boundaries of economics.
>
> *Wassily Leontief, 1959*

Joel Mokyr has pointed to 'the refusal of inventions to obey the fundamental rules of arithmetic, which is the source of their stubborn defiance of quantification attempts'.[11] Many writers have argued that there is a strong stochastic or random element in the production of *inventions*. On the other hand, several have claimed that the process of *innovation* may be understood in broad socio-economic terms. Thus it has recently been reargued that major innovations of a Schumpetarian kind (transforming or creating industries) appeared in clusters during major trade depressions, as they did in 1825, 1886 and 1935. The explanation is that innovations are induced because investors become desperate and are willing to assume the risks associated with basic development of technique. As a result, there is no *consistent* time relationship between inventions and innovations – the former occur randomly while clustering of innovation appears as a result of the trade cycle.[12]

To understand the contrasts in approach it is necessary to distinguish clearly between invention and innovation. The gap between the two is filled by time, money and initiative. When James Watt hit on the idea of a separate condenser for the atmospheric steam engine in 1763 he created an invention. This only became an innovation after certain technical difficulties and scaling-up problems had been solved, and this required his time and Matthew Boulton's money. Furthermore, the case may be extended to show that the process of technical *development*, which is the measure of this gap, may itself generate incremental inventions *and* innovations which are eventually embedded in the final technique. Thus, by 1785 the standard Watt steam engine was a superior technology not only because of its incorporation of a separate condenser, but because the introduction of a flywheel allowed the production of a circular motion whilst the use of a governor created an elementary feedback system which reduced fuel inputs. The sources of invention are not necessarily the causes of innovation. In a series of quite separate studies of 240 significant *inventions* of the first half of the twentieth century, it was found that only 20 per cent originated from within major firms in their respective industries, most of the remainder emanating from a variety of individual inventive efforts.[13] However, if we once more define *innovation* as the expression of a sustained development effort, Freeman and others have found that corporate R&D looms far larger. In the present age, successful innovation results from large size, strong R&D, attention to potential markets (including education and assistance to users), strong entrepreneurship, firm links with general science-technology networks and an astute use of patents to protect property rights and promote bargaining power.[14] Firm size is not universally important but is particularly so for innovations needing large initial investments or serving fairly concentrated markets. Such results suggest that, whatever the source of invention, twentieth-century innovation is a function of forces endogenous to relevant industries. The first generation of computers emerged in several loci. In the second phase of transistorised computers (from the late 1950s) an American technical lead was established. This was the result of the attainment of industry leadership qualities in such areas as management, learning by doing, financing, marketing, government agency relations and components supply.[15]

Focusing on the invention development level, Schmookler emphasised the *multipurpose* nature of modern scientific knowledge, which was thus 'amenable to development at almost all points'. More specifically, Carter and Williams have concluded that 'invention is now often the result of a planned and cooperative research and development programme'. Reviewing the field, an economist has recently collapsed the whole subject into the formula that a shift in the production function (technological progress) is 'a result of semi-autonomous forces of scientific engineering and corporate research and development'.[16]

All positions may hold. Prior to the 1940s it was generally true that both invention and innovation often emerged from an array of related technologies stemming from the general trend of scientific advance, e.g. the metallurgical and electrical industries of the late nineteenth century, the internal combustion engine and the development of synthetic fibres. On the other hand, then as now, smaller modifying or perfecting inventions and innovations were more likely to come from within industries as a reflection of the attainment of specific skills and support mechanisms.[17] This is even more true of the development work required to convert significant invention into innovation, or to successfully establish a new process or product after initial scaling-up to commercial levels.

Around the year 1900 some 80 per cent of US patents were granted to individuals; by 1957 the proportion had fallen to 40 per cent. Enterprise research and development (R&D) had emerged. The late nineteenth-century beginnings of formalised R&D in private enterprise occurred in the USA, at General Electric, du Pont, Eastman Kodak and Bell Telephones. Kendall Birr has argued that this awaited the fulfilment of three preconditions; the 'maturation' of science itself, the emergence of positive business attitudes, and the establishment of new *organisational technologies* within enterprises themselves.[18]

The software techniques of management are almost certainly subject to learning-by-doing (i.e. scale and time) effects, but overt organisational planning may well telescope the innovation process at this level. In Japan, Sweden and elsewhere a new 'scientific management' has moved away from assembly-line type factory layouts towards assemblage by autonomous groupings and the establishment of independent product shops. The notion is to return to an organisational technology of a parallel-groups type, a (very!) loose analogue to the putting-out system or proto-factories of the eighteenth and nineteenth centuries. Figure 1.2 gives a schematic outline of the new form of organisation.

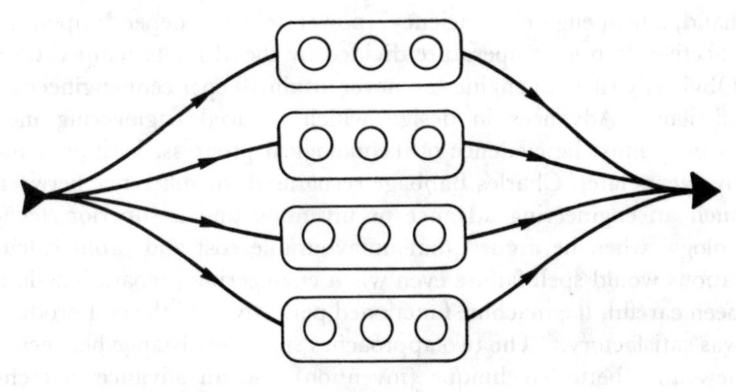

FIGURE 1.2 Assembly line and parallel groups

Here 'problems such as technical disturbances . . . can often be handled within the unit and need not to any great degree influence the production system as a whole'.[19] A new *technology* is evidenced in reduced downtimes, lower costs for quality control, improved balance of work loads and reduced personnel turnover. Thus the organisational technologies of the Lombe or Ford type are increasingly seen as themselves inefficient and unable to yield the enterprise-specific technical innovations required for productivity growth. Newer organisational forms are seen as superior technologies and as measures which induce process and product innovation from within the firm without great increases in formal R&D expenditures.

1.4 THE IMPORTANCE OF INVENTION AND INNOVATION

> The greatest invention of the nineteenth century was the invention of the method of invention.
>
> *A. N. Whitehead, 1946*

In the early nineteenth century Sadi Carnot published *On the Motive Power of Heat* in which he advanced a thermodynamic concept of 'efficiency'. Working on the power of overshot waterwheel technology he calculated that the power delivered to the mill depends primarily on the height from which water is dropped. On the other hand, a heat engine's 'efficiency' (power delivery) depends upon the effective drop in temperature divided by the absolute temperature. Obviously such an engine can never attain 100 per cent engineering efficiency. Advances in design which reduced engineering inefficiency must be evidence of technological progress. Writing some five years later, Charles Babbage recognised the difference between such an engineering advance or invention and a superior 'technology' when he argued that unfavourable cost and profit calculations would spell failure even when engineering preparations had been careful, the machine functioned perfectly, and the end product was satisfactory.[20] The two approaches span the distance between a new and 'better' technique (invention) and an advance in technology (realised in innovation). It is quite possible for an invention to represent both an engineering or design advance and an inferior technology, if it may only be commercially utilised at a total cost above that of other, existing techniques. For instance, Carnot efficiency may be increased by a switch from one fuel to another, yet the cost of the latter fuel may reduce the actual technique to less-than-best/average practice in terms of commercial utilisation. In this sense, inventions are not necessarily of great economic importance. On the other hand, a superior technology may be created out of a novel juxtaposition of existing machine techniques – as in the case of the Lombe mill (above).

Most inventions never become innovations if we define the latter as incorporating an extended development phase and a period of commercial usage. In market economies the distinction is hard to quantify, e.g. patents often represent innovations as much as inventions. In the Soviet Union invention is confined by definition to the provision of technical specifications and perhaps working drawings in the process of registration. State payments are made only after an invention has been used in the economy and are based on the economic returns which it yields. Table 1.1 suggests that only 16 per cent of all inventions registered during the 1960s emerged as at all significant innovations.[21]

More importantly, even major inventions and significant inno-

TABLE 1.1 Inventions and innovations in the USSR (1960s)

	1960	*1965*	*1966*
Inventions registered	10 485	13 158	16 648
Inventions taken up	4 500	12 752	13 427
Inventions for which return calculated (innovations)	965	3 098	2 516

vations do not necessarily coalesce to create a noteworthy increase in economic efficiency. A series of small, sustained improvements within an industry or enterprise may be more fundamental in economic advance, if only because they are necessary to reduce the establishment or operating costs of a spectacular technical break-through, e.g. modification of the shape and size of new furnaces, better lubrication, improved metals and alloys in construction, etc. As J. L. Enos found from his study of change in petroleum-refining, 'in terms of almost any measure other than originality – reduction of costs, saving of resources, expansion of output – the improvements made in the processes subsequent to their initial application were as significant as the innovations themselves'.[22]

For such reasons, we would argue that the history of technological progress is a history of adaptation, adoption, diffusion and transfer of organisational and machine technologies. Furthermore, the boundary conditions of such processes are not necessarily those associated with invention and innovation.

1.5 THE ADOPTION AND DIFFUSION OF TECHNOLOGIES

It has often been said that the ancient world failed to progress technologically because it possessed superabundance of cheap slave labour. Such an argument does not claim a lack of invention or even innovation. Rather, it implies that the adoption or diffusion of superior technologies was held back in so far as such new technologies tended to replace inexpensive labour with expensive capital or organisational structures.

In contrast, the economic history of medieval Europe was dominated by the *diffusion* of major, basic technologies – the water wheel and windmill, the heavy plough, the three-field system, the horseshoe and harnassing. What was new in the growth of Europe was not invention – many of the technologies had been long known – but their diffusion and assimilation, probably induced by an increase in demand in the agricultural sector, combined with natural interrelatedness – the adoption of one technique encouraged the adoption of another. Many of the new technologies originated in China, but did not diffuse there. This might be explained in terms of underlying economic structures. In Europe, new technologies could often be applied to a *variety* of existing productive pursuits. For instance, in China the water wheel was used principally for blowing bellows in metal works. In Europe the same wheels were increasingly used for not only grinding but for the preparation of malt, the fulling of cloth, the sawing of wood and the production of silk thread. Similarly, the Chinese and Europeans became acquainted with the Persian (vertical axle) windmill at approximately the same time, the twelfth or thirteenth centuries. Although the windmill was not commonly used in China, in Europe it was *adapted*, first by mounting the sails on a horizontal axis, and secondly by developing the tower structure. Adaptation led to adoption. The reason for this disparity of experience almost certainly lies in the fact that in Europe the windmill satisfied a multitude of growing demands – for silk, ribbons, fulling and calendering cloth, rolling copper plates, and so on.

Much the same can be said of the diffusion of the shipping, navigational and printing improvements of early modern Europe. This is not to deny that such adoptions did not then effect European cultures in a broader sense.[23] Nor would we deny that there existed important facilitating mechanisms, e.g. the frequent movement of key personnel from nation to nation for gain or from persecution. But in scholar-gentry China, people moved too.

By displaying the factor inputs for a fixed level of output, and restricting the analysis to two inputs only, economists can draw so-called 'isoquants' for an industry, which show productivity changes in a simple manner, as in figure 1.3 below. Lying upon this isoquant are an array of *techniques*, each combining labour and capital in different ways to produce the same output at the same total cost. The isoquant represents the *technology* for the industry or main-product. With *technological progress*, that is, the *adoption* of a superior technology in the industry, the isoquant shifts nearer to the origin, illustrating the reduced consumption of factors of production

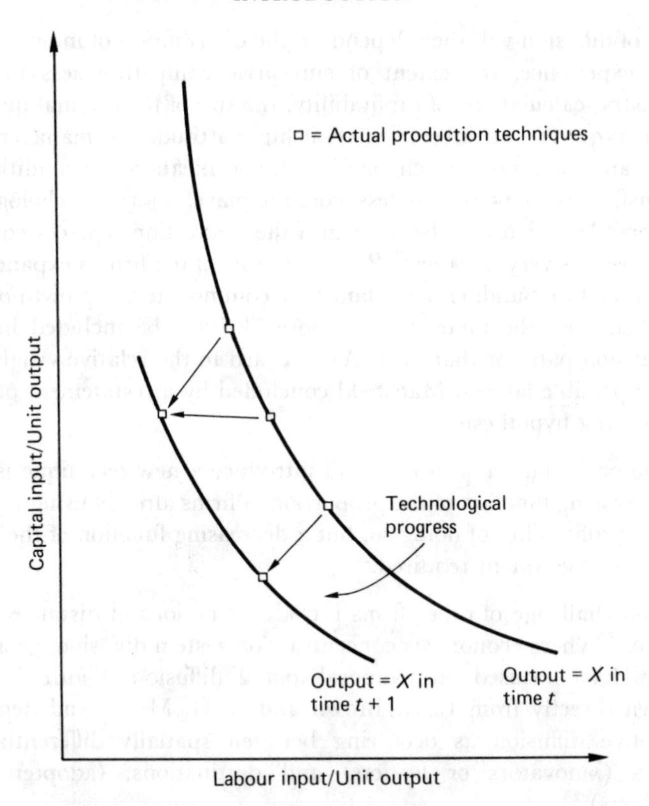

FIGURE 1.3 Techniques of production and technical innovation

which results. All these new techniques collectively represent 'best practice' at a time, even though they may look very different from an engineering or organisational perspective. A number of economists, including Mansfield, Nelson, Rosenberg, David, Heertje and Salter have investigated the conditions under which a new isoquant (i.e. adoptive diffusion) might be created.[24]

Of course, as many parts of the present volume indicate, improvements in the technology of production do not in fact occur simultaneously or even systematically over the whole best-practice front, for they are localised around particular existing techniques.

Focusing on the realistic assumption that movements of the industry isoquant in fact occur as a cluster of imitation-adoptions around a particular new technology, Mansfield has argued that the rate of diffusion will then depend on the distribution of information and experience, the extent of enterprise competitiveness in the industry, calculations of profitability, the size of the original investment required (an inverse relationship), attitudes of managers to risks and of workers to change in job specifications or conditions. Mansfield also points to less commonplace, more psychological factors. Adoption may be delayed if the innovation replaces equipment that is very durable.[25] Related to this, if the firm is expanding and therefore building new plant to accommodate the growth of its market, then the innovation is more likely to be included in an expansion path for that firm. With regard to the relative weight of these possible factors, Mansfield concluded by a restatement of his organising hypothesis:

> the probability that a firm will introduce a new technique is an increasing function of the proportion of firms already using it and the profitability of doing so, but a decreasing function of the size of the investment required.[26]

The challenge of other firms introduces notions of distance and space.[27] Where economists concentrate on system diffusion, geographers have focused on temporal-spatial diffusion. Figure 1.4 is drawn directly from L. A. Brown and E. G. Moore and depicts adoptive diffusion as occurring between spatially differentiated nodes (innovators or leaders) and destinations, (adopters or followers).[28]

Both cultural and economic geographers have stressed that a diffusing item – whether a machine or a culture trait – may be at once a stimulus to further innovations and itself subject to *adaptation* as it spreads from its physical point of origin. In Hagerstrand's work, adoption of an innovation is the outcome of a learning process, and because diffusion is across space, factors relating to the *effective flow of information* become paramount. It is expansion of specific information which reduces resistances to adoption.[29] Information comes from other adopters and from more generalised means of transmission, with the first taking priority. This follows from the findings of both sociologists and economists that interpersonal communication is more influential upon adoption than are mass media.[30] The network of interpersonal contacts which is required if *specific* information is to flow freely is conditioned by territorial distance and by social distance. Social and geographical factors

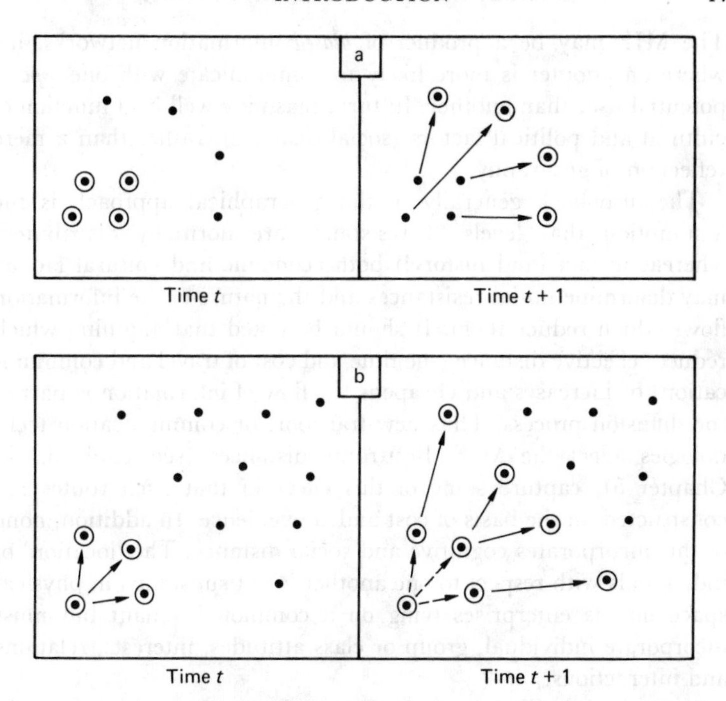

● – Indicates a potential location of the diffusing phenomenon.
◉ – Indicates an actual location of the diffusing phenomenon during the time indicated.

FIGURE 1.4 Diffusion graphs – (a) relocation diffusion and (b) expansion diffusion

dictate the cost of an information network and Hagerstrand does not fully consider the perhaps varying economic benefit (above) of an innovation as an element in the recognised variations in resistance to adoption.

The notion of an information net may be expanded to that of the mean information field (MIF), which provides a statement of the probabilities of an individual communicating with other individuals at different locations. In Chapter 3 we argue that in England and Scotland in the late eighteenth century the probabilities of relevant information spreading to potential users were higher than anywhere else at that time, especially so when social distance is incorporated.

The MIF may be a product of *biased* information networks, i.e. where an adopter is more likely to communicate with one *type* of potential user than another. In turn, bias may well be a function of cultural and political factors (social distance), rather than a mere reflection of geography.

The drawback generally of the geographical approach is the assumption that levels of resistance are normally distributed, whereas in fact (and history!) both economic and cultural factors may determine initial resistances and the nature of the information flows which reduce them. It should be noted that anything which reduces effective distance (the time and cost of travel and communication) or increases and cheapens the flow of information is part of the diffusion process. Thus new transport or communication technologies affect the MIF. Best-route distances (see Table 5.1 of Chapter 5), capture some of this effect in that such routes are constructed on the basis of cost and convenience. In addition, none of this incorporates cognitive and social distance. The 'location' of individuals with respect to one another is not subsumed in physical space nor as enterprises lying on a common isoquant but must incorporate individual, group or class attitudes, interests, relations and interactions.

Normal market forces may be quite sufficient to encourage a ready diffusion of a superior technology. This is especialy the case when there exists within an economy a thriving sector engaged in the specialised manufacture of machinery, i.e. a capital goods sector. Thus in nineteenth-century America, the Lowell Mills in Massachusetts emerged when initial user firms became suppliers of equipment to other firms. This suggests that a minimum internal demand within a sector or sectors is required prior to specialisation in machine production and that this is an effective precursor of a capital goods sector which concentrates on making *general* machine tools (i.e. in this case, lathes, planes, boring machines) for a variety of industries. Such a sector becomes a key to the diffusion of technologies within the nation – a resource to be drawn upon by a variety of manufacturers.[31]

The price of coal conditioned the diffusion of the Watt engine in England, just as the inexpensive availability of Swedish iron may have held back the diffusion in England of new processes of coke smelting and puddling.[32] In Pennsylvania during the 1830s the canal system penetrated into the anthracite coal fields, which only then allowed the diffusion of a complex of iron and steam technologies that had been otherwise available for many years.[33] Electric power generation diffused rapidly in Japan during the first three decades of the twentieth century primarily because of its small initial costs and the high demand within the production structure for small power units.[34] In such examples, market forces were predominant.

But markets may fail to diffuse superior technologies for various reasons, including lack of funds and information, an inability to capture all benefits of an innovation, miscalculations of risk and so on. This may be countered by government intervention in general economic policy where it is felt that the potential social gains from diffusion are being lost as firms focus on private returns only. Chapters 4–7 show clearly that the history of the nineteenth century abounds with examples of government intervention in this sphere. If an adopter is primarily a risk-taker, then anything in the external environment which acts so as to reduce risks by providing information must induce technological change. It has been argued that in modern Japan such organisations as the *Kenkyu Kumai* (research cartels), the Ministry of International Trade and Industry, the Trade Council and so on provide both market and technical information to aid diffusion in this way.[35]

Douglas North has pointed out that the most efficient production technologies may not be followed in a world of positive but variable *transaction costs*.[36] The latter embrace all the costs of specifying and enforcing the contracts that underlie all exchange. A lower production-cost technology may not diffuse because the transaction costs associated with its efficient introduction and utilisation are higher than for otherwise 'less productive' techniques. This notion effectively links the conventional approach with a more institutional perspective.

As a nation's techniques of production become more complex, sectors dealing with reduction of transaction costs will develop within firms (e.g. legal, marketing and personnel departments), within government and within specialist service organisations (e.g. insurance). Furthermore, a large but not measured proportion of transaction costs is composed of information costs.[37]

Finally, it is quite possible for old technologies to revive and survive even when the effectiveness of new ones has been demonstrated. The efficiency of a canal system may improve if it is challenged by railways.[38] A good example of this effect is provided in the history of the British heavy chemical industry in the late nineteenth century. The Leblanc process was the established one for the manufacture of soda (alkali) and it remained as a representative technology in the face of the seemingly superior Solvay process.[39] Although this has been seen as a failure of vision, it may be interpreted as the outcome of rational decisions on the part of British inorganic chemical producers. Firstly, Leblanc technology represented a heavy and durable investment associated with a stream of incremental *improvements* (above). System innovations included the improved recirculation of waste heat, the use of revolving furnaces, the development of other uses of caustic soda in cotton treatment and paper manufacture and the recovery of sulphuric acid – a key product – from tank wastes. Secondly, Leblanc producers maintained their profitability by reducing their output of soda and increasing their output of bleaching powder and other products. Thirdly, technical advance *outside* the industry – electrolysis – inexpensively provided the very chemicals which Leblanc relied on. A period of response and adjustment led to a situation of roughly comparable cost structures between the two processes.[40] The importance of technologies is that they are at once competitive and adjustable.

1.6 THE TRANSFER OF TECHNOLOGIES

> It may be seriously argued that, historically, European receptivity to new technologies, and the capacity to assimilate them whatever their origin has been as important as inventiveness itself.
>
> *Nathan Rosenberg, 1982*

In the late 1750s the Smithfield chemist, Humphrey Jackson visited Russia in order to discover how isinglass (used as a substitute in brewing) was made there, and in March 1760 obtained a British patent for its manufacture. Technology transfer is the movement of a technology or product from the context of its original invention and diffusion to a different economic context, and is normally conceived of as occurring between nations.[41] Obviously enough, many of the inducements and barriers associated with technological diffusion within an industry are common to such transfers. However, the process of transfer generally does involve a variety of transfer mechanisms and environmental considerations not normally associated with industry-based diffusion processes. Together, diffusion and transfer convert potential techniques of production at one locus into available techniques elsewhere. Transfer mechanisms have embraced a great range of phenomena, and are dealt with in detail in Chapters 2 and 4–7. It is often argued that at the present time transfer mechanisms generate a learning process between nations which allows 'latecomers' to speedily catch up with early adopters.

Technology is not simply a collection of hardware or a set of manuals, 'it is these things plus people and the experience and skills they bring to the process, be they operatives, craftsmen, or managers'.[42] The role of key personnel has always been considered vital, from the transfer of technology into the silk and woollen manufactures of Bologna in the thirteenth century to the mechanical engineering and textile machinery transfers to America from Europe in the early nineteenth century. British industrialisation has been interpreted as predicated upon the importation of great waves of skilled peoples – from Elizabeth's employment of German and Italian experts in mining, metallurgy and glass manufacture, through the protestant migrations from the Low Countries and France to the Dutch contractors involved in the draining of the fens. Study of the transfer of Western technology into the USSR in the twentieth century has concluded that greatest successes were gained when systems of equipment and plant were employed alongside key engineering and management personnel. Blueprints do not transfer easily.[43]

The *trading* of technologies or the products of technology has at all times been a feature of technology transfer. In the work of Nurkse and others the value of trade depends upon both its income and structural effects.[44] Table 1.2 shows that the period of 1840–70 was a key one for trade expansion.[45] In Chapters 4–7 and Chapter 10 we show that this coincided with an increase in technology transfers. Major transfers were to North America, Sweden, parts of Germany and Belgium. If anything, transfer accelerated in the years 1870–1900, and this suggests the importance of the new institutional and political factors described in Chapters 4 and 5. Britain was both the workshop of the world and its greatest trader: during the 1870s domestic exports represented 23 per cent of GNP and were produced within a range of industries whose technologies were fairly readily transferable – textiles, pig iron and steel, with a growing

TABLE 1.2 World trade and industrial production, annual average percentage growth, 1720–1977

Years	*World trade*	*World industrial production*
1720–80	1.10	1.5
1780–1820	1.37	2.6
1820–40	2.81	2.9
1840–60	4.84	3.5
1860–70	5.53	2.9
1870–1900	3.24	3.7
1900–13	3.75	4.2
1913–29	0.72	2.7
1929–38	–1.15	2.0
1938–48	0.00	4.1
1948–71	7.27	5.6
1971–7	5.30	6.1

percentage of exports of machinery, transport equipment and chemicals. By 1899 textile production contributed 40 per cent of all world trade in manufactures.

If it had not been for the vagaries of trade protection and colonialism, trade might have continued as a major transfer agent into the early twentieth century. Whatever its motivation, colonial policy which centred on financial sufficiency did tend to focus on the growing or mining of export staples using *European* enterprise and cheap, indigenous labour. Technology transfer was not encouraged in a context of low opportunity costs for indigenous labour, plantation systems and cultural confusion or dissipation. At the same time the financing of transfers into the indigenous economic systems became sensitive to international market forces and, in particular, to the terms of trade. The coverage of India and China in Chapters 8 and 9 shows that colonialism was not conducive to meaningful technology transfer.

Even if mechanisms for transfer are viable, success in transfer depends on what Solo has termed 'the social capacity to assimilate advanced technology'.[46] In such a vein Sawyer has explained the success of transfers into America in the nineteenth century in terms of social structures – the absence of class and craft rigidities, a lack of hereditary constraints on work processes and sumptuary standards, a high focus on individualism and competition, 'and the mobility, flexibility, adaptability of Americans and their boundless belief in progress'.[47] Chapters 5–7 and Chapter 10 deal with such matters more broadly.

Binswanger and Ruttan isolate three classes of 'obstacles' to success in technology transfer:

(a) the costs of acquiring information and screening technologies (see Chapters 2, 5 and 7);
(b) the environmental sensitivity of new technologies;
(c) inadequacies of local research capacity where technological adaptation is required (see Chapters 4, 8 and 11).

But 'obstacles' may be surmounted.[48] It is also allowed that institutional as well as technological change may be produced in receiver nations, and that there may be positive feedback between the two. In a crudely-fashioned nutshell, if change in factor prices or demand may induce technological transfers, then they may also encourage appropriate institutional responses as the less-developed nation moves towards 'development'. This is a view of a 'big bang' theory of development. In such terms we argue in Chapters 7–9 that formal colonial relations prohibited such responses. As Benjamin Higgins has summarised: 'The most damaging colonial policy . . . was the failure to promote the spread effects [of industrial development] in the secondary and tertiary sectors of the colonial economy.'[49]

But if we are to more fully comprehend the problems involved in the assimilation of transferred technologies, we must be prepared to go beyond the isolation of transfer mechanisms or the heritage of colonialism. The very idea of indigenous technological capacities must be seen as conditional upon the physical character of imported technique itself, and this has changed through time. Galbraith has isolated the following as the commanding technological imperatives of modern techniques:

(a) Lengthening of gestation periods; the time lapse between the setting-up and commercial production of technologies has lengthened, and may continue to do so. This creates a potential threat of debt, as the receiver may be contributing to repayments long before income accrues.
(b) Capital commitments and intensities; the size of and relative importance of fixed investments in the new technologies has increased with time.

(c) New technologies are increasingly specialised as both process and products, and may therefore be designed to satisfy a level of demand which has not yet been reached in the receptor nation or industry.
(d) Both in construction and usage modern technologies now involve a high level of management and technical skills, and these may be unavailable and costly to create.
(e) Complex hardware demands a sophisticated software technology. 'More even than machinery, massive and complex business organisations are the tangible manifestations of advanced technologies.'[50]
(f) As a result of all this, successful settlement of advanced technology requires a command over time and a commitment to the future.[51]

If such imperatives mean that adoption awaits adaptation of technique, this process itself requires substantial engineering and management skills, attributes which have accumulated but are forever changing in 'donor' nations. Because of this latter tendency, more than specialised skills may be required during a prolonged period of transfer. During the life of a major project the ability to adapt skills, procedure and even goals may be of great importance. Official support mechanisms may help. In the 1880s, the Meiji government of Japan sponsored a conference of private Japanese investors to discuss the viability of the use of Western technology in the nation's cotton industry. The major problems recognised included the large scale of initial investments, the high cost of motive power and transport, and a lack of worker training. The tactic adopted by the government was to set up its own demonstration plants, in which the costs and experience of introduced technology were absorbed, and to disperse the subsequent, relevant information through a variety of institutional mechanisms.[52]

Under what conditions does today's advanced technique transfer from a TNC into the indigenous sector of an underdeveloped nation? If a more appropriate technology is not readily available, the tactic of an affiliate or agency may well be to adapt essential features (i.e. imperatives) of its technology to the given characteristics of the host nation or industry, e.g. to account for a lack of skills. Here transfer is effected by adaptation within part of the donor environment. An alternative vision is that the more general institutional environment must change in order to accommodate the given imperatives of advanced technologies. Frank Bradbury has pointed out that:

> It is noteworthy that any multinational company in transferring its technology to new centres around the world strongly emphasises standardisation of parts and specifications. In other words, its selection environment is a uniform pan-world one, which damps out local variations.[53]

In summary, it is history, the character of transfer mechanisms and the changing nature of advanced technologies which have together given rise to the central problems of development and development policy. At what cost and by whom can advanced technologies be converted into appropriate technologies? Do alternative technologies exist? Does a 'world technology' demand a 'world culture' (i.e. the 'convergence' of value and institutional systems)? Is cultural revolution to be avoided only at the expense of continued underdevelopment in poor nations? Are there different 'models' or paths of development, each with its own appropriate socio-cultural system? May failures in the market place be met by government planning or control? To what extent has economic modernisation through technology transfer *always* been a matter of ingenuity, exploitation and conflict?

1.7 SCIENCE, TECHNOLOGY AND THE SOCIAL MATRIX

> One may come by many routes to appreciate the uses of history. A thorough indoctrination in the modes of thought formalised by the now-dominant neoclassical tradition of economic analysis, however, does not happen to be among them.
>
> *Paul A. David, 1975*

> Institutional inertia is not *necessarily* to be regarded as a pathology. Institutions provide the framework for economic life. A completely flexible framework is a contradiction in terms.
>
> *R. C. O. Mathews, 1986*

T. H. Huxley once gibed that Herbert Spencer's idea of a tragedy was a deduction killed by a fact. The use of theory is to illuminate the particular, historical event or series of events. If theory does not do this or if it does not indicate how data relating to events may be somehow improved, then the theory should be discounted. The historian tries to protect the integrity of the individual case.[54] Robert Solow has concluded that because modern economics has been constructed as an axiomatically-based science, 'economic history is as much corrupted as enriched by economic theory'. Although economic activity itself is 'embedded in a web of social institutions, customs, beliefs and attitudes', economic analysis conventionally proceeds in implicit denial of this, 'as if economics is the physics of society'.[55] Paul David too has noted that economists conceive of the market system as a mechanical analogue in which time is not, in fact, chronological, but rather the expression within which causal sequences must be placed, e.g. the simple phrase 'supply adjusts to demand'.[56]

In fact, departures from natural science occur because the axiomatic system of economics is very complex and, therefore, throws up competing hypotheses whose veracity depends not on reproducible experiment but on existing evidence.[57] As Robert Solow has noted, the trouble with history is that the evidence base is often so general as to allow a fit for all such competing hypotheses! The result may be a series of 'virtuoso finger exercises'.[58] These abound particularly in micro-economic theory, much of which assumes that firms operate efficiently and respond to circumstances (e.g. isoquants move towards their origins). Yet Cyert and George find that 'firms do not operate as efficiently as assumed in economic theory'.[59] More to our point, Leibenstein has explicitly argued that 'firms do not always introduce technical change when available and profitable', a consideration somewhat different to those met in Sections 1.5 and 1.6 above.[60] Although Leibenstein has tended to explain such inertia in terms of conventional market failure (e.g. the existence of monopoly), there is good reason to believe that it may stem from lack of information and high transaction costs as well as from wider institutional and attitudinal forces. Outside of the firm, historians are constantly faced with a series of institutions which impinge upon the activities of firms or groups but which exist only in terms of law, control and the nation-state – usury laws, navigation acts, limited liability, partial inheritance, royal monopolies and internal tariffs.

It seems that, if for today's natural or social scientists, the empirical world is annoyingly recalcitrant, then to the historian the world with which he or she deals is far more problematic for it is partial and often preselected. Many tests of data cannot be adopted with propriety, subjects may not be interviewed, attitudes may not be confidently derived and, therefore, values may not be identified with any intellectual ease. It is for such reasons that in history both great theories and great men loom too large.

However, we have already shown that several economists have departed from what David Hamilton has recently schematised as the tripartite division between institutional, technological and supply–demand (natural science) approaches. Innovations are themselves part of existing *institutions*, and the routines of the latter may constrain or encourage their innovative activities. Some firms may respond to the competition of a rival's technique by advertising or pricing policy. Others may immediately turn to research and development which is as close to established routine as possible. Yet another enterprise might venture out into the technologically unknown. Such variable responses to a common problem may relate to firm size, the extent of R&D facilities, the prevailing opportunity and transaction costs of innovation for the particular enterprise (transaction costs may be variable between firms in an industry) and other characteristics of the enterprise as an institution. But non-enterprise institutions also provide information, transactions-servicing and technologies. In the view of Thomas Sowell an array of economic, social and political institutions can be viewed as means to develop and transmit information, itself central to the choices (above) which firms or individuals make.[61] In a polyarchical market system of the Stiglitz type market forces may sufficiently provide individuals with signals (information) enabling them to correct faulty decisions.[62] On the other hand, other institutions may transmit the accumulated wisdom of generations. R. C. O. Mathews has distinguished between institutions which define and confer property rights, those which stabilise conventions of economic behaviour, and those which circumscribe the operation of markets for labour, capital and so on. It is quite possible that institutions which provide 'sets of rights and obligations affecting people in their economic lives' may be understood in terms of general economic laws.[63] But it is more obvious that there exist many institutions in the cultural, schooling and communications areas which are not particularly sensitive to economic circumstances, and, thus, to economic analysis. It may be argued that some institutions emerge as a *result* of market forces. The emergence of the factory has been interpreted in terms of rising transaction costs in earlier forms of enterprise organisation. But such institutions may in turn induce technological or other organisational changes which, in a larger view, increase economic growth in a sustained fashion. The later development of the multi-unit enterprise has been interpreted in such a manner.[64] On the other hand, it is argued in Chapter 3 that the rise of innovative enterprise during the British industrial revolution depended on the existence of a complex of *cultural* institutions and processes, which through a multitude of informal contacts and information flows reduced the costs of technological change and of transactions. On a grander scale, David Landes has pointed to a number of non-market institutional factors which encouraged the innovative activities of the European bourgeousie. In nineteenth-century Europe the arbitrary actions of the state were minimised, protection of property was maximised, contracts *thereby* replaced force and status. Within this environ, Britain was perhaps the most polyarchical of nations, with a relatively high degree of social mobility and decentralisation of power. But we may acknowledge that the relevant social matrix is not easy to identify. Joseph Needham would argue that China represented a more than viable alternative setting, e.g. its literacy, education and meritocracy.[65]

The spread of new knowledge within a national setting depends on individual motivation, literacy levels, social exchange and mobility and some willingness to assimilate new ideas or procedures and to break with custom. All societies are endowed to some degree with a potential supply of inventors, innovators and risk-takers, but nations are vastly unequal in the supply of facilitative institutions which serve to transmit information or reduce risk. At the same time, learning-by-doing takes time, and in the Arrow approach is closely associated with the formation of new capital goods.[66] Formal education or cultural institutions may well both advance the flow of information and strengthen the ability of an economic system to absorb new physical capital. Thomas Schultz has calculated that the stock value of education in the USA increased by approximately 850 per cent between 1900 and 1956, compared to an increase in physical capital of 450 per cent, and that the economic returns to the former were at least as high, if not higher, than returns to reproducible capital.[67]

Four final points are worth noting. First, in periods of relatively speedy, hitherto unexperienced economic change, very little institutional adjustment may take place. There will always be some degree of cultural lag. Second, those transfers of technology which comprise imperatives which demand certain market conditions, may illustrate the inappropriateness of particular institutions to economic development. But this is another side of a coin which defines such technologies as inappropriate. Third, the chain of causal regress which relates technologies to institutions is of indeterminate length and is composed of links which are difficult to negotiate. At what point in an analysis of change can the economic contribution of *particular* institutional changes be separated out from other institutional change and from technological progress? Fourth, institutions are extremely complex and serve a multitude of purposes.

Such matters are of importance to historians. For instance, much of this book focuses on the institutions of the state itself. Such institutions varied enormously through time and place and at many points impinged upon the play of market forces or, indeed, served as substitutes for them. At what cost? With what benefit? Historians have hardly begun to address such questions.

1.8 SCIENCE, TECHNOLOGY AND UNDERDEVELOPMENT

> The economic histories of the advanced countries provide important guidelines rather than complete formulae for the developing countries today.
>
> *Benjamin and Jean Higgins, 1979*

Experience since the 1960s has shown than even those poor nations which achieved a high rate of economic growth did not generally enjoy 'development' in the sense of real improvement in welfare and security for the masses. Indeed growth of national incomes at times occurred alongside growth of unemployment, inflation, malnutrition, ill-health, illiteracy, economic inequality and political repression. As Benjamin Higgins has noted, such evidence suggests that societies are not seamless webs in which changes in one quarter will bring corresponding, empathic changes in others.[68] Though pleasing waves and webs may be discernible over the very long run, in the time period in which people starve and governments fall, economic growth can occur alongside a dynamic process of underdevelopment, a process identified in Chapters 6 and 8–9 below. Thus the replacement of traditional by modern technologies need not launch a process of pervasive economic modernisation. In certain locations and historical contexts (e.g. Japan as in Chapter 7) new technologies may settle amidst little-changing national institutions and mores. In other settings new technologies may never successfully transfer if they fail to accord with prevailing economic, social and cultural givens, change in which is problematic, difficult to control and frequently disastrous. A decade-and-a-half ago Adelman and Morris wrote that:

> The frightening implication of the present work is that hundreds of millions of desperately poor people throughout the world have been hurt rather than helped by economic development. Unless their destinies become a major and explicit focus of development policy in the 1970s and 1980s economic development may serve merely to promote social injustice.[69]

To the historian there is little new in this. Late nineteenth-century India showed several signs of economic modernisation, as did China in the early twentieth century, but for both these major nations the true history of development begins sometime after 1948.

The problem of modern underdevelopment reflects a dramatic application of elements already discussed in Sections 1.5–1.7. The drama is a partial result of the fact that the technologies available for transfer and subsequent diffusion are changing year by year, both in their scientific content and in the nature of their commanding economic imperatives. Such changing imperatives have a decisive impact upon the probability of the economic success of transferred technologies in the social and cultural context of receiver nations. History cannot repeat itself, but does illustrate the variety of development paths which have been trodden, the importance of technological progress widely defined, and the need to create independent, innovative patterns of response to changing circumstances. Whether change should occur in technologies or in the institutional frameworks which harbour them remains debatable. Economic development in nineteenth-century Europe illustrates that success may result when there is a coalescence of appropriate technologies, transfer mechanisms and institutional adaptations. The history of China since 1949 suggests that institutional reformation may be a precursor of successful development through technology transfer. A problem is that *appropriate technologies* are not easily identifiable until time has passed. For instance a skill-and-capital intensive transferred technology may at a later stage in its life cycle employ large numbers of abundant unskilled or semi-skilled workers directly and employ a range of available resources through its demands on ancillary servicing sectors.[70] At the same time the *improvement* engineering developed in order to modify a relatively inappropriate technology may well generate skills and expertise and facilities which will produce growth bonuses at a later time.[71] R&D improvements may increase the absorptive capacity of the economy as well as improve the quality of new, original technological choices. But some modern technologies are relatively unadaptable. Modern nuclear power stations now employ huge vessels, heat exchangers and pumps the construction or emulation of which are well beyond the technological capabilities of most nations – manufacture is indeed confined to nations such as America, the USSR, France, West Germany, Britain and Japan. With such technologies large and essential fragments of expertise remain untransferable.[72]

Chapter 2 is fully centred on the eighteenth century, but serves to illustrate many of the more abstract themes already sketched. A scientific culture diffused throughout much of Europe but not all of Europe then developed; parts of Europe developed later, some European nations did not achieve an industrial revolution prior to the Second World War. Technologies transferred by various mechanisms, but at only a few locations did new technologies transform industries, and only rarely were whole economies significantly altered. More often, new imported techniques represented a net cost to the importer nation, a cost defended by ruling elites whose prime purpose was the retention of prestige and power through a military-linked industrialisation process.

2. Mental Capital – Transfers of Knowledge in Eighteenth-Century Europe

2.1 THE SCIENTIFIC ENTERPRISE PHASE (1) – COMMUNITIES AND ACADEMIES

Savants and students should participate as much as they can with other people and in the world.

G. W. Leibnitz, 1669

FROM the time of the publication of his *Memorandum on the Founding of a Learned Society in Germany* in 1669 to his death in 1716, G. B. Leibnitz held fast to his view that the growth of a scientific enterprise in Europe depended on increased communion between *savants* in different nations and between *savants* and other social groups within nations. The study of the natural world should be a practical and wordly, as well as a theoretical and abstract pursuit. Leibnitz carried theory into practice with his calculus and his calculating machines, his mechanical pumps for use at the Harz mines, his plans for a universal language for scientific discussion, and in his demand that the provincial German scientific society, the Leopoldine Academy, become a focus of a united national effort in scientific, technical and commercial research. His cultural entrepreneurship led to the foundation of the Berlin Academy of Science, which began publishing its transactions in 1710. Although much more gifted than most, Leibnitz was but one of several projectors of a post-Newtonian scientific enterprise for all Europe.[1]

Newtonianism served to at least partially dampen distinctions arising from different national traditions. Newton was popularised before he was published. The *Principia Mathematica* of 1687 was not available in English until 1729. But prior to both dates, in 1686, B. le B. de Fontenelle published his *The Plurality of Worlds*, which not only epitomised much of the Newtonian corpus in malleable form, but also hammered out at least some of the implications of the new science for the world view of the European intellectuals. By inference, induction was carried into the European religious and political system.

Writing solely of France, Barber has claimed that 'the *philosophes* and their writings became for many relatively unsophisticated bourgeois the symbols of secularised thought'. Frenchmen continued to attend church and kept their philosophic and deistic opinions to the fireside or the provincial *salon.* Furthermore, the high esteem in which such intellectual pursuits were held promoted social mobility, as the natural theologians became the interpreters of a new *Weltenschauung*.[2] The development of 'natural theology', whereby the study of nature was to be used as a base for the understanding of the existence and benevolence of God, was what brought the scientific enterprise into general prominence. Thereafter, *savants* were to gain patronage from not only wealthy devotees or State institutions, but also from an audience of officials, professionals, journalists and businessmen, who received their message through a plethora of journals, periodicals, public lectures and meeting places.

The general notion that the revelation of the Scriptures might be merely a supplement to the revelations of nature took many specific forms. In Europe's outposts deistic conclusions seemed the most logical. In the American colonies Newton was to become for intellectuals what God had been for the natural world. Newport town became a centre of deism, whilst Cotton Mather communicated to the Royal Society his opinion that 'Newton is the perpetual dictator of the learned world'.[3] In contrast, at least for some time, civic virtue in Haarlem kept Dutch Newtonianism in the mercantile, utilitarian mould. In Germany, Leibnitz grew deistic and at a later date both Kant and the Scottish school of philosophers maintained that it was the theological implications of the study of nature that made such study 'useful'.

In this, perhaps, such elements were following Newton himself, as well as those such as Thomas Burnet, John Ray and William Derham who in England maintained the need to consider both Scripture and nature with equal finesse. In England, the Boyle lectures for the diffusion of Newtonian natural philosophy became also a vehicle of natural theology.[4] William Whiston, a Boyle lecturer in 1707, was removed from his fellowship and professorship at Cambridge (and his living at Lowestoft), when the Colleges pronounced him guilty of Arian heresy. Although his subsequent energetic speading of Newtonianism in public lectures in London, Bristol and elsewhere appears to have excluded his brand of primitive Christianity as 'foreign to the mathematics', by the 1730s the scene was set for a *popularisation* of 'science' in London which seems to have owed much to its link with theology.[5]

One of the most public forums for intellectual debate in London during the 1730s was The Oratory, a debating house situated at the corner of Lincoln Inns Fields, near Clare Market. Much puffing was in evidence: 'Letters to the Oratory from Madrid, Paris, and Rome will shortly be answered'.[6] Amidst exhaustive lecture courses on scientific subjects by Theophilus Desaguliers, George Gordon, John Clarke, Peter Shaw and Francis Hauksbee, the Oratory mounted debates under such headings as 'The Works of God, delineated in the most important Branches, the Heavens, Stars, Earth, Vegetables, Dominion of the Heart etc.'.[7] By this date, 1731, Whiston's own public lectures in London, delivered in courses of 21 lectures at the Amsterdam Coffee House, were well-publicised and seemingly well-attended.[8] In February 1732 the Oratory advertised a debate on 'a Comparison between the Sense of Dr. [Conyers] Middleton, who is a Naturalist in Religion, and Dr. [Zachray] Pearce, of St. Martin's, who is for revealed Religion'.[9] In April 1732 the Oratory debated 'The Nature of God, discernable by Reason, in view of all that is *Scientific* on that subject', [my emphasis].[10] In June the same forum attacked Whiston with its sermon on the Divinity of Christ 'in opposition to Whiston, and the other Haresians [sic] against the Trinity'.[11] From what may be discerned indirectly from the more popular of London's newspapers, the disputes over natural theology were attracting greater public attention than were the more demanding courses of 21 or 30 lectures on Newtonian science.[12] It now comes of no surprise to learn that in February 1735 'Mr. Whiston's desire to be Assistant at the Oratory is refused'.[13]

Urban provincialism spawned science everywhere. Writing of intellectual systems more generally, Edward Shils has argued that their development requires the existence of 'institutions which, although apparently only ancillary, are indispensable conditions for the effectiveness of the central institutions'.[14] In France, the Academie prospered against the backdrop of a foundation of no less than 100 academies between 1700 and 1776, including such distinguished centres as Bordeaux (1712), Rouen (1716, 1735, etc.), Dijon (1740), Lille (1758) and Mulhouse (1775). In Britain, the provincial movement took the form of small, informal coteries and the 'Literary and Philosophical' societies of the second half of the century. Similarly, the work of the Berlin Academy was boosted by the formation of a series of provincial academies in each of the German states. Between 1692 and 1792, 11 towns in Italy formed scientific academies, including that of Turin (1759), instigated by J. Lagrange (1736–1813), Professor of Mathematics in the town's royal artillery school. It is of note that most of the cultural entrepreneurs were relatively young men. By 1790 there may have been some 220–250 major academies for discussion and research in the natural sciences and mathematics. The metropolitan centres did tend to attract the provincial talent. Utilitarian pursuits and the encouragement of officialdom proved a powerful combination at Berlin. The Academy was directed by Leibnitz in its early years – of the first 60 articles published in its transactions, 12 originated with him. Statistical methods and political arithmetic were joined by technical laboratory research. Under A. S. Marggraf (1709–82) the Berlin laboratory succeeded in recovering sugar from beetroot in 1746 and demonstrated that a component of alum, alumina, is found in clay.[15] But throughout Europe the 'ancillary' academies played important roles. Here *savants* were actively engaged in collecting, surveying and in experimental programmes. The emphasis upon communication of knowledge was marked: 75 per cent of the academies published proceedings, and nearly all of these devoted their pages to translation, summarisation and popularisation of the advances in knowledge. Several forums were more specialist; in England the Botanical Society was formed in 1721, the Society for the Promotion of Natural History in 1782 and the better-known Linnaean Society of London in 1788. The producers of knowledge did not yet own it, nor could they contain it within national frontiers.

Within a pattern of differences, it was probably in England more

TABLE 2.1 Scientists, technologists and higher education in England, percentages

	Oxford/Cambridge		*Edinburgh*		*Other*		*None*	
	Sci.:	*Eng.*	*Sci.:*	*Eng.*	*Sci.:*	*Eng.*	*Sci.:*	*Eng.*
Pre-1700	47	34	3	4	19	5	30	57
c. 1700–60	30	8	19	10	13	5	38	76
c. 1760–85	20	11	17	27	11	9	51	53
Average	36	18	12	12	15	6	36	63
Number	242	43	85	30	105	15	248	152

than anywhere else that the growth of a scientific enterprise arose from the socio-cultural reflexes of a relatively open society. Table 2.1 suggests that amongst both the 'scientists' and the 'engineers' of eighteenth-century England, training in elite institutions was not of paramount importance. There existed a consequent dependency on science within the urban culture itself.[16]

Only 36 per cent of 680 scientists and 18 per cent of 240 engineers were at any time connected with either Oxford or Cambridge, and throughout the century the tendency was downward. Of more significance, as time went on scientists were decreasingly likely to have attended any higher education institution, and the prelude to the industrial revolution was associated with the emergence of a group of knowledgeable engineers who owed little to any university education at all.[17] Within a nation the conduits for the diffusion of scientific knowledge were the journals, transactions, newspapers, and academies, much of which lay far outside any centralised guidance, and many of which were hardly affected by restrictive legislation imposed by religion or state.

The emergence of two distinctive tendencies within the culture of European science remain to be considered more fully. The first of these relates to the changing nature of the mechanisms whereby scientific knowledge was diffused *within* nations. The second concerns the strengthening connections between the *production* of knowledge and its *application* to industry and agriculture, which by the end of the period were such as to forbid any rigid distinction between the two.

Amongst the earliest compendiums of the new science were dictionaries, devoted both to learning and its several trade-based applications. Prior to the beginning of the century appeared Thomas

Corneille's *Dictionnaire des arts et de Sciences* (published as a Supplement to the Dictionary of the French Academy in 1694, the third edition of which was revised by his nephew, Fontenelle), and John Harris' *Lexicon Technicum* of 1704.[18] Frequently such efforts were associated with a public lecturing programme, as in the case of Harris's free lectures at the Marine Coffee House in London, sponsored by Sir Charles Cox MP. During the 1730s this continued with the *Dictionarium Polygraphicum* and a series of publications which, particularly in England, were more precisely aimed at increasing the technical basis of existing trades – *The Instructor*, the *Nature of Fermentation*, *The Builders Dictionary* or the *Method of Learning and Drawing in Perspective* were representative titles.[19] From that decade was issued Zedler's *Universal Lexicon* in 64 volumes. For Europe's *savants* the tendency was headed by the French *Encyclopédie*. Its publishing history has been exhaustively explored by Robert Darnton.[20] It was the less expensive, somewhat slighter versions of the Diderot edition which achieved greatest circulation: the Geneva quarto (from 1777) sold 8525 copies at a price which fell from 384 livres to 240 livres in 1781; the Lausanne–Bern octavo (1778–82) sold perhaps 6000 copies at the lower price of 225 livres. Of the known subscribers to the last of these, 80 per cent were residing outside France. Darnton writes of a 'Frenchified cultural cosmopolitanism that remained [in most cases] restricted to a tiny elite'.[21] However, outside France, the *Encyclopédie* filtered down to the professional groups more generally, perhaps particularly so in Italy, the Netherlands and Britain. It is noteworthy that this Europe-wide cultural enterprise was state-sponsored. Most of the Enlightenment figures of France, contributors to the project, were in fullsome receipt of State appointments, pensions and privileges.[22] Indeed, the whole production and diffusion of knowledge in France was controlled by privilege as much as by patronage. As Voltaire wrote in his *Lettres Philosophiques* (1734: Letter 24), 'A Seat in the Academy in Paris means to a geometrician, to a chemist, a small secure fortune; on the contrary, in London it costs money to belong to the Royal Society'.

From the cultural coteries emerged the journals and transactions. Priority disputes over scientific advancements had been building up. A function of the eighteenth-century journal was to establish intellectual priority.[23] But general scientific publishing incorporated encapsulation, repetition and popularisation. These elements did reinforce initial published claims, but they served other functions as well. Of 501 substantive scientific journals that have been surveyed, it was found that the majority in fact contained much in the way of confirmation and repetition of existing findings. Moreover, by far the greatest proportion came from Germany, a 'Nation' which did not rank foremost in Europe as a centre of creative investigation. We might judge that publications served more to communicate than to advance knowledge.[24]

It will be noted that several of the titles so far noted conflated science and the arts. This is also true of Benjamin Martin's (1704–82) *Bibliotheca Technologica* and of Ephraim Chambers, *Cyclopedia or General Dictionary of the Arts and Sciences*, a direct precursor of the better-known encyclopedia of the Unitarian Dr Abraham Rees. The growing number of manuals and guide books for mechanics and engineers contained detailed information on systems of measurement, the drawing of technical plans, the calculation of velocities and basic findings in mechanics, hydrostatics and hydraulics. A good example was Francis Walkinghame's *The Tutor's Assistant* of 1751, which by 1783 had passed through 18 editions each of which had sold 5000 to 10 000 copies, a total very far in excess of the various editions of the *Encyclopédie*.

Cardwell refers to such prominent inventors as Newcomen, Smeaton, Watt and Wedgewood as 'scientific technologists'.[25] In doing so he captures the spirit of the eighteenth century. The word 'technology' was seemingly first introduced by Johann Beckmann (1739–1811), who after an extended tour of Western Europe and Russia initiated public lectures on agriculture, mineralogy and commerce at Gottingen University under the new title.

In a careful survey Hahn has uncovered some 50 or so 'Societies of Arts' established in Europe and elsewhere during the eighteenth century.[26] The society formed in London in 1753–4 was composed of 'Noblemen, Clergy, Gentlemen and Merchants'. That established in Paris in 1726, the *Société Academique des Arts* met in the Comte de Clermont's palace, limited its membership and refrained from dissemination of its findings. What such varied societies had in common was the emphasis upon application of formal knowledge and technique to the betterment of practice, and the award of prizes, gifts and medals for the advancement of best practice. Much further removed from the artisans themselves, the learned institutions also refused, or were unable, to draw a distinction. At Cambridge during the 1760s Richard Watson's lectures compounded theory with discussions of mordant dyeing, metallurgy and assaying, gave demonstrations and illustrations of industrial processes for production of green vitriol, nitre, common salt, sal-ammoniac, plate glass and enamels, and outlined Marggraf's method for the extraction of sugar from roots.[27]

It becomes fairly obvious that scientific knowledge and its applications were equally integral to the rise of the European scientific enterprise during the eighteenth century. New scientific specialisms arose on the back of application. Table 2.2 gives a deceptively quantitative summary of an essentially qualitative process. Drawn from the 435 'major advances' in chemistry which are listed by the Clows, the table does serve to roughly depict many of the points emerging from our account.[28] Scotland and England dominate throughout, even though French chemistry was far ahead of British in the late eighteenth and early nineteenth centuries. This is because the British chemical enterprise was also composed of initial applications of knowledge and their subsequent commercial exploitation. The study of chemistry owed much to institutions, and to regulations designed to monitor or encourage the products of chemical enquiry.[29] Defined in such a way, chemistry gained its spectacular rise in the years 1781–1830 (column B), a period which in simple numerical terms saw the applications of knowledge growing at a faster rate than the production of knowledge.

2.2 THE SCIENTIFIC ENTERPRISE PHASE (2) – SCIENCE IN CULTURE, 1780–1800

There was nothing traumatic about the year 1780. But from about that date, and particularly in England, the movements of the scientific enterprise towards popularisation, provincialism and specialisation all accelerated, and the links between science and both utility and political ideology were strengthened. The journals and academies of the earlier years had at least partially taken over the role of colleges and universities. Throughout Europe the work of leading intellectuals owed less to formal institutions. The information dispersals of the first phase *permitted* (i.e. they did not necessarily cause) the entry of men of humble background into the second phase of the scientific enterprise. In itself, this was likely to generate a form of

TABLE 2.2 The chemical enterprise of the eighteenth century

	A	B	C	D	E	F	G	H	I	J	K	L
	Total entries	*Total entries per decade*	% *which were Scot. & Eng.*	% *Intell. advance*	*Total D entries per decade*	% *Applic. of knowl-edge*	% *Instances of Ind. Com. Dept.*	% *F + G*	*Total F + G per decade*	% *of Institu-tions/In-stitutional*	% *Instance of Soc/ Legal Regulations*	% *Total (cols. D–K)*
Seventeenth century	36	3.6	78	30.5	1.1	36.1	13.9	50	1.8	5.5	8.3	94
1700–1780	107	13.4	72	29.9	4.0	37.4	24.3	61.7	8.2	1.9	6.5	100
1781–1830	183	36.6	63	31.1	11.4	49.7	10.4	60.1	22.0	4.4	4.4	99
1831–1856	107	42.8	71	26.2	11.2	49.5	8.4	57.9	24.8	8.4	4.7	100

TABLE 2.3 Patented technique in Britain, 1689–1829

		Total patents	*Patents per decade*
William and Mary	(1689–1701)	104	80
Anne	(1702–13)	26	21
George I	(1714–26)	91	70
George II	(1727–59)	253	76
George III	(1760–80)	536	255
George III	(1781–1819)	3124	801
George IV	(1820–29)	1355	1355

scientific enquiry closely related to practical pursuits.

Science production accelerated. Of a total of 1052 scientific journals and transactions identified by Kronick for 1665–1790, 20 per cent originated in the decade 1770–79, 40 per cent in the decade 1780–89. In this measurable acceleration the formal associations were of importance; of 307 *Proceedings* and *Memoirs* published between 1670 and 1790, 43 per cent were produced during the 1770s and 1780s.[30] The latter years of the century were also characterised by an increase in the quality and continuity of journals designed to provide general scientific information at reasonable cost. To the *Botanical Magazine* (from 1787) and *Der Naturforscher* (Halle, from 1774) were added the *Journal der Physick* (1790), the *Repertory of Arts and Manufactures* (1794), *Nicholson's Journal* (of Chemistry, 1797) and the *Philosophical Magazine* (1798). In these publications a tendency towards specialisation is clear, and was embellished with the *Annales de Chémie* (1789), the *Neue Annalen der Botanick* (1790) and the *Zoologisches Archiv* (1790). The growth of specialist societies supported such publications – in London alone the years witnessed the foundation of the Linnaean, Horticultural and Geological Societies, forums for the study of astronomy, meteors and zoology, and the Institute of Civil Engineers.

The production of discrete technologies accelerated in similar vein. Table 2.3 illustrates the growth of patented technique in Britain between 1689 and 1829, a subject returned to in detail in Chapter 3.[31] The late eighteenth century watershed is very clear.

We may hypothesise that prior to circa 1780 scientific knowledge moved between regions and across nations with relative ease, but was *mostly* confined to 'upper-class' social groups. (The definition of 'upper class' varies between nations). With a few exceptions, downward filtration of knowledge occurred via publications. From around 1780 societies and academies contained *within* them a greater diversity of social groups. Significant advances were now coming from the lower professions, merchants, business groupings and artisans. Whether this was primarily a result of the earlier development of scientific publication, or whether it was more part of a wider process of social change occurring after 1780 is yet a matter for debate. The fact that the opening-up of scientific knowledge to wider social groups occurred far more in some nations than in others, does at least suggest that social changes within nations may have been of paramount importance.

Britain and France offer vivid contrasts. The case of Britain suggests that the application of science to industrial pursuits did not wait upon its intellectual 'maturity' but upon its more widespread *diffusion*. In the metropolis such small groupings as the philosophical society of the Chapter Coffee House gathered together some of the most active popularisers of science in England. To Richard Kirwan and Edward Nairne were added the scientific entrepreneur J. H. de Magellan (1722–90), the public lecturers George Pearson, Adam Walker and William Babington, the political activists and eventual exiles Thomas Cooper and Joseph Priestley and the industrial entrepreneurs Matthew Boulton and Josiah Wedgewood.[32] The explicit sub-culture of the Dissenting Academies, whose science provision reached levels above that of the English universities, flooded through to most urban provincial centres during these years.[33] To this Unitarian-dominated dissenting culture of England itself was added the flow of Scottish engineers and scientists, directed to not only London but Manchester, Liverpool, Newcastle and other urban centres. The New College of Arts and Sciences, founded at Manchester in 1783, numbered amongst its lecturers Thomas Henry FRS, on chemistry, and John Dalton on natural philosophy and mathematics.[34] Similarly, at nearby Liverpool at the turn of the century a Society for the Encouragement of Arts, Science, Trade and Commerce was formed, instigated by the instrument-maker Hugh Williams.[35] John Money has recently illustrated the similar acceleration of science provision in the local schools of Birmingham during the years 1781–93.[36] From 1798, at Dublin, William Higgins (1762–1825) offered his ambitious 40-lecture chemical courses from the elaboratory of the Dublin Society, where he was Professor of Chemistry and Mineralogy. The first ten alone were

> appropriated to the investigation of the Philosophical principles of chemistry, and by the application of which many striking phenomena in meteorology, aerology, vegetation and animalisation will be illustrated and explained. . . . [later lectures will be] applied to the development and explanation of the various processes employed in different branches of manufactures and the Arts [Trades], particularly such as are indebted to chemistry for their past progress and dependent on it for their further advancement.[37]

Such selected examples may be, and have been, added to almost indefinitely. Of more significance, it has been suggested that the rising culture of science and technology in Britain owed little to established social groups – whose earlier patronage was fast giving way to the security offered by a large public audience for science – but originated in the dissenting and radical sub-culture of the urban provinces. Joseph Priestley is perhaps the best-known of the scientific radicals. It might be remembered that the first provincial conflicts which were *explicitly* anti-philosophic *and* counter-revolutionary were the 'Priestley riots' in Birmingham during 1791, at which time the *savant's* property, library and scientific equipment were targets of 'Church and King' mob violence.[38] In his account of the social position and cultural significance of Joseph Priestley, Edward Thompson summarises the English version of 'rational enlightenment' as a composite of Unitarianism, science and political reform.[39] G. A. Williams refers to the riots as 'an establishment coup against the powerful dissenting interest', and pictures the object of the attack as a 'nucleus of Dissenters, intellectuals and radical industrialists'.[40] It is certain that in provincial England during the early 1790s there existed a large group of scientific intellectuals at odds with establishment society and institutions. Members of provincial philosophical societies in Liverpool, Derby and Sheffield published their support of Priestley.[41] Scientific dissenters felt culturally marginalised and at risk.

It seems fairly clear that much of the provincial scientific culture of England in the last years of the eighteenth century revolved around social groups in conflict with existing social structures, groups legally victimised in such legislation as the Test and Corporation Acts. Remnants of eighteenth-century elitism existed alongside a more thrustful provincialism. In France, the revolution of 1789 and the subsequent revolutionary wars created a contrasting environment for science.

Gillispie has shown that by the later eighteenth century French science cannot be understood without reference to the close nexus of science, industry and the state.[42] Science remained civic and self-consciously elitist, and gained through its increasingly professional relations with the state, e.g. in its procurement of service contracts. Under the new regime, science was even more 'politicised'. We might suggest four approximate and overlapping stages. In the first, a general policy of cleansing led to the disestablishment of science. The scientific work of the Bureau de Commerce was closed off in 1791, the Académie met the same fate in 1793. This first phase is therefore one of discontinuity. A second phase saw the reimbursement of science. Although such *ancien régime* figures as Lavoisier and Condorcet were sacrificed, the phase saw a certain democratisation of the scientific enterprise and the foundation of the Lycée des Arts and the Société d'Historie Naturelle. Technology was encouraged through the establishment of a patent system which offered 15-year protection, reform in technical education and state activity in machine collection and exhibition. In the third phase, science was politicised once more. Remnants of the earlier regime survived in Carnot, de Morveau and Fourcroy. Under official auspices were founded a mining college (1793) and the well-known Polytechnique (1794). The work of the Academie reappeared with the Institut de France. Both the latter institutions were directed to serve military and imperial functions. The Ecole Polytechnique attracted a large portion of scientific talent and mounted research programmes in formal mathematics, chemical nomenclature, the thermodynamics of the steam engine, the theory of light, electricity and the turbine.[43] The needs of a military state were met in research upon hydrogen production for ballooning, the manufacture of saltpetre and the organisation by French chemists of soda and gunpowder manufactures. With its own distinguished journal and its training links with the Ecole Normale, the Polytechnique represented the professionalisation of the scientific enterprise in France.

Phase four, during which French science regained its European dominance, was associated with the founding by the Directory of the Conservatoire des arts et métiers, with new industrial exhibitions and lycées, and the formation in 1802 of the Société d'encouragement pour l'Industrie nationale de la France.[44]

The specialist literature yet debates whether science under the revolution, under Napoleon (from 1799) and under the imperial system (from 1804) benefited or suffered from the vagaries of state interference.[45] Bearing in mind the major features of the British scientific enterprise during these years, we might suggest four generalisations.

First, the post-revolution educational institutions produced such scientists as Gay-Lussac (1778–1850), Thenard (1777–1857) and Fresnel (1788–1827), individuals whose creative work was to maintain France in its leading position in chemistry and other sciences for another 30 years.

Secondly, the war years promoted a state-instigated search for strategic supplies of such products as salt and sugar, and thus encouraged and rewarded laboratory research. However, the work which resulted in the most significant commercial breakthrough, that of Nicholas Leblanc (1742–1806) in 1789–90, originated prior to the revolution itself in an *ancien régime* programme for making sal amoniac from urine and salt and soda from salt, and was transferred and modified as best technique not in France itself, but in Britain. (See Chapter 3.) Perhaps a greater impact of the war economy was to instill a bureaucratic character to French science and an imperial (rather than an industrial) direction to research programmes.[46]

Thirdly, whereas British science was incorporating a wider range of social groups at both the creative and 'audience' levels, in France the payment and politicisation of scientists created a new elite, a scientocracy. Although the new institutions rewarded merit, those who actually succeeded in Napoleonic science had little *need* to disseminate their knowledge to the 'public' or to apply it to industrial purposes outside of imperial pursuits. Except for perhaps one very short period, the institutions of French scientific research and discourse remained exclusive.[47]

Finally, the fact that science was in many ways formally valued (i.e. paid for) meant that it increasingly resembled any other *private good*. When knowledge is priced through generous rewards, patents, salaries and pensions, it is less likely to be freely disseminated. The 'audience' for science becomes the state itself. Thus, French science could be creative and professional as well as elitist and independent of the demands of the growing industrial middle classes.

2.3 EIGHTEENTH-CENTURY IDEA-MERCHANTS: THE TRANSFER OF SCIENCE

For many years of the eighteenth century France and Britain were at war. Yet there is really little evidence that this resulted in any blockage in the flow of scientific information between the two nations. During wars, ships which could claim some scientific purpose were immune from hazard. Groups such as the international masons could cut across the lines drawn by international conflict.[48] When France and England were struggling over possession of America, the English chemist Joseph Priestley was swelling in Paris 'in a bag wig, a sword and coloured clothes. So much does a commerce with the gay world induce a man to relax from the strictures of his principles'.[49] Indeed, the channels through which scientific information flowed were of a complex character, and may only be summarised here under a few sub-headings.

Random Walks

In the eighteenth century, individuals mattered, even those of humble origin and status. The London instrument maker, John Cuthbertson, gained much from his residence in Amsterdam during 1769, 1782 and 1793, his fellowship of several European scientific societies and his cooperative work with Van Marum during 1787. Following a course of electrical experiments at Amsterdam; 'On my return to London, after delivering courses of lectures upon electricity, I once more returned to the subject of [the production of metallic oxides by electricity].'[50] At the other end of the social spectrum, when the Conde des Penaflorida returned to Spain from a tour of France in 1746, he immediately set about the creation of a *salon* for mathematical, scientific and political discussion. In 1767 this became the locus for the establishment of the Sociedad Bascongada de los Amigos del País. Both Louis XIV of France and Peter the Great of Russia, through their informal information network, learned of a small New Mathematical School, which had been founded in London in 1673. From 1683 to 1702 the two imperial monarchs were active in the adoption of the new set-up in their own countries, and the impact of this relatively humble establishment may be also found in Berlin, Silesia and Serbia up into the 1780s.[51]

Because of its comparatively open social structure, Britain acted as an *entrepôt*, within which the cultural goods of the idea-merchants were refined and re-exported. In 1741 the Prussian Ambassador in London attended the public scientific lectures of Desaguliers and immediately 'engaged almost all the foreign ministers to be of the party'.[52] J. T. Desaguliers (Oxford MA 1712, FRS 1714) was himself the son of a French Huguenot immigrant, a curator of the Royal Society and a freemason. His public lectures were eventually published in 1724 as *A Course of Mechanical and Experimental Philosophy*, the vernacular problem being solved by alternate learning in English, French and Latin!

Publications

The function of the eighteenth century scientific journal was to monitor and digest, and by abstraction and translation to reduce the time-consuming process of information search. That is, scientific publications became vital to the *assimilation* of knowledge generated in other nations. The *abstract* was vital. Sixty per cent of all eighteenth-century abstract journals were published from Germany, with the prime aim of providing discrete information in simplified or corrected form. Good examples of books with the same function were the 3-volume version of Lorenz van Crell's 12-volume *Die neuesten Entdeckungen in der Chemie* and Christian Adelung's 6-volume *Mineralogische Belustigungen*. Translation of important or timely works often involved simplification or alteration 'to render natural philosophy plain to the meanest readers'.[53] On such a basis Thomas Dale translated and condensed Regenault's *Philosophical Conversations, or a New System of Physics*, and Timothy Dallowe, a former pupil of Boerhaave, translated the latter's *Elements of Chemistry* the second volume of which boasted '227 Chymical Processes'.

Academies and Centres of Learning

By 1785, Benjamin Franklin, though situated in America, was a member of some 20 scientific societies throughout Europe. Franklin's famous course of electrical experiments may be traced to Adam Spencer's lectures, first delivered in the colonies at Boston in early 1744. The traveller William Black reported that Spencer lectured on electricity at the Library Company, State House, Philadelphia on 29

May, and from this date began the 'Philadelphia Experiments', in electricity of Franklin and others, published in 1751. To the Philadelphia Academy, founded in 1749, came the English lecturer D. J. Dove in 1750, and Lewis Evans followed in 1751 and 1752.[54]

Such fairly informal flows between central and peripheral institutions were natural to nations sharing a common cultural tradition. On the other hand, in Russia flows of knowledge were dictated by State policy. Saint Petersburg and its Academy soon became Peter the Great's 'window into Europe'.[55] At the Academy the Swiss, Daniel Bernoilli began a research programme in physics, mechanics and medicine, whilst another, Leonard Euler, promoted investigations into optics, astronomy, hydrodynamics and the calculus of variations. Here was very strong State patronage, although at times poems to the Czar might gain more notice and more roubles than did reports on research. Patronage has many effects.[56]

Mass Movement

The most famous of the mass migrations of the period was that of the French Huguenots. As a result of the revocation of the Edict of Nantes (1679–85) some 80 000 Huguenots settled in England, 50–75 000 in the Dutch United Republic, 30 000 in Germany, 25 000 in Geneva and Switzerland, perhaps up to 10 000 in Ireland.[57] Again, it was in England that the Huguenots were most readily assimilated. The revolution of 1789 witnessed a new surge of French emigrés to Britain, arriving not only in London but at Staines, Petersham, Richmond and Norfolk. Absorbed as a group, the contribution of Huguenot and French intellectuals to the transfer of ideas and information is probably impossible to measure. Certainly, intellectual gatherings and influence are identifiable. In Fleet Street, London, the Rainbow Coffee House served as a forum for Huguenot intellectuals. Therein met De Moivre (mathematician, FRS), Halley, Newton and De Saint-Hyacinth (translator of *Robinson Crusoe*). John Chamberlayne and Ephraim Chambers (both FRS and Huguenots) translated the papers of the French Académie into English and the second published his famous encyclopedia in English in 1728.[58]

The State

All European states were in some way involved in the production and transfer of scientific information. But the character of involvement varied greatly. In England little was done directly, the State being more involved in creating a trading, colonial and legislative environment in which knowledge and technologies could be exploited. However, even in England pensions and rewards could be gained through proven achievements in areas which related to the needs of a mercantile state, e.g. the award of £20 000 to John Harrison in 1772 for his invention of a technique for measuring longitude at sea. More direct state *tutelage* occurred in nations of relative economic or industrial backwardness. Thus the Russian government ran such instruments of scientific and technical modernisation as the Institute of Mining (1773), the Military Academy and Medical Academy (1799) and the Forestry Institute (1803). In contrast to the promotion of aristrocracy at Oxbridge, in Russia the University Statutes of 1787 encouraged meritocracy and demanded the study of natural science.[59]

It may be fairly summarised that the movement of scientific knowledge in eighteenth-century Europe was a highly individualistic, and somewhat random affair. What of the transfer of technology?

2.4 CHANGE AGENTS: THE TRANSFER OF TECHNOLOGY

As with science, neither war nor restrictive legislation could effectively prohibit the movement of techniques across national frontiers. French rulers exerted hegemony over the Spanish economy throughout the early years of the eighteenth century and actively discouraged French entrepreneurs and technicians from introducing new industries into Spain. Although the French established a sophisticated system of controls, this did not halt a flow of expertise into the Spanish woollen, glass, iron, gunpowder, paper and silk industries. In Britain, the formal restrictions on the movement of skilled workers were lifted only in 1825, whilst those on the export of machinery continued in force until 1842. Such legal restrictions did not prevent the inventor John Kay, frustrated by the English patent system, from removing himself and his new spinning machinery to France, where he commenced business by utilising equipment already smuggled out of England by John Holker.[60] It is true that Samuel Slater, who had moved to Rhode Island, America in 1790 after employment at the famous textile mill of Arkwright and Strutt (best technique), could not get blueprints or models of the technologies out of Britain. But he solved the problem, as did so many others, by carrying precise information in his head as *mental capital*, quickly used for the establishment of mechanical spinning in New England.[61] Although there were some similarities with the movement of scientific information, the transfer of technology was also bound up with the expansion of trade and the penetration of foreign capital and financial interests into new, developing regions.

Science and Technology

When considering the eighteenth century, there is little insight to be gained from an artificial separation of interests in natural science and interests in technique – the same publications and the same congeries were involved.[62] In Britain, the Lunar Society of Birmingham has become the best-known of the societies whose research and discussions embraced science and technology and whose personal connections were international in scope. But in many locations, scientists with specific knowledge of industrial processes were concerned in the establishment of new technologies in new environments, and several entrepreneurs were trained in natural science in other nations. Ambrose G. Hanckwitz, FRS, learned of a method of manufacturing phosphorus from urine whilst assistant to the scientist Robert Boyle. In 1710 he established his own business along those lines at the same time as he was chemical analyst to the Royal Society. One Huguenot intellectual, Lewis Paul, invented in England the roller-spring, whilst another, the engineer Charles Lebelye introduced the *caisson* method for building underwater foundations when erecting the first Westminster Bridge. The French scientist de Gensanne supervised the erection of furnaces at Hayange to smelt iron with coke in 1723.

Recent technological breakthroughs were reported in scientific journals almost as frequently as in specific, industry-orientated outlets. The first international publication of R. J. Eliot's invention for smelting iron from black magnetic sand, demonstrated at Killingworth, Connecticut, was in the famous *Transactions* of the Royal Society of London in 1762. Entrepreneurs seeking new methods could be profoundly influenced by the availability of specific scientific information. Thus the knowledge gained by James Keir when translating P. J. Macquer's *Dictionary of Chemistry* inspired him to set up his alkili-manufacturing works at Dudley.[63] It is patently obvious that any analysis of the industrialisation process of the eighteenth century which divorces 'science' from 'technology' is illegitimately modernist and fraught with danger. An artisan inventor of 1750 or 1790 who read a scientific journal also imbibed explicit knowledge of new techniques.

Britain as Receptor

A German born in England, Benjamin Huntsman (1704–76) first produced pure and hard 'steel' in the northern town of Sheffield. The Dutch were essential to the improvement of British drainage and canal systems, the French in civil engineering. M. I. Brunel was a French refugee, and Manchester's engineering community included the Swiss, J. G. Bodmer and, at a later date, the Americans J. C. Dyer and Jacob Perkins. From the seventeenth century, Germans settled at Keswick and elsewhere in order to mine or process copper and other minerals. In fact, the London-based German merchants of the 'steelyard' acted as a local committee for the British interests of the Hanseatic League. Such connections carried not only specific knowledge, but were also important in the spreading of foreign language sources as well as the founding or running of institutions, e.g. the work of the German mathematician John Müller at Woolwich Academy from 1741. Of most importance was the cumulative long-term effect of such early transfers upon the creation of a widespread, interlocking culture of science and technology in Britain. This hybrid culture was the immediate backdrop for the technological changes associated with the industrial revolution.

Transfer Episodes – (1) the Rouen Factory

Transfers of technology into French textiles took the form of a fairly integrated, government-instigated development effort. For instance, John Holker gained government assistance in the form of expenses, salary, subsidies and the granting of a royal 'priviledge' (patent of monopoly) when he established a factory at Rouen, mostly concerned with the weaving of woollen and cotton materials using British machinery and artisanal spinning skills.[64] Of 86 skilled artisan employees at the Rouen works in 1754, the most important positions were held by 20 skilled British workers, who acted with the aid of a full time interpreter. The Rouen factory may be characterised as a *development project* for the following reasons.

The technology and organisation of the factory was directly emulated and established in other places – Vernon, Elbeuf and Pont de 'lArche and a number of French employees received specific formal education at the Rouen factory and later became foremen at works in other places. In addition, several British employees of Rouen were later builders of textile machinery in France and some British employees even set up their own independent factories elsewhere in France – Danuel Hall at Sens, James Morris near Rouen. These were copied by French employees including Pierre Fouguier at Bernay and Thomas Leclerc at Bourges.

The original Rouen factory diversified into other areas of production. For example, in 1763 Holker ran a cotton velveteen branch and his cousin, James Morris, organised a calendering establishment which received royal privilege in 1766. Holker offered advice to the French authorities on how to avoid British restrictions upon machine exporting (by using Rotterdam as an entrepôt) and also stimulated recruitment of skilled workers from Scotland and from Irish regiments stationed in France. When he was appointed by the French authorities as Inspector General of Factories his salary was boosted to 8000 livres per annum!

At a different level of diffusion, as Inspector, Holker obtained government subsidies for other British technological ventures at Bourges, Lyons, Amiens and elsewhere and enticed British technicians to come to France to teach the use of the spinning jenny and the manufacture of spinning machinery. He also encouraged the use of British skills in the establishment of factories for muslin, cloth, cotton, iron, armaments, glassware and chemicals.

Lastly, there was continuity across generations. John Holker Jr (1745–1822), after employment by the French government in England inspecting Hargreaves' and Arkwright's textile technologies, became the French Consul-General at Philadelphia. His son, J. L. Holker (1770–1844) at Rouen and Paris developed a procedure for continuous combustion in the manufacture of vitriol (sulphuric acid) which he established on a commercial scale.

This example suggests that in the eighteenth century the *successful* transfer of technologies involved a complex 'package' of ancillary elements or 'software' – entrepreneurship, specific skills, government finance or patronage, as well as an adequate commercial demand. When such conditions were satisfied, initially transferred technology could *spread* not only from one location to another but from one project or industry to another.

Transfer Episodes – (2) Russia

In Russia, the role of the government was of even greater importance than in France in setting up the chain of circumstances required by the transfer process. It was statist fear of Poland and Sweden which stimulated the employment of key foreign nationals in the leading sectors of the economy, especially in the heavy area of iron manufacture.[65] Perhaps the best illustration of the Russian model of transfer was the industry and machinery partnerships of Charles Baird, Charles Gasgoyne and Francis Morgan during the 1790s.[66] Baird, trained in Scottish industry, was invited to Russia in 1786 in order to establish an iron foundry and machine works.[67] In 1790 in partnership with Gasgoyne and Morgan he began an enterprise which by the turn of the century was producing at an annual value of 130–140 000 roubles. By 1825 the Baird works had constructed over 140 steam engines, and Baird himself had diversified into advising the government on military, civil engineering and machine construction elsewhere in Russia. Apart from direct orders, the State aided the enterprise considerably. Of 133 employees in the machine works at the turn of the century, 67 were State-paid apprentices. In 1804 the College of Mines decided in Baird's favour when he applied to purchase peasants for factory labour. In 1806 Baird obtained permission to employ up to 100 Russian workers from other existing manufacturing enterprises as five-year-apprentices at his works. When Baird's son, Francis, was ennobled by the Russian government in 1852 he employed no less than 442 Russian peasants. In essence, the Baird enterprises in Russia seem to have represented a continuing *technological project*, akin to that of Holker's project in France.

The State

At this juncture we may posit a relationship between economic backwardness and state intervention in technology transfer. The more relatively backward the economy, the more evidence there is of fundamental state activity. Throughout the eighteenth century, most of the 'factories' within which modern technologies were settled were engaged in the production of State needs – dockyards, arsenals, armaments, metallurgy and mining and transport systems. After all, most nations were at war. Russia is the extreme example, but in relatively backward Spain a combination of local State governments, nobility and the Church promoted a programme of technology 'improvement' centred on the Sociedades Economicas de Amigos del País – Friends of the Country. Thus in 1774 the Fiscal of the Council of Castile, the Conde des Campomanes published his *On the Development of Popular Industry*, of which some 30 000 copies were distributed to local officials and clergy throughout the nation. The success of Prussia in Silesia prior to 1786 stimulated a programme of technological transfer under the paternalistic guidance of Frederick the Great. The first stage was policy which encouraged internal improvements – the erection of State shipyards, the establishment of agencies for the marketing of Silesian iron ore, etc. At the second stage, Prussian officials in Silesia organised technique transfers into the region from France, Britain, Belgium and Switzerland.

Foreign Trade

Through trade, machine or chemical technology could be imported directly for military or related needs. Treaties which opened up foreign competition were at times *designed* to promote greater technical efficiency. Thus it has been suggested that the 1786 commercial treaty between France and Britain, which exposed the French market to British cottons, acted to force industrial modernisation in France. Likewise, Russia's penetration of the British market for iron, together with the emergence of coke-smelting and puddling in Britain, induced the several Swedish efforts to revamp that nation's capabilities by direct transfers.[68]

Human Capital: Skills

I have suggested that the transfer of a technology often depended on much more than initial change agents or formal knowledge, mental capital. Work skills were required to settle techniques, to get them going and to maintain, adapt or repair them. We have seen that Holker in France depended on skilled artisans from Britain. One such was Michael Alcock, a metalworker, who from the 1750s acted as a French recruiting agent for the metal industries. For placing British mechanics in the armaments factory at Saint Etienne, Alcock received a fee of 2400 livres. Similarly, in the 1790s English artisans operated mule-jennies at Bernhard's cotton-mill in Saxony.

Starved of European skills, Russia once more offers the prime example of dependency. Artisans attracted to Russia could become registered members of guilds, which under State legislation included formal apprenticeship training. By 1766 the Petersburg craft-guild community contained 777 German and 562 Russian master craftsmen. In the silk-producing area of Astrakhan, foreign artisans soon became entrepreneurs in their own right. So, from Hamburg, in 1764 came A. Verdiev and Jean Roux, who received a grant of 6000 roubles for the erection of sericulture and silk-weaving establishments, as well as grants of land and the obligation to train *Russian* apprentices. They were later successful in selling their thriving business to Russian merchants.

2.5 CONCLUSIONS: THE CONDITIONS OF SUCCESS

We must begin by admitting that highly qualitative and selective material cannot yield very firm results. The overall economic *impact* of even successful transfers remains in most cases problematic. Even highly successful projects may have absorbed as much in the way of scarce economic resources as would alternative indigenous techniques, especially given that the agents of change were not necessarily operating upon strictly commercial principles.

Initial Transfer

Eighteenth-century evidence seems to indicate that the first-stage transfer of technology involved a relatively simple set of procedures.

The relatively effective spread of scientific and technical information and the needs of the State combined to promote frequent transfers. It seems that in nations where the receptor State was particularly concerned (e.g. Russia, Prussia), the transfer process was likely to be swift and to cut across barriers erected by legislation, war or market forces. Given the latter, then transfer was not *necessarily* a primarily commercial concern.

Appropriate Technology

In well-tempered cases, an initial import of machinery and personnel gives way to some capacity for maintenance and repair. During this learning process indigenous skills are revamped and redefined. This may well culminate (e.g. in France in our period) in the ability to *construct* and *modify* the initially transferred archetypes, and this ability may spill over into other regions or industries. The stark *inappropriateness* of the original, imported technique may very well inhibit this linear progression. Boulton and Watt steam engines were speedily transferred to a variety of nations, but long lags in the ability to *construct* efficient copies at a commercial level are evident. Again, early attempts at diffusing better methods of woollen production failed at Guadalajara, in Spain. The scale of production appropriate to the *technology* was far too great for the intended market, and the Madrid warehouses were soon overflowing. At the same time, the goods produced were both inferior to those of the woollen industry of Barcelona and more expensive than the high-grade production of Abbeville.

Diffusion of Transferred Technology

Costly, capital-intensive and skill-intensive technologies were often quick to transfer but slow to diffuse within receptor nations, e.g. Russia. That is, as illustrated in Figure 2.1, the arrow marked 1 tended to be increasing in strength (effectiveness) and breadth (frequency) and shortening (length of time) as the eighteenth century progressed.

But just as technologies improved (block X), became more capital- and skill-using, and more knowledge-specific, arrow 2 became longer (took greater time) and slimmer (is less frequently evident in the historical material available to historians). Tech-

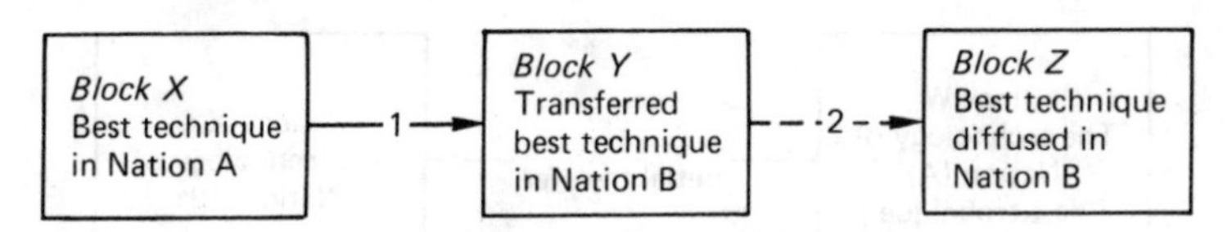

FIGURE 2.1 Transfer and diffusion of technologies

nology may not diffuse effectively (block Z) because it is too costly, supplies an insignificant market in the 'host' nation, or a product which does not fit the 'tastes' of consumers or producers. In addition, diffusion is inhibited if transfers produce products which compete with readily-available internal produce (e.g. that of handicraft industries) or cheap imports, or which utilise raw materials which are not available internally, obtainable at only great cost or risk, or available in forms (e.g. particular coal types or iron ores) which do not lend themselves to effective, commercial utilisation. It would seem that the *technologically* successful textile complex at Yamburg, Russia, inspired by both foreign entrepreneurs and the State, failed commercially (i.e. did not *diffuse*) because of a combination of such factors. The result of diffusion failure is, of course, *enclave* development, nodal points of technological modernity set in hinterlands of traditionalism.

Transfer, Diffusion and the Wider Context

During the eighteenth century in Europe, under what conditions was a transfer of technologies from nation A to nation B most likely to result in diffusion within B? Figure 2.2 charts a feasible series of relationships and conforms to most of the qualitative evidence above.

As in Figure 2.1, it is block Z which is of crucial concern. But our new block Y represents a new complication. Assume that Nation A is late eighteenth-century Britain, whilst Nation B could represent any of, say, France, Spain or Russia. In this scheme the newly-created best technique at block W, filters into the highly-developed institutional network for science/technology diffusion at Y. This internal diffusion process involves *adaptation* of the original W technology, which is now 'available' to Nation B via flows 1 and 2. The capacity of Nation B to absorb the technique at X has been vastly enhanced, because no technology monopoly has been created and

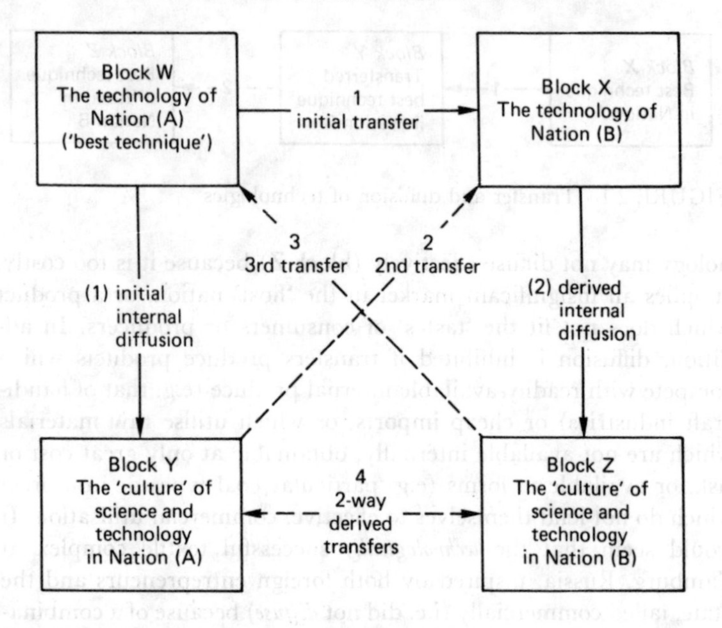

FIGURE 2.2 Feedback mechanisms in technology transfer

advanced technique is now available in several forms, at least one of which may be especially *appropriate* to the economic or other needs of Nation B. But if Nation B (e.g. Russia) possesses little in the way of a comparable science-technology culture (block Z), or if *market* forces in B are insufficient to commercially induce replication or modification, then technology in block X, whilst transferred, may yet remain *enclavist*. However, if Nation B (e.g. France) can boast a reasonable capacity and a strong S–T information net, technology diffuses into Z, and further adaptation may even mean its feedback transfer to Nation A (via arrow 3). Continued feedback occurs when the S–T systems of nations become approximately equivalent, for then transfer arrow 4 develops and a sustained process develops. This schema posits that the existence of *general* scientific and technical information systems helps not only in the diffusion of 'best practice' within that *nation* but also increases the likelihood of a even flow of information or of technologies between leading nations. Cumulative causation then emerges, but nations whose cultures or institutions forbid or inhibit the existence of blocks Y or Z are henceforth left beyond the pale of sustained technological advancement. For purposes of *historical* analysis we might now hypothesise that such a dynamic process was emerging in the years prior to 1780 and accelerated thereafter.

A careful reading will show that this chapter has not in fact attempted in any way to argue that the transfer of scientific knowledge, or of machines and skills, determined the industrialisation of receiver nations in the eighteenth century. Holker may have been relatively successful, but most transfers failed at some level. What we have argued is that the *relative ease* of transfer of techniques or the discrete knowledge upon which they were based or improved, illustrates that the mere *possession* of a technology or set of technologies is not a sufficient explanation of the industrialisation process. Technologies transferred. The nation which *industrialised* was the one which diffused, modified and improved existing techniques, brought them out of their origin in enclaves and into the market place. The explanation of the diffusion process must lie in a complex of cultural, social and economic factors, many of which are addressed in Chapters 3, 5, 6 and 7. The reasons for the *failure* of transfers and diffusion in the nineteenth century are explored in Chapters 8 and 9.

3. Science and Technology in the British Industrial Revolution

3.1 INTRODUCTION – DEFINITIONS

> Closer acquaintance reveals that industrialization in Britain was by no means a single, uninterrupted, and unitary, still less a nation-wide process.
>
> *Sidney Pollard, 1979*

> The economic transformation of Britain from 1780 to 1860 is strange and important . . . The evidence of the experiment is elusive.
>
> *Donald McCloskey, 1981*

IT is possible to unearth the medieval roots of the Industrial Revolution.[1] But for the purpose of this chapter the British industrial revolution is defined as the combination of structural, efficiency and social changes which occurred between the years 1780 and 1850. Removing the industrial revolution from the face of British history, a popular exercise for historians at the present time, requires faith in the relevance and authority of aggregate data (i.e. estimates of national income, the capital stock, and so on).[2] Dampening out the notion of industrial revolution also neglects the social and institutional changes of the period, which are difficult to assume away and impossible to disassociate from fundamental economic change. Finally, removing the industrial revolution may simply lead to boredom.[3] As yet there seems little reason to disallow David Landes his flourish:

> The Industrial Revolution began in England in the eighteenth century, spread therefrom in unequal fashion to the countries of Continental Europe and a few areas overseas, and transformed in the span of scarce two lifetimes the life of Western man, the nature of his society, and his relationship to the other peoples of the world.[4]

This book traces Landes' ramifications into the twentieth century. In the present chapter we highlight some of the major features of the industrial revolution, the links between scientific and technical knowledge and industrial production, and the nature of technological and institutional change in the years prior to 1851. The latter year, the year of the Great Exhibition, represents a symbolic break between the period of original industrialisation and the phase of its greater diffusion within the 'workshop of the world' and transfer to other parts of Europe and elsewhere.

As such, industrialisation is normally associated with the rise of a manufacturing sector which absorbs resources from elsewhere and then releases resources, as technical progress, economies of scale and falling transaction costs (improvements in institutions) yield significant growth in output and even more significant growth in efficiency. Because *later* examples of industrialisation (e.g. in Germany or Japan) were linked to previous development in other economic systems (e.g. movement of capital into Germany from France and Britain, movements of skilled personnel and ideology into Japan from everywhere), the onset of industrial modernity is more marked there than in the classic British example. In Britain, industrialisation was a partial and haphazard process, whose origins may be reasonably traced to the workshop and artisan culture of the eighteenth century and even earlier. Where in Europe or Japan an explicit, often State-orientated *industrial drive*, centred on a fairly narrow range of manufacturing, transport and military industries, then gave way to a broader process of industrial revolution, in Britain the Industrial Revolution was a direct product of many happenings 'from below'.

It would seem that the industrial and commercial sectors of the British economy doubled in their real annual output between 1760 and 1800, at which time agriculture increased its output by perhaps 40 per cent. The share of industry and services in total GNP was around 54 per cent in 1770, 68 per cent in 1801 and 78 per cent in 1841. Such structural change is what we might expect, but it should also be observed that in 1770 Britain was not merely an agricultural nation.[5] What is of great interest is the association between structural change and an acceleration in total economic growth. Growth in real output per annum between 1780 and 1800 occurred at twice the rate of growth of the years 1740–80. But annual average *industrial* growth between 1780 and 1800 was three times that of the years 1740–80.[6] Whilst agricultural production almost certainly grew in efficiency in the early eighteenth century, the accelerating efficiency of the economy after 1780 was indisputably associated with the growth of manufacturing industry and its related service sectors, e.g. transport.[7]

We would argue that the very partiality of the industrialisation process, centred as it was on certain regions and certain products (outstandingly, cotton textiles and metal production and working), testifies to its importance. If the fast growth of output and efficiency in limited parts of a system can significantly effect *aggregate* measures of the system's performance, then this is a statement of the power of the new processes and organisations over that of the old. The industrial revolution in Britain was characterised by fast growth in certain sectors of the economy: measurable increases in the *efficiency* of those sectors, which to an extent spread into other industrial, service and, finally, agricultural areas; and an introduction of both new technologies and new institutions which together created most of the efficiency improvements, the latter arising not so much from the immediate impact of initial, new 'breakthrough' innovations (e.g. the steam engine, enclosures, the factory) as from the impacts of a longer period of adaptive diffusion.

Sixty years ago H. L. Beales noted the 'smoothing' effects of aggregation:

> It is sometimes maintained that a quantitative examination of the leading features of the period of so-called revolution tones down the high colours and produces a softened picture . . . But if it is argued that the social and economic changes of the industrial revolution were in any final sense changes merely of degree and not of form, changes of quantity not of quality, one must dissent.[8]

But it is not simply the 'social' or institutional which is neglected. With selected aggregations, the regional and sectoral bias of economic change is also downplayed. Growth in the aggregate may appear less than breathtaking, but the explanation of such growth, and the reasons for its acceleration or its spatial diffusion at a *later* date (say, in the years 1840–80) may lie in the quite surprising transformations exhibited in a cluster of interrelated industries. Here, at the levels of 'cause' and 'process', change may, indeed, be revolutionary.

As Clapham claimed, and as Pollard has demonstrated more recently, the growth of manufacturing industry did not sweep over the entire face of Britain.[9] Because certain regions and their industries possessed specific advantages of natural and human resources, location or access to transport, communications and markets, change occurred at nodal points on the map. Because of institutional backwardness and resource immobilities such nodal growth did not automatically 'spread' throughout the existing, complex agrarian, commercial and industrial hinterlands.[10] Industrialisation was not osmotic. The interfaces between modernity and backwardness were not semi-permeable membranes but were often barriers constructed out of fractious materials – accepted and entrenched ways of doing things, lack of information and scarcity of resources. Commercial institutions were by no means always instantly available to move surplus finance from one advanced region to another in 'need' of growth. At the same time, the initial advantages of particular regions (e.g. London or Lancashire) were increased as skills and physical resources moved into and built up within them. Indeed, it has been argued that the early industrial revolution (circa 1780–1820) involved an increase in economic differentiation between regions in Britain and that this in turn impacted widely over a range of social and institutional processes.[11] A Marxian proletariat was observable in some areas, but not in others. Political radicalism arose in Birmingham, but that city could not convince the country.

All too often, the continuing existence of small workshops and hand labour have been seen as relics or anachronisms. In contrast, we might hazard that the British industrial revolution was characterised by a three-sector economy. These sectors are perhaps best thought of as being separated on the grounds of differences in output and productivity growth rates, rather than in purely regional or final-product terms. A final product may be the result of a series of discrete processes involving both the factory and the workshop, steam power and hand power, skilled and unskilled labour.

The fastest-growing sectors of the economy (cottons, metals, engineering, transport) were all growing in efficiency – costs of production per unit of output were falling – at a greater rate than in other industries. These were the sectors more likely to exhibit the phaseology of a period of resource inputs from outside the industry (e.g. from trade or land or from other manufactures), followed by a release of resources due to efficiency increases. New institutions, (the factory) and new technologies, (the steam engine) were most clearly associated with the development of such a leading sector complex. A second industrial sector, very much concerned with manufacturing, remained at a smaller scale of organisation (this being distinct from their increasing size of output *in toto*, e.g. building, leather). Evolving in various orgnisational forms, from putting-out, to artisan workshops to proto-factories, this sector employed the great variety of artisanal skills which had accumulated over a considerable period, but which were transformed by circumstance.[12] This was a labour-intensive sector, whose size and importance may well have meant that, overall, technical change in the industrial revolution was labour-using, not labour-saving. Many improved tools increased rather than replaced the demand for skills. Through subcontracting into this sector, labour-saving machinery in factories created a demand for labour in workshops or cottages. Factory spinning required, at some times, domestic and workshop weaving. At the cutting edge, coalmining may have passed through little change, but the Savory and Newcomen engines allowed deeper shafts and demanded a new organisation of labour. Much new equipment (that for stamping of metals, riveting equipment in shoemaking) was employed in workshops, not factories.[13] The third sector of the economy – agriculture – had already yielded efficiency increases to the economy in the eighteenth century, captured further improvements through organisational changes until the 1840s, and from then was the recipient of both chemicals (fertilisers) and machinery (reapers) from the industrial sectors. All three sectors were 'serviced' by industries ranging from banking to transport and therefore themselves varying greatly in organisation and technique.

Fixed capital did not loom large in manufacturing.[14] Several historians have argued that the Industrial Revolution must have been associated with an acceleration in the rate of investment or capital formation.[15] But within our framework, the financing of new forms of production could obviously arise from transfers of existing resources between regions and industries or between sectors. When large capital was required, as in the scaling-up of factory manufactures in the 1830s or the railway boom of the 1840s, it appears to have been in ready supply as a result of the high profits generated by industrial production itself. New techniques could be brought into production as part of replacement and conversion investment, small equipment changes might be financed out of income or the selling of stocks. In addition, much of technical change (especially in sector 2 above) was 'disembodied', or required very little setting-up capital, and great efficiency increases arose from organisation change. Thus the rate of technological change, in reality a function of the diffusion of existing best techniques – was by no means necessarily dependent upon the rise of capital formation. Even in mid-nineteenth-century Britain, most fixed capital lay in areas quite divorced from the equipment, machinery or power needs of the manufacturing sectors, i.e. most capital was held as structures (bridges or buildings) or stocks.[16]

The evidence is partial but reasonably clear. Chapman's revision of Colquhoun's estimate of the capital value of *all* cotton mills in Britain in 1787 yields a figure of £500 000.[17] Fifty per cent of such capital was located in Lancashire, Derbyshire and Nottinghamshire. James McConnel and John Kennedy set up their textile machinery establishment with an initial capital of £1700. Honeyman shows that *conversion* of buildings and equipment from one product (e.g. silk) to another (e.g. cotton) reduced the establishment costs in new industries substantially.[18] A small-scale cotton mill (jennies, mules, capstans and building) of the 1780s might require £1000 to £2000, and even a purpose-built factory of the Arkwright water-frame type with 1000 spindles might only cost £3000. As such, fixed capital loomed small. Samual Oldknow's original mill of 1787 required £90 to convert, £57 for machinery (durable capital) and £261 in stocks; i.e. 65 per cent of the total investment was in the form of circulating capital. Chapman's analysis of the insurance policies of 1000 early textile entrepreneurs shows that (a) much capital was converted from one function to another (which added to its 'value') and that setting-up required modest expenditures only; (b) most entrepreneurs held the bulk of their assets in houses, buildings, land and tenaments, not as machinery; (c) most building was of a semipermanent type, cheap and versatile.[19] Much of the fixed capital investment associated with the Industrial Revolution occurred in canals, mines and roads at a very early 'prerequisites' stage – by 1790 some £1.5 million had been spent on the canal system alone – or was absorbed in the massive railway investments at the end of the Industrial Revolution period. Although individuals at times faced financial embarrassment the Industrial Revolution as a system was not short of funds.

3.2 INVENTION, TECHNOLOGICAL CHANGE AND INDUSTRIALISATION

> The leading characteristic of English inventions has not been their ingenuity.
>
> *Alfred Marshall, 1919*

> Few countries are equal, perhaps none excel, the English in the number of contrivances of their machines to abridge labour.
>
> *J. Tucker, 1757*

Donald McCloskey has concluded that 'Ordinary inventiveness was widespread in the British economy 1780 to 1860 . . . indubitably Britain from 1780 to 1860 ate a massive free lunch.'[20] The growth of the economy is not explainable in terms of increased amounts of physical inputs which went into the production process. Nick Crafts views 'inventive activity', 'better machines', and 'improvements' in organisation as the keys to the growth of the British economy in the later eighteenth century.[21] Appendix I on pages 304–6 shows that many of the 'breakthrough' technologies of the Industrial Revolution were released in the years before 1780. However, the thrust of the argument must be that it was from around 1780 that the economic impact of such major technologies was felt, and that this lag is a measure of the time taken for inventions to become adapted and adopted innovations, a process analysed in Chapter 1, sections 1.2 and 1.4; and that several important inventions in sectors 1 and 2 of the economy emerged after 1780.

The years 1780–1850 were literally crammed with incremental improvements which underwrote the diffusion of both older and newer breakthrough technologies, and the Industrial Revolution was replete with organisational changes, from the diffusion of the factory to the better establishment of property rights, to the creation of an informational system, all of which abetted the cost-reducing impact of technological change.

Using admittedly 'arbitrary' estimates of the shares of profits, wages and rents in the economy, Crafts estimates that 30 per cent of the growth rate of the economy between 1740 and 1780 came from increased productivity, whilst 80 per cent of the faster growth rate of the years 1780–1800 may be so explained.[22] Prior to this, most phases of economic growth had in all probability arisen from increases in the availability of the conventional factors of production organised within improved commercial and social institutions. But from 1780, by far the greater part of the fall in production costs of cotton textiles derived from technical change rather than from a fall in the price of factor inputs.[23] Similarly, technical change was fundamental to cost reduction in various parts of the chemical industry, transport and metallurgy.[24] McCloskey's approximate calculation is that, of the total growth in productivity of the years 1780 to 1860, cotton textiles contributed 15 per cent (all textiles 23 per cent), the major transport sectors (canals, railways and shipping) contributed 19 per cent, and agriculture 10 per cent. The other areas of the economy, mostly but not entirely captured in our notion of sector 2 above – contributed 46 per cent of the nation's growth of productivity during the Industrial Revolution.[25] Thus efficiency improvements were of great importance in the 'non-steam' part of the economy, which employed an array of traditional and revamped skills, workshops and factories, rural and urban production facilities. It was this phenomenon, technological progress over a wide range of processes and products, which yielded the 'change in quality' within the process of economic growth. It is this which must be emphasised by those historians who wish to retain the 'industrial revolution' as a major organising principle in their work. Until stark regionalism was reduced during the 1830s and 1840s, the process of economic change may be visualised as a series of 'neighbourhood effects' of the Hagerstrand type outlined in Chapter 1, section 1.5. Regional efficiency increased as a result of the increasing comparative advantages in natural resources, infrastructure and institutional endowments. As Pollard emphasises, regions soon developed their own 'distinct technological traditions', but because regions varied and encompassed all sectors of the economy, 'ordinary inventiveness' became the hallmark of the British economy.

Most historians acknowledge the ultimate power of incrementalism. Landes' concern with spectacular change does not inhibit him from also emphasising that 'the many smaller gains were just as important as the more spectacular initial advances'. At the heart of bringing great artefacts into commercial production was incremental invention, in 'the articulation of parts, transmission of power, and the materials employed'.[26] Beales had noted the same thing: 'The revolution in industry comprised a series of improvements rather than a series of startling inventions. The elaboration of large scale manufactures necessitated . . . the successive conquest of processes and parts of processes by machinery'.[27] Raphael Samuel argues that the major process of mechanisation itself 'advanced by small increments rather than by leaps'.[28]

Given that many major technologies were also available to other nations at that time and that they transferred between nations, historians often resort to a 'demand' argument when attempting to separate British experience from that of the Continent.[29] Whilst increases in demand for key products – especially that stemming from depletion of old natural resources or from disjunctions within an improving sector, as with weaving versus spinning – may indeed have led to a search for new applications, this argument is an uneasy one when applied to the total economy. It ignores the fact that applications, when sought, are seldom found, and it fails to properly acknowledge that such demand was rising in a wide range of locales throughout Europe, and, indeed, elsewhere. At several points David Landes begins to erect 'supply side' arguments for British technological change. The social structure may have allowed greater opportunities for social mobility and experiment, may have offered greater rewards to invention, or may have involved less in the way of institutional barriers (e.g. guilds). That is, the social system may have encouraged or at least permitted technological change.[30] As with Sidney Pollard, the social and the regional are then combined in the hypothesis, 'Was it not only that the English atmosphere was more favourable to changes, but also that special experience in certain areas provided unique facilities for training?'[31] Drawing on the early results of Musson and Robinson, Landes points to specific micro-environs of skill, adaptation and innovation, which favoured further change.[32] But in the end 'the pressure of demand . . . called forth the new techniques in Britain'.[33] Similarly, Crafts suggests that demand may have been a central element in prompting the process whereby better *knowledge* of technique was translated into better technologies.[34] But knowledge, or information, may be regarded as a factor of production (one which expands with use), the supply of which itself must be explained. Although we have hints and useful indications, it seems that we still need to know why it was that any increasing demand (presupposing this) did not result in a high price for both knowledge and goods and a *lower* rate of economic growth. What if the supply of both information and inventions was *particularly* forthcoming in Britain? And what if this increased supply had little directly to do with the needs of large capitalists or the rate of investment?

3.3 CONNECTIONS: SCIENCE AND TECHNOLOGY

> Invention is mainly a practical problem of science and engineering technique
>
> *Friedrich Rapp, 1982*

> If anything, the growth of scientific knowledge owed much to the concerns and achievements of technology.
>
> *David Landes, 1968*

> Science had little direct influence on the development of technology in the early stages of the Industrial Revolution, even though it was later to become all-important.
>
> *Colin Russell, 1982*

The three quotations above appear to define an arena of real disagreement. In model-building vein, Rapp posits a causal nexus between science and technology, one which has been energetically explored over a host of events by Musson and Robinson.[35] Russell negates the claim whilst Landes joins others in reversing it.[36] Taking the long view, Kurt Mendelssohn quite vehemently asserts that 'the industrial revolution was not hatched in the workshops of artizans but in the meetings of learned societies'.[37] Is it possible to reconcile the seeming confusion? One possibility is that within certain fields of endeavour, invention related to the formal application of scientific knowledge. Thus Russell, a chemist by training, gives a good summary of his claim that at least the chemical industry owed a considerable debt to science, a statement barbed with detail in the work of Haber.[38] Contrariwise, the older claim that natural science knowledge was fundamental to the development of the improved steam engine is now disavowed, although this is not true of its later history.[39]

Another possibility is that the availability of specific scientific and technical information was important in creating the host of incremental and adaptive innovations which in many instances followed upon important inventions, as in section 3.2 above. Although it may prove 'extraordinarily difficult to trace the course of any significant theoretical concept from abstract formulation to actual use in industrial operations', this may not be the most representative link between knowledge and production in the British industrial revolution.[40] The availability of packets of scientific and technical information, transmitted through a network constructed of scientific institutions, journals and associations may well have been an important factor in contributing to empirical technical research and invention, and the gross availability of such information amongst diverse social groups and between diverse localities may be what distinguishes the British case from that of any other.

Amongst the several 'intricate force conjunctions' of the industrial revolution, the availability of specific scientific and technological information and the pattern of its diffusion has been neglected by economic historians.[41] Whilst most analysts now recognise the expansive power of commercial information, they yet fail to feel the similar power of organised knowledge more generally. As Thackray has suggested, even that knowledge which is later shown to be incomplete or ill-constructed might have generated insights useful to production.[42] Adam Smith, who above all emphasised the generation of innovative skills in a learning-by-doing-and-specialising framework, also acknowledged the influence of more formal, extraneous knowledge in the process of technological advancement. Smith recognised feedback mechanisms between work and extraneous knowledge:

> There is scarce a common trade which does not afford some opportunities for applying to it the principles of geometry and mechanics, and which would not therefore gradually exercise and improve the common people in those principles, the necessary introduction to the most sublime as well as the most useful sciences.[43]

The juxtaposition of industry or work-specific knowledge and formal knowledge and information may, then, have been a very common feature of the natural history of British industry, to be found to a far lesser extent in any other nation at that time.

In 1799 William Henry of Manchester published his lectures on 'chemistry and its application to Arts and Manufactures'.[44] In this book Henry was at pains to emphasise the value of formal knowledge of chemistry as a prelude to the advancement of the chemical and other industries. His grounds were that,

> a talent for accurate observation of facts and the habit of arranging facts in the best manner, may be greatly facilitated by the possession of scientific principles. Indeed, it is hardly possible to frame rules for the practice of a chemical art [application], or to profit by the rules of others, without an acquaintance with the general doctrines of the science. Were chemical knowledge more generally possessed, we should hear less of failures and disappointments in chemical operations; and the artist would commence his proceedings, not as at present, with distrust and uncertainty, but with a well-grounded expectation of success.[45]

But Henry then went beyond this general statement to claim, with such writers as Dugald Stewart (*Elements of the Philosophy of the Human Mind*) that creativity only arose from a merging of work-specific, practical knowledge with more generalised, abstracted knowledge. For Henry, what was happening to industry in Manchester could not be understood other than in terms of a general diffusion of the research findings of such foreign chemists as Scheele and Berthollet. The advantage of formal knowledge was that the investigator could use it to predict the value of a large-scale operation from the basis of small-scale experimentation. So:

> If the Chemist fails in perfecting an economical scheme on a large scale, it is either because he has not sufficiently ascertained his facts on a small one, or has rashly embarked in extensive speculations, without having previously ensured the accuracy of his estimates.[46]

Chapter 2 has suggested that information about science and technology was diffused both more widely (spatially) and more broadly (socially) in Britain than in other nations during the late eighteenth century. Sections 3.4 and 3.5 below show that the years of the industrial revolution involved a far greater increase in the production and distribution of useful knowledge. Prior to 1851 no other nation built this kind of an information system.

3.4 ACTIVISTS – THE SOURCES OF SCIENTIFIC AND TECHNICAL INFORMATION

> We cannot take the stock of knowledge for granted . . . we must have some theory of its production and accumulation.
>
> *Simon Kuznets, 1965*

Although the main locus of industrialisation was England itself, the sources of information upon which inventors and industrialists drew included Scotland and Europe. Table 3.1 below itemises the educational background of 498 notable applied scientists and engineers of the period 1700 to 1859.[47]

Noteworthy is the large number of such individuals who boasted no higher education whatsoever, a feature understandable in terms of the argument in sections 3.5 and 3.6 below. Of the remainder (34 per cent), most had been educated in the protestant universities of Scotland or of Europe. Although a highly partial sample, these figures do at least suggest that most of the well-known applied scientists and engineers of the period did not come from the establishment culture as represented by Oxbridge, that the large proportion of university technicians were from the dissenting religious sects, and the majority of technicians were from relatively humble backgrounds.

From what has gone before, none of this is surprising. England was clearly open to the influence of various sources of knowledge, whose production lay outside the older hegemony of land and church. Chapter 2 has shown that information transferred indiscretely across national borders throughout the eighteenth century. Chapter 4 describes an institutional framework which boosted such transfers in the nineteenth century. Most of such formal scientific and technical institutions were the product of the later nineteenth century. The Industrial Revolution in Britain, therefore, depended for its information base on a number of less formalised mechanisms for knowledge diffusion. In that sense, the Industrial Revolution was a product of the eighteenth century.

TABLE 3.1 University education of 498 applied scientists and engineers

Period of birth	*None*	*Oxford and Cambridge*	*Edinburgh*	*Other Scottish*	*Foreign*	*Total*
1700–79	106	15	27	8	3	159
1780–99	68	5	9	2	2	86
1800–19	73	13	18	5	3	112
1820–39	48	5	8	7	16	84
1840–59	34	12	4	3	4	57
TOTAL 1700–1859	329	50	66	25	28	498

(A) Transfers: Europe as a Source

Chapters 2 and 5 explore in some detail the role of Europe as a source of scientific and technical information. Basically, the Industrial Revolution itself increased the attraction of England as both a place of employment and as a social setting for knowledge production. The early function of Britain as a shelter for intellectual and religious refugees was transformed as industrialisation opened up employment to skills and professional expertise of a new kind. The availability of capital as well as an established system for the registration and protection of intellectual property rights meant that Britain became host to a number of foreign patentees and businessmen.

(B) Intellectual Colonisation: Scotland as a Source

The role of Scotland as a supplier of human capital and information to the English industrial revolution requires greater elaboration. In 1784 a group of Scottish schoolmasters penned a memorial which advocated new legislation relating to their payment and status. In doing so, they captured the image of Adam Smith's North Britain:

> Scotland, or North Britain, struggles with many natural disadvantages. The climate is cold, the sky seldom serene, the weather variable, the soil unfruitful, the mountains bleak, barren, rocky,

> often covered with snows, and the whole appearance of the country very forbidding to strangers; yet, by an early attention to the education of youth, to form good men and good citizens, has been always deemed an excellent nurse of the human species, and furnished, not soldiers only, but divines, generals, statesmen and philosophers, to almost every nation in Europe.[48]

A firm historiography has established the notion of an intellectual meritocracy in the eighteenth century, the tip of which has been termed the 'Scottish Enlightenment'.[49] Innovative parish schools, numerous charity endowments, and universities sensitive to both the 'new philosophy' and to their immediate urban environs have been interpreted as the bases of a social and geographical mobility based on human capital formation. More recently, Houston, Smart and the Bulloughs have provided more systematic material which combines to question the intellectual meritocracy theme.[50] Thus, the social-class background of the Bullough's sample of Scottish 'achievers' is found to be very similar to that of the English eighteenth-century achievers as analysed earlier by Joseph Schneider.[51] Possibly the earlier conviction about an 'achievement from below' mechanism stemmed from the high visibility of what was in fact a small minority of humble highlanders who entered the universities, combined with a failure to recognise that most of those of humble background who did enter the universities did not obtain a final status beyond that of the lower ranks of the law, parish schoolmaster or parish minister.

What, then, explains the flow of North Britons and their influence into England during the Industrial Revolution? Of course, there were obvious pull-factors, but the cultural and institutional effects of socio-economic peripheralisation *within* Britain may lie at the heart of the matter,[52] as in British colonies, secured at a later date, a relationship of metropolis and province may give rise to a philosophy of 'improvement' in the colonised region, in effect an ethos or ideology of emulation and response to economic subservience deliberately created by a national elite. This may be an important dynamic explaining the drain from the North. Thus, as Nicholas Phillipson has stressed, the origins and sustenance of the Scottish enlightenment lay less in the 'democratic intellect' than in an ideology of improvement.[53] The institutional and psychological ramifications of the Act of Union included a search for economic

TABLE 3.2 The Scottish production of British medical graduates, 1701–1850

Period	*All graduates Practising*	*% of graduates Scots-Trained*
1701–50	1408	29
1751–1800	3040	85
1801–50	8291	96

growth, civic renewal and rejuvenation of status within Britain, rather than a movement to political independence.[54] The growth and character of the University of Edinburgh, the centre of intellectual change, may be interpreted as an institutional expression of the needs of a relatively small Scottish elite.

During the Edinburgh-based enlightenment the city was, in fact, suffering a loss of population and of economic power. In Edinburgh intellectual and cultural capital was in oversupply, and this led to a flow of highly-skilled personnel into England.[55] To this we should add the reverse flow of English dissenters, seeking a protestant, scientific and utilitarian education in the North.[56] The natural philosophy, mathematics and medicine which they studied was imbued with the spirit of improvement. The colleges at Aberdeen rejected all elements of scholasticism and disputation in favour of 'employing themselves chiefly in teaching those parts of philosophy, which may qualify Men for the most useful and important Offices of Society [with teaching] founded on an induction of particulars'.[57] Oxford and Cambridge, hampered by their own histories, could simply not compete with this, nor were the English dissenters allowed entry to the establishment institutions. Medical training became the great vehicle of scientific and technical information-gathering. Robb Smith provides the details of the importance of Scottish training in England shown in Table 3.2.[58]

The impact of Scottish training or of Scots professionals and intellectuals in England has yet to be fully investigated. The list of Scots MDs who had a significant role in disseminating scientific and technical information in London would include Robert Willan, David Pitcairn, Matthew Bailie, Henry Halford, James Hope, Neil Arnott, James Copland, Thomas Bradley, A. J. C. Marcet, James Curry and George Birkbeck. This is an impressive list which omits William Hunter, who began advertising as a disseminator of science in the *London Evening Post* in 1746 and who planned and petitioned for a technical and scientifical academy in the English metropolis. Again, medical men trained in Scotland exerted a quite disproportionate effect in setting-up scientific and technical information systems in the English provinces, e.g. Rotherham at Newcastle or Currie and Bostock at Liverpool. James Hutton, James Keir and John Roebuck were all taught by Andrew Plummer at the University of Edinburgh, and this triumverate had an undeniable impact on English industrialisation. James Watt and his 'steam intellect associates' were Scottish, as too were Thomas Telford, John Macadam, David Mushet and J. B. Neilson. Scotsmen of note in the English textile sector included John Kennedy and George and Adam Murray. The list would be very long.

(C) Reproduction: Social Marginality and Urban Information Networks

In 1947 T. S. Ashton had little doubt about the close relations between information, invention and entrepreneurship in the British Industrial Revolution:

> Physicists and chemists, such as Franklin, Black, Priestley, Dalton and Davy, were in intimate contact with the leading figures in British industry: there was much coming and going between the laboratory and the workshop, and men like James Watt, Josiah Wedgewood, William Reynolds, and James Keir were at home in the one as in the other. . . . Inventors, contrivers, industrialists, and entrepreneurs – it is not easy to distinguish one from another at a period of rapid change – came from every social class and from all parts of the country. . . . Clergymen and parsons, including Edmund Cartwright and Joseph Dawson, forsook the cure of souls to find out more efficient ways of weaving cloth and smelting iron. Doctors of medicine, among whom were John Roebuck and James Keir, took to chemical research and became captains of large-scale industry. Under the influence of a rationalist philosophy, scholars turned from the humanities to physical science, and some from physical science to technology.[59]

It is of some importance to draw a distinction between an interest in science and technology and the resulting supplies and flows of information on the one hand, and the 'demand' arguments on the other. Quite simply, information did not become available merely because it was in *economic* demand. The price of information did not rise, if anything it fell. The alternative proposition is that social forces generated information, that the scientific and technical networks were constructed for social purposes, and that this represented an information pool (stock) and flow which could be drawn upon by producers of goods and technologies. Musson and Robinson have demonstrated the strength of the networks and the frequency of information–production interactions.[60] Arnold Thackray has argued, cogently, that the composition of such networks was socially inclusive because they represented a locus for the regional status-gains of marginal individuals. Individuals entered such networks for a multitude of reasons, but the particular attraction was that they captured 'a particular affinity between progressivist rationalist images of scientific knowledge and the alternative value system espoused by a group peripheral to English society'.[61] Thackray is subtle, and almost certainly accurate. Entry into such scientific and technical information networks gave dissenting and radical provincials a location within which and from which they 'affirmed their commitments to high culture, announced their distance from the traditional value systems of English society, and offered a coherent explanatory scheme for the unprecedented, change-oriented society in which they found themselves unavoidably if willingly cast in leading roles'.[62] Even such small centres as Derby, so proximate to the earlier factory industry, boasted viable institutions for information dispersal, from the Derby Philosophical Society of 1783, through the Derby Literary and Philosophical Society of 1808 to the Derby Literary and Scientific Society of the 1840s.[63] Urban centres of the new industries or services were also the most involved in the provincial information network. But in all urban centres, the activists were intent upon information dissemination and vehemently publicised its attainment:

> from this establishment more learning has been diffused throughout the town, than could ever have found its way along the private channels of a few rich individuals. Into this room, as into a focus, I would endeavour to collect every ray of intellectual light, and from this place, as from a centre, I would have it emanate to every point of the neighbourhood.[64]

The civic science of the 1830s and 1840s directed enquiry and information-gathering towards the service sectors. Indeed, activists explicitly forged correlations between civic science and reform in municipal government and utilities.

> All questions relating more particularly to the Public Health, such as the removal of nuisances, and the prevention of deposits or works which may vitiate the atmosphere in a way that may be detrimental to life . . . and other sanitary objects, as my mind may be supposed to be professionally habituated to them, will naturally be expected to engage my peculiar care. . . .[65]

In summary, the supply of scientific and technical information both to and amongst English industrialists appears to have derived from a combination of three sources. England, in a sense, derived its 'free lunch' (McCloskey above) from three distinct social systems. Chapter 2 has shown that much of the knowledge stock generated in Europe was under the auspices of the State. In Scotland, economic peripheralisation led to an elitist ideology of improvement which generated institutions and attitudes which, through one channel or another, flowed into the English industrial system. Finally, in England itself information was disseminated within a provincial, urban 'culture' which was in many ways at distinct odds with established society. Neither the State nor the English elite were, at that time, prime movers of the new information system. The flow of information from all sources was quickened by the proximity of and improved communications between urban centres.

3.5 AUDIENCES – THE DIFFUSION OF SCIENTIFIC AND TECHNICAL INFORMATION

> A thing not yet so well understood, is the economical value of the general diffusion of intelligence among the people.
>
> *John Stuart Mill, 1848*

Of some 1020 associations for scientific and technical information operating in England and Wales when Mill was writing, 72 per cent were specifically designed for members of the middling class, artisanry and labour aristocracy.[66] A very minimal membership figure for this latter group of institutions is 200 000 individuals, but there is every reason to argue that this estimate may be doubled.[67] In industrial and commercial centres the facilities for information-dispersal were highly developed by the 1830s but had been increasing in both size and social inclusiveness from the early years of the century. The city of Liverpool, the largest in Britain after London and Dublin in 1851, provides a good example. The academies and discussion groups of the mid-eighteenth century, which emphasised lectures and experiments on 'Natural Philosophy' gave way to more formal associations. By 1780 Richard Postlethwaite could insert a very wide range of scientific subjects into the regular teaching of his School for Mathematics: astronomy, conic sections, pneumatics, hydraulics, geometry, altimetry, mensuration, gauging, navigation, dialling, gunnery, architecture, algebra, chemistry, mechanics, calculus, optics, hydrostatics and magnetism were all taught in addition to the material at his regular high school.[68] The evolution of evening and part-time facilities illustrated the commercial focus of the new developments. The cost of scientific, commercial and technical instruction was falling throughout the period 1750–1800. In June 1804 the local success of the applied chemical lectures of John Stancliffe was such that in August a Liverpool School for the Encouragement of Arts, Science, Trade and Commerce was instigated by the broker and instrument-maker Hugh Williams. Designed to sponsor and finance inventions, annual admission was set at 10/6d, this allowing each member 'Authority to sign 10 notes of admission to Mechanics'.[69] This was the effective backdrop to the formation of the Liverpool School of Design (1812), Liverpool Marine School (1815), Liverpool Apprentices and Mechanics Library (1823), Liverpool Mechanics' Institute (1825) and the host of scientific and technical forums which then followed. By 1850, some 39 of such forums existed in Liverpool, some of which claimed audiences of over 500 for their lectures and discussion meetings.[70] Associations were especially vital in sub-regions of manufacturing. Thus the Liverpool Northern Mechanics' Institute was formed to provide for

> the vast number of the working class congregated in this Northern part of the town and the numerous establishments connected with steam machinery in all its branches. . . . we must admit that such an institute is most particularly needed.[71]

To many such forums admission was free for apprentices, whilst fees for artisans might range from 4/- to £1 per annum. In others, as in the Clarence Foundry Mutual Improvement Society or the Edge-Hill Mechanics Club, membership was mostly confined to groups of skilled workers from particular factories or workshops at perhaps 1d or 2d per meeting. Almost all such associations had library, lecture or classroom facilities, whilst several provided museums and laboratories. All of the early and some of the later associations for artisans and other industrial workers were instituted by the dissenting intellectuals and industrialists of the scientific and technical networks.

A case study such as this may be multiplied many times across the face of urban Britain, and there is much evidence of the diffusion of information into areas of low population density also.[72] By the 1830s the membership of the Doncaster scientific and technical association was composed of approximately 46 per cent professionals and service occupations, over 10 per cent merchants and manufacturers and nearly 45 per cent small businessmen, skilled tradesmen and other working-class groups.[73] There is no doubt that by the 1820s and 1830s the early networks of dissenting academies, lecture societies and discussion groups had spawned a far larger information system, one which incorporated a goodly range of social groups.

3.6 APPLICATIONS – THE SOCIAL LOCATION OF TECHNOLOGICAL CHANGE

> The Industrial Revolution did not, in fact, develop in an unlikely place. . . . New inventions were, so to speak, in the air, the environment was favourable to industrial progress.
>
> *H. L. Beales, 1928*

Such trends appear of greater importance after examination of the social location of the culture of invention itself. Eric Hobsbawm has calculated that the 'labour aristocracy' (a combination of the older artisans and the new skill groups in spinning and engineering) represented some 15 per cent of the working class during the Industrial Revolution. W. H. Chaloner has argued for a specific culture of the artisan, forged in apprenticeship, trade clubs, friendly societies, educational institutions and common interests. Table 3.3 illustrates the grouping of the artisan class in evening instruction facilities, some of which provided scientific and technical facilities.[74]

TABLE 3.3 Evening instruction and the artisans, 1850–51 (percentages in brackets)

Categories	*National*	*Cheshire*	*Lancashire*	*ER Yorks & City*	*NR Yorks*	*WR Yorks*
Total scholars	39,783	1643	9687	149	85	7785
Scholars known	28,686	1367	7204	91	53	5627
Artisans	14,405(50)	800(58.5)	3440(48)	61(67)	27(51)	4936(88)
Agric. labourers	6709(23)	139(10)	565(8)	22(24)	11(21)	417(7)
Factory hands	4418(16)	224(16)	2705(37.5)	0(0)	0(0)	25(.5)
Dom. servants	1317(5)	14(1)	124(2)	6(6)	15(28)	171(3)
Soldiers/sea	431(1.5)	0(0)	12(0)	0(0)	0(0)	0(0)
Weavers & knitters	334(1)	187(14)	70(1)	0(0)	0(0)	38(.5)
Lead etc. miners	324(1)	1(0)	103(1)	0(0)	0(0)	0(0)
Clerks & office	251(1)	0(0)	185(3)	0(0)	0(0)	0(0)
Labourers	211(1)	0(0)	0(0)	0(0)	0(0)	0(0)
All others (25)	413(1)	2(0)	0(0)	2(2)	0(0)	40(.5)
TOTALS	(100)	(100)	(100)	(100)	(100)	(100)

Of nearly 29 000 individuals whose occupations are known, 50 per cent were artisans, although this figure reached 89 per cent, for the West Riding, where artisan skills in metallurgy and related occupations were especially concentrated. In Lancashire in particular, the artisan was joined by the factory hand. It was these working-class groups who made up the bulk of the membership of mechanics' institutes and polytechnic societies. As Chaloner stresses, many of these individuals were upwardly mobile and entered the ranks of small businessmen: 'It should never be forgotten that during that Industrial Revolution economic mobility and openings for the new entrepreneur were greater than ever before, and inside the working class the artisans, both of the old and the new groups, were the best fitted to make the most of their opportunities'.[75] Many of these businesses, particularly in such places as London, Sheffield or Birmingham, originated in invention and patenting. In Birmingham, and through their inventive efforts alone, Henry Clay and William Whitemore moved from journeyman status to their own engineering firms, Edward Thomason passed from the trade of buttonmaker to the position of manufacturer, and others moved upwards from journeyman status to engineers, brass and iron founders and merchants.[76] Fairly typical of London was Junius Smith, whose patents for washing, cleansing and whitening cottons and linens became the basis for his large steam washing company at Phipp's Bridge in Surrey.[77] In 1835 the engineer William Newton – himself a major figure in the information system – saw such individuals as working at the hub of incremental technical change, whether this arose in the steam economy or the artisan workshop sector.

> Their province is to improve in detail, to give a finish to the detached parts of the extensive combinations formed by superior minds, and to fill up the chasms that occur frequently in the plans of the greatest inventors. Happily, this class is immense, being spread thickly over the whole body of mechanics, from the manufacturer and engineer down to the lowest workman; such men constitute expert mechanicians, who are never at a loss for expedients for overcoming the practical difficulties of detail that occur in the business, and are perpetually making trifling inventions which they require for immediate application.[78]

A very distinguished engineer, John Farey, made the same point in more detail in his evidence before a Parliamentary Committee of 1835:

> by the operation of patents, the making and using of patent articles (which have merit enough to sell) is multiplied and accumulated into considerable trades, which would never have arisen to any such extent without patents; because no individuals would have devoted themselves to have created such trades, if others could have supplied the demand as freely as themselves. . . . That is the origin of a number of considerable trades at Birmingham, Sheffield and in London.[79]

By such a date the artisan inventors were firmly joined with such activists of the scientific and technical information systems as Birkbeck, Babbage and Newton in pressure-group forums like the Museum of National Manufactures and of the Mechanical Arts, established for the dissemination of technical information and for reform in the educational and patent systems.[80] It was little wonder that in 1829 Farey could give evidence of 'the very expressive nature of the technical language that is used amongst all our artisans; also the established habit that the English have, more than any other people, of associating themselves into bodies and societies to act in concert to effect a common object'.[81]

Table 3.4 goes back into the earlier period of technology diffusion by presenting the social location of 602 patentees of the 1790s.[82] Of the total patentees, 48 per cent were skilled tradesmen. The division of these into five groups is based on the sums of money required to set up as a master in that trade, group A paying the highest amount.[83] This measure of social location has been taken as superior to premiums for apprenticeship, as patentees very often used their inventions to move into independent manufacture and because it is a measure divorced from skill or education *per se*. Many of the small businessmen (group 6) and engineers (5) and a smaller proportion of professionals (4) would have themselves have emerged from a process of apprenticeship. Intuitively, many of the smaller manufacturers would have been in transition from the tradesmen groupings, and would at times fall back into their trades. We may hazard that up to 50 per cent of all patentee-inventors were at some time apprenticed and are likely to have been exposed to the information-dissemination system described above. Table 3.5 is also of some interest in illustrating that 'steam intellect' was not especially locked

TABLE 3.4 The social location of British patenting in the 1790s

Social groups	*Number of occupations in each group*	*Number of patentees*	*% of patentees*	*% of London to total patentees*
1 Nobility and gentlemen	2	138	23	42
2 Manufacturers group A	8	45	7.5	9
3 Merchants	4	25	4	36
4 Professionals	5	35	6	54
5 Engineers/mach. makers	3	56	9	42
6 Manufacturers group B	10	14	2	14
Sub-Total for groups 1–6	32	313	52	37
Skilled tradesmen				
7 Group A	8	18	3	40
8 Group B	11	70	11.5	57
9 Group C	8	37	6	65
10 Group D	33	124	20.5	55
11 Group E	8	25	4	40
12 Others	6	15	2.5	30
TOTAL (1–12)	106	602	100	45

into the larger industrial centres at this time but was located in London and the provincial hinterlands. The culture of invention was geographically as well as socially widespread.[84]

Patents may not be dismissed as unrelated to the growing industrial sector of the 1790s. Table 3.6 shows that patenting occurred across a wide range centring on industry (Appendix 2 on page 307 gives further information on the categories).

Although some major inventions were never patented, most were. But here we suggest simply that the process of incremental invention is reasonably illustrated in the patent material. A final view from John Farey is of interest:

> in fact it almost always happens that the inventions which ultimately come to be of great public value, were scarcely worth anything in the crude state in which they emerged from secrecy; but by the subsequent application of skill, capital and the well

TABLE 3.5 The geographical location of British patenting in the 1790s

Location	*1790*	*1791*	*1792*	*1793*	*1794*	*1795*	*1796*	*1797*	*1798*	*1799*	*Total*
All Patentees	73	55	85	44	55	49	47	31	77	83	599
% Patentees in London	48	51	49	43	47	45	34	39	53	48	47
% Patentees English provinces	48	47	46	48	47	55	55	55	41	41	47
All other %	4	2	4	9	6	0	11	6	6	11	6
Number of patentees											
Manchester	0	0	0	2	1	0	2	0	0	2	7
Birmingham	6	1	3	0	3	2	4	0	7	2	28
Sheffield	1	2	3	1	4	2	1	0	3	0	17
Leeds	1	1	0	1	0	1	0	0	0	2	6
Liverpool	0	0	3	0	1	0	1	0	0	0	6
Newcastle	1	0	0	0	1	0	0	1	3	2	8
Nottingham	1	1	0	1	0	0	0	0	0	1	4
7 Cities as % of total	14	9	11	11	18	12	17	3	17	11	13%
Number of Patentees											
Bath	0	1	0	0	0	1	0	1	0	0	3
Bristol	5	1	0	0	1	2	1	2	4	0	16

TABLE 3.6 Approximate subject typology of British patenting in the 1790s

	Category by function	*% of total*	*% of 7 cities (Table 3.5)*
1	Industrial processing of raw materials	9	11
2	Motive power (generation, conversion)	8	6
3	Machine processes for producer manufs	8	6
4	Machine processes for consumer manufs	1	9
5	Improved fabrics/fabrication	7	6
6	General machinery improvements	3	1
7	Lifting and moving machinery	7	16
8	Machine tools	4	1
9	Agricultural machinery	3	1
10	New chemical processes for new manufs	2	3
11	New chemical processes for old manufs	2	3
12–24	All other	46	37
TOTAL 1–24 %		99%	99%

directed exertions of the labour of a number of inferior artizans and practitioners, the crude inventions are with great time, exertion and expense, brought to bear to the benefit of the community . . . such improvements are made progressively, and are brought into use one after another, almost imperceptibly, in consequence of the experiments and alterations made in the cause of extensive practice.[85]

3.7 CONCLUSIONS – INFORMATION, INSTITUTIONS AND INDUSTRIAL REVOLUTION

the statistical and the literary approaches to history are not so very far apart.

Donald McCloskey, 1981

He who seeks conclusiveness should flee from British economic history in the eighteenth century.

Nick Crafts, 1981

The Industrial Revolution may have been the product of a marriage between the frankly mundane and the fairly dramatic. McCloskey, in his survey paper, goes on to emphasise that 'Nothing very definite can be said about the "preconditions" for British growth, which leaves room for unrestrained speculation on what they were.'[86] This is true also of the dynamics of the growth process after 1780. We have suggested that broadly 'social' or 'institutional' factors may not be safely relegated as a background feature of the Industrial Revolution but should be brought forward into our essential understanding of its timing and character.

If we consider information as an input into industrialisation, then we may also go on to postulate that, in itself demand does not conjure up supply. Increased demand (at this point we remain agnostic) may create a search for information which is not successful, or it may encourage the application of existing information or inventions to the process of production itself. If information is unavailable because it is either non-existent, spatially distant or socially constrained, then an increase in demand for products may not result in an efficient increase in their supply. During the Industrial Revolution there is little evidence of a shortfall of information in any of those sectors of the economy which grew in efficiency at the fastest rate. Within and between any of the three major sectors of the economy, information was carried by such mechanisms as mechanics journals, patent specifications and a panoply of more academic scientific and technical literature, and flowed through a great complexity of institutions. These were the means whereby the social interests of the 'activists' became the information system of the 'audience'. Information expands with use. Because information was, as it were manufactured for social and ideological reasons, it was, if anything, oversupplied during the years of industrialisation. As Chapter 4 outlines, this profusion became part of the comparative advantage of the British economy in the later years of the century. In other places, States or enterprises had to create their own information systems through the development of universities, technical schools and other such more formalised mechanisms.

There remains the possibility that there developed another dynamic linkage between techniques and institutions. Douglass North has suggested that a key to the dynamics of the British case lay in the relations between factory organisation and technical change: 'With the better management of individual contributions came reduction in the cost of devising machines to replace men's and women's hands. . . . The industrial revolution came about as a result of organisational changes to improve the monitoring of workers.'[87]

Although there is something in this, it should be remembered that 'putting-out' and other earlier organisational forms very often failed to lead to the factory and for a long period coexisted with it.[88] Many of the early factories which merely increased the scale of production and control over the workforce were abject failures. In addition, control over the workforce *per se* could be gained and monitored within a variety of organisations.[89] Possibly of importance was the emergence of true factories, central power sources which produced a continuous *flow* of goods rather than a series of batches, represented in such 'breakthrough' institutional innovations as Lombe's silk mill and Arkwright's cotton mill.[90] The function of such organisations may have been to convert techniques into better technologies. Fundamentally, the separate components of the Lombe mill were discrete techniques which in most case had been established in Italy and elsewhere for a hundred years and more. Yet even the smaller-scale mills showed an improved profitability, machines did replace labour and the purchase price of the finished product was reduced by around 30 per cent. The arrangement of power, building and machines was the new *technology*, even though each technique, from Sorocold's undershot waterwheel, to the 78 winding engines, 8 spinning mills and 4 twist mills, were all established techniques. But flow production created pressure for increased efficiency, with each component effecting improvements in the other. Soon the spindle speeds in the mills achieved 3000 revolutions per minute, four to ten times the speed of exactly the same French spindles, this alone reducing production costs by 50 per cent. The sophisticated flow required precise stop-motions in the winding process. Similarly, the roller spinning machine of Lewis Paul remained but a technique until Arkwright placed it in sequence in his mill, i.e. reorganised it into a superior technology. By so doing he gained a six-to-eight-fold increase in labour productivity in spinning over Crompton's mule of 1780.[91] Given that factories did not require vast capital, then ingenuity, incremental technological change and organisational innovations do appear to have been at the heart of the British economy.

4. The Scientific Enterprise: Institutions and the Diffusion of Knowledge in the Nineteenth Century

4.1 INTRODUCTION: SCIENCE, TECHNOLOGY AND INSTITUTIONS

the sole aim of our thoughts and our exertions must be the kind of organisation most favourable to industry . . . including every kind of useful activity, theoretical as well as practical, intellectual as well as manual

Saint Simon, 1817

Man has mounted science and is now run away with

Henry Adams, 1862

AT a meeting of the Glasgow branch of the British Society of Chemical Industry in November 1885, its chairman, J. Neilson Cuthbertson, offered yet one more comment by British industry upon recent European and American innovation in the organisation of science. Cuthbertson noted the splendid provision for science and engineering at the newly erected University of Strasburg, its 92 professors and large laboratory facilities. His measure of scientific improvement was manifold and included its contribution to industry, its institutional formalisation, its support by and of the State:

> I might refer to the great success that has attended Continental science in different ways; in organic chemistry, in the introduction of the ammonia process in the alkali trade, in the ventilation of deep mines, in the application of water power, in the invention of the dynamo machine, in the construction of roofs and bridges, in the printed cottons of Mulhouse, in the woolen yarns of Verviers and in the ribbon trade of Basle . . . all the principal machinery that is now at work on the Continent has been derived from England, or, at all events, is the direct offspring of suggestions made in Great Britain . . . the beginnings of the modern industrial systems are due in the main to Great Britain. . . . This was no doubt the occasion of the great efforts put forth by Continental nations to cope with this country, and they began by instituting Polytechnic Schools. There is l'Ecole Centrale at Paris, and there are the Polytechnic Schools of Germany and Switzerland; and the first step they took was to get men from England and Scotland to start them.[1]

Even this brief section from Cuthbertson's speech captures much of the argument about the organisation of science during the nineteenth century. Prior to 1914 public advocates of the scientific enterprise commonly resorted to international comparison in their efforts to stimulate industrialists or governments into financial involvement. In doing so they normally merged the notions of science and technology (or pure and applied research) in the common understanding that improvement in the production and organisation of knowledge led to improvement in the production of commodities and services, particularly those associated with industrial manufactures. Through brief consideration of just four leading scientific and industrial nations (Britain, France, Germany and the US) and one industrial newcomer (Japan) the present chapter highlights the institutional context of that surge of technological and industrial development which is the subject of Chapter 5. The transfer of knowledge and machinery *between* industrial nations in the years prior to 1914 was facilitated by the prior or contemporaneous development of institutional structures which hastened the creation and diffusion of knowledge *within* nations.

Very frequently, the successful applications of knowledge in industry (e.g. Germany or the USA) have been associated with superior organisations for science, rather than with the character of the *demand* for science. Until relatively recently, this institutional supply-side argument has concentrated on the superstructure of the scientific enterprise – upon elite scientific programmes, universities, societies and publications – and has neglected its infrastructure, those institutions concerned with education, training, diffusion, adaptation and application of knowledge. Usually it is to the universities and specialist research institutes that historians turn when illustrating the economic value of the scientific enterprise. Here we suggest that an emphasis on the higher learning leaves many problems unremarked and neglects major aspects of the support for nineteenth-century science and technology.

A cause of this somewhat partial treatment may be discovered in the nature of public rhetoric. Thus when R. B. Haldane argued for support of science in his 1905 pamphlet *The Executive Brain of the British Tongue* his plea was for the establishment of a *corps scientifique*, which could act as a vanguard for national efficiency. A focus on higher learning and research was equally found in France, Germany and the US, and emanated in the main from the academic and scientific establishment. When addressed to the goals of national prestige and security, civic efficiency, cultural superiority or intellectual progress this commanding rhetoric was appropriate enough. But whether a large movement of public or private resources into such areas would in fact have improved overall industrial or economic performance is quite another matter.

In their discussion of the scientific enterprise, Hilary and Steven Rose agree with those writers who see science and technology as 'interacting terms', a view of especial relevance to the years before 1914. In this chapter, two of the Roses' 'legitimate' uses of the term 'science' are collapsed as one: 'science' includes, 'the whole field of research and development, that is, both science and technology . . . [and] the social institutions within which the activity is carried out'.[2] Sections 4.2 and 4.3 below deal very briefly with institutions concerned with the creation and diffusion of knowledge, section 4.4 considers the role of private enterprise and section 4.5 the role of the State. As a prelude to Chapter 7, section 4.6 introduces the institutions involved in the establishment of western science and technology in industrialising Japan. Finally, section 4.7 offers a synthetic treatment of the relationship between institutions, technological systems and industrialisation in the years prior to the First World War, a theme which is developed over a far wider framework in Chapter 5.

By the middle of the nineteenth century the support for the scientific enterprise derived from any one or more of three major sources. The first of these was *socio-cultural*. This, of course, was the traditional source of funds for science, whether through direct patronage of individuals, or by the indirect support given to scientific intellectuals by a larger audience for science – through societies, associations, educational institutions or publications. The second source of support was private, profit-seeking *enterprise*. This was very much a phenomenon of the late nineteenth century and has normally been associated with the spectacular industrial emergence of Germany and the USA. The third source of support was the *government*. Just as French science has been seen as hampered by premature state interference (see also Chapter 2 above), British science has been visualised as wasted by lack of government support, and German and American science have been interpreted as major beneficiaries of state activity. Sections 4.2–4.5 below indicate that varying national trends in scientific organisation might be interpreted as direct reflections of the different sources from which the national scientific enterprises gained their support. An almost infinite variety of such national traditions may be generated, for each source was in fact a complex range, and no source in *itself* dictated the exact direction which institutionalisation would take. Individual patronage could be directive or liberal, could endow a scientist, or scientific programme, a university (the British and American traditions) or a research institute (Germany). Private enterprise might encourage government activity through public pressure (the probably ineffectual British mode), or operate through external institutional endowment (the USA) or through the internalisation of knowledge production into the structure of the firm itself (Germany is the normal example). Similarly, the institutional impact of provincial or municipal government support was quite different from that of highly centralised government activity.

But despite national differences, the development of the scientific enterprise during the nineteenth century was associated with certain new or revamped characteristics. The most obvious features were the two related ones of institutionalisation, and a more formal linkage between higher scientific activity (knowledge creation) and induction into the scientific enterprise (training or education).

The concept of 'polytechnical' training which emerged more or less in all four nations was visualised as generally enlightening and radical as well as the mathematical and mechanical basis of engineering practice. At times, as in Germany, a mandarin ethos might have thrust 'applied' science outside the universities, but in all four nations the specifically created technical institutions did develop 'pure' teaching or research components. From the beginning, the new wave of institutions supported scientists directly . The Ecole Polytechnique employed Monge, Fourier, Laplace and Berthollet and later Ampère, Guy-Lussac, Arago and Fresnel and was a forum for such visitors as Volta, Rumford, von Humboldt and Liebig. Here praxis and laboratories met with meritocracy and examinations, and from it was published the *Journal Polytechnique*, a symbol of the new institutionalised nexus between knowledge creation and knowledge diffusion. Previously, the newly created knowledge of one institution had to pass through other institutions prior to direct transmission to a wider audience of *savants*, active amateurs or technical innovators. The political power of academics and professionals in Germany ensured that scientific knowledge was first institutionalised on a significant scale in new rather than old universities. The technological universities of Berlin, Breslau or Bonn and the *Technisches Hochschulen* linked teaching with research and harboured specialised research institutions which were often especially designed to be of relevance to local industrial or civic concerns. From the new institutions of the higher scientific teaching emerged laboratories and increased specialisation, perhaps best symbolised in Justus Liebig's chemical laboratory at Giessen from 1825. Through Liebig's students the model spread – A. W. von Hoffman set up laboratory teaching at the Royal College of Chemistry in London and larger research groups formed at Göttingen, Munich and elsewhere. All featured the academic leader, programmatic research and guided teaching laboratories.

In promoting special expertise the new formats also stimulated the growth of professionalism and the formal scientific career. In major industrial nations, universities and institutes trained specialists who were later employed in government departments and in a variety of industries – chemistry, metallurgy, mining, electricity, all of which were applied in sectors requiring quality control technologies. The movement of the industrial state into social control and welfare (see Chapters 5 and 6) also demanded the services of a variety of scientists and engineers, from factory inspectors to public analysts.

Specialisation and professionalism were developing aspects of the scientific associations. In Britain the tripling of the number of scientific professorships between 1810 and 1860 took place alongside a rapid growth of specialised scientific forums for the presentation, dissemination or popularisation of knowledge. In Germany, the Society of German Naturalists and Natural Philosophers from 1822 had proclaimed expertise by restricting its membership to those who had published. The British formula was to allow for expertise within the sections of its more general scientific associations, the classic compromise of the British Association for the Advancement of Science from 1833. In most nations this occurred against the background of a growing number of specialised professional (i.e. occupation-based) societies devoted primarily to solidarity, pressure-group politics and knowledge dissemination. Such differing formats do not disturb Mendelsohn's generalisation that 'ultimately the specialist in science became the normal man, and the machinery of every society which made a pretext of harboring a scientific tradition became geared to the needs and used to the language of specialised science'.[3]

A last general feature of the scientific enterprise of the later nineteenth century was the breakdown of any clear distinction between knowledge and its applications. Engineers employed the concepts, mathematical techniques and experimental procedures of the physical sciences. It is this which justifies our broad approach to the scientific enterprise. The increased application of mathematics to civil engineering in the later eighteenth century, associated with the work of Coulomb, Poisson, Navier and Cauchy had given way to the developments in physics and chemistry. Technology was emerging from research grounded in the law of conservation of matter or founded on an improved understanding of thermic processes and electrical phenomena. Research breakthroughs in chemistry and electro-chemistry were of almost immediate importance to an increasing range of new industries at the end of the century. J. Neilson Cuthbertson's identification of the progress of science with its applications was powerful rhetoric because it was well-based collective wisdom.

4.2 THE ORGANISATION OF SCIENCE: NATIONAL TRENDS

There has been too great trust in the energy and hard-working qualities of our race and a consequent neglect to activate the brain power of the nation.

Yorkshire Union of Mechanics' Institutes, 1879

Many of the older accounts of the modernisation of scientific organisation directly or implicitly contrast Britain and France on one side against Germany or the USA on the other. In the former the forces of tradition cramped the development of science; in the latter, new modes of thought ushered in a new age of formalism, professionalism, specialisation and large-scale science. A not-too-violent caricature of such distinctions might result as shown in Table 4.1.

This schema combines or is associated with at least six underlying postulates, which may be summarised as follows:

(a) France and Britain did not significantly modernise the organisation of their science prior to 1914.
(b) The reasons for this are complex but appear to lie in a failure of vision which itself resulted from a mixture of social and econ-

TABLE 4.1 Dichotomies in the organisation of science

Causal factors	*Organisational backwardness (France and Britain)*	*Organisational innovativeness (Germany and USA)*
Vintage of organisation	Early	Recent
Educational system	Amateur and polite	Professional
Industrial structures	Failing and fragmented	Growing
Political economy	Smithian/Malthusian	Expansionist/interventionist
Pressure groups	Agrarian and colonial	Industrial and commercial
Social preferences	Stability and protection	Growth and competition
Social institutions	Burdensome	Adaptive

omic attributes (e.g. the difficulty of scrapping well-established institutions, the inability of industry to finance the supply of applied science etc.).

(c) This organisational failure was an important component of the relative decline of the British and French economies prior to 1914 and, conversely, in Germany, Japan and the USA institutional reformation assisted industrial revolution.
(d) Organisational innovation is a function of supply-side factors rather than of demand.
(e) An increase in the production of well-trained and well-financed scientists and engineers is necessarily beneficial not only for intellectual or moral reasons but for socio-economic ones as well.
(f) The twentieth century scientific enterprise is the child of the German–US organisational mode rather than the traditions of France or Britain.

Postulates (e) and (f) are not of any particular concern to us here. This chapter suggests that (a) is at best correct if the meaning of 'scientific enterprise' is confined to numbers engaged in higher science and to their immediate support structure (training and employment), and that therefore postulate (b) requires modification to allow for the possibility of a breakdown in the implicit ranking (Germany, USA, France, UK) of different national tendencies. Postulates (c) and (d) appear to be unfounded.

In the new German Reichstag of 1881, 44 per cent of deputies were academics, lawyers or administrators, and only 13 per cent were industrialists or engaged in trade.[4] This power of academia may well have had something to do with the continuities in Germany's stress upon the higher learning and upon theory in the universities, a trend cutting across the economic transitions from proto-industry to industrial revolution (Chapter 5). Table 4.2 summarises the position of natural science in the German university system between 1875 and 1910.[5] It might be noted that as early as 1864 the universities employed 260 professors in scientific subjects.

It is clear that the tremendous growth of science in the German universities was an outcome of growth in the total system. But in addition to the universities, with their Humboldtian characteristics of a disciplinary rather than a collegiate ethos, specialisation combined with speculation and research with teaching, there were also the technical institutes or Technische Hochschulen, which in Prus-

TABLE 4.2 Natural science in the German university system (student numbers)

Year	*Natural Science*	*Index (1900 = 100)*	*% Natural Science to total students*	*% Natural Science to Law students*	*% Natural Science to Theology students*
1875	1710	35.6	10.4	38.1	74.4
1880	2815	58.7	13.3	53.8	93.8
1885	2820	58.8	10.4	58.3	51.5
1890	2298	47.9	8.0	34.4	41.2
1895	2821	58.8	9.9	36.8	63.5
1900	4796	100	14.2	49.3	121.4
1905	5444	113.5	13.2	46.0	140.3
1910	7276	151.7	13.6	67.5	172.4

TABLE 4.3 Staff and students in chemistry at Berlin and Hanover *Technische Hochschulen* (1885–1900)

	1885	*1890*	*1895*	*1900*
Professors	8	10	10	12
Lecturers	5	3	7	10
Private lecturers	5	4	15	16
Assistants	11	18	22	26
Students	153	253	363	563
Staff/student ratio	1:5	1:7	1:7	1:9

sia by the 1870s were producing more graduates in science and technology than were the universities: at their peak in 1900 they enrolled twice as many students in science and engineering than did the universities. Although beginning as a distinct tendency, by the later nineteenth century the technical institutes had developed a sturdy research component (for instance, the research institutes at Munich from 1875, Darmstadt from 1882) and had become heavily involved in theoretical training in mathematics and science.[6] Table 4.3 provides details of academics employed in chemistry only at two of the major technical institutes during 1885–1900.[7]

Specialised chairs were established in organic and inorganic chemistry, chemical technology, electro-chemistry, photo-chemistry and metallurgy. Staff also included specialists in food chemistry, mineral oils and naptha products, the design of chemical works and plant, and special laboratories were devoted to photo-chemistry, metallurgy and chemical technology from 1884 onwards. At the same time the science of the German universities had developed strong applied components. At Berlin there were no less than 11 professors in chemistry alone, several of whom were linked to the Berlin Technical Institute, to the Imperial Patent-Office, to the Royal Department for Testing Explosions and so on. Within the university were four chemical institutes specialising in organic and inorganic chemistry, chemical technology and pharmacy. Between them, the annual expenditure of these four labs on apparatus, instruments and chemicals was in the region of £3000 to £4000 by the turn of the century. The erection cost of the first of the institutes (1860–70) totalled some £45 000, whilst its new building (1897–1901) required an initial expenditure of £70 000, to which was added another £7500 for instruments and apparatus.

Such concentration of both physical and human capital attracted large numbers of students from elsewhere in Europe and America and at the turn of the century, from Japan, whilst in Germany itself there is some evidence of an oversupply of educated manpower in scientific and other fields.[8] Most of the foreign students were post-graduates, and it was the German graduate schools which greatly influenced the organisation of science in the United States. The impact of German refugees from the early 1840s together with the public persuasions of such key figures as H. P. Tappan led to the lab-postgraduate emphasis of such major institutions as the Lawrence Science School at Harvard or the Sheffield Science School at Yale. From the 1870s there was a general increase in the numbers of undergraduate scientists and engineers and a spectacular development of the PhD degree, again associated with the German model; by 1900 some 5700 graduate students had gone through the system; in 1918 alone 562 PhDs were awarded.[9] The laboratory science of Johns Hopkins or Stanford stimulated the large private endowments of the US system, e.g. Rockefeller at Chicago, which continued the important trend established by Stephen van Rensselaer (with the foundation endowment of Rensselaer Polytechnic Institute in New York in 1824). A large boost to the college scientific system came with the land grant schemes from the 1870s, which encouraged the growth of science and technology based colleges, numbering 70 in 1872, 126 in 1917. By 1897 some 10 000 students attended colleges of technology.[10]

The emphases upon research, organised programmes, professionalism and the large scale were supposedly resisted in Britain, where scientists were not as closely involved in the political process as in France or Germany. In fact, similar tendencies were evident, but they were on a smaller scale, provincial and *ad hoc*. For example, professionalism and specialisation were demonstrated in the large number of learned scientific societies in Britain (500 were officially registered in 1883), of more importance in British scientific organisation than in Germany or the US. Omitting the enormous number of associations and institutions concerned with the diffusion of knowledge (see 4.3 below), the numbers involved in national scientific societies may have reached over 20 000 by the later 1860s. Again, the early development of engineering training and laboratory work lay outside the universities, in such private or bureaucratic venues as the East India Company (Addiscombe, 1807), the School of Naval Architecture (1811), Putney College of Civil Engineering (1839), the Royal School of Mines and the Royal School of Naval Architecture (1864). Chemical labs were established at the endowed Royal Institution and at several private institutions prior to the first wave of university establishments, from William Thomson's lab at Glasgow (1866) to that of James Clerk Maxwell at Cambridge (1874).[11]

From 1826 the colleges of London University took up the German example. During its first decade this new university appointed 17 chairs in medicine, science and engineering. But the new university of Durham had no professor in a natural science, only one professor of mathematics, a reader in natural philosophy and a lecturer in chemistry and mineralogy. In the undergraduate provision no prior scientific training was assumed, and a 'popular course of lectures on the history of the sciences' was followed by the survey courses on mechanics, hydrostatics, optics and astronomy: The 'application of science to manufacture' was relegated to a short and entirely optional 'popular course'.[12]

This was not the stuff of which the German model was made. At the time of Durham's establishment, there were perhaps 20 or so professors or readers in the natural sciences in England, and university provision of science was brought up to a reasonable level only by the Scottish universities.[13] Overall, the early part of the century (to around 1860) witnessed a fall in the proportion of the population at universities. Although there was some reform at both Oxford and

Cambridge following a Royal Commission of 1850–52, the real break with tradition came with the scientific and engineering emphases of the newly-founded provincial universities of the later part of the century.[14] Under the auspices of London or Oxbridge examinations, the new university colleges at Manchester, Birmingham, Liverpool and elsewhere were faculty rather than college based, emphasised both science and technology and responded to local industrial and civic concerns.[15] They also opened up formal higher education in science in England to a wider social intake.

In the aftermath of the revolution (see Chapter 2), innovation in the organisation of science had come first in France with the foundation of such important institutions as the Ecole des Ponts et Chaussées (civil engineering), the Ecole Polytechnique (engineering, mechanics and chemistry with a greater industrial orientation), and the Ecole Normale (more associated with research training in science). Although this more formalised and bureaucratised mode diffused throughout Europe and transferred to the USA, and created a whole generation of French scientists, the later years of the century have been seen as a phase of decline, ushered in by militarism and imperialism, an overemphasis on praxis and a prematurely bureaucratic and centralist ethos.[16] More recently – alongside changing views of the French economy – the system has been seen as adaptive.[17] The emphasis on the centralism of the national institutions has neglected the enormous importance of provincial science and the associations geared to the civic and industrial needs of very diverse localities.[18] By 1886 there were some 655 *Sociétiés Savants* producing 15 000 volumes of memoirs and transactions. These were highly localised in terms of both the nature of the science pursued (often fieldwork, as in the natural history societies in Britain) and in the emphasis on local application.[19] Again, the *Ecole pratique des hautes études* was a central institution, but its object was to fund laboratory work in the provinces as well as in Paris, and to diffuse science through publication.[20] Lastly, aspects of the German model developed in the combined operations of local government, central authority and private finance. Research institute science in France was very strong and, as at times in Germany, often served to forge formal links between universities and industry. In 1894 the municipal council of Marseilles paid 50 per cent of the cost of a professorship in industrial physics and the departmental government voted funds for a thirty-year period to finance a chair in agricultural botany. During the 1890s several public figures were claiming that science in France was too greatly devoted to practical rather than to theoretical pursuits.

The differences between nations in the way in which national scientific enterprise was organised is clear enough, but none of the four nations was entirely isolated from observed tendencies in the others. At the same time, although contrasting trends at the level of the higher learning may have been of real importance to scientists and educators, industry and the state benefited from (and were benefactors of), developments elsewhere in national scientific and technological systems.

4.3 HUMAN CAPITAL: THE DIFFUSION AND APPLICATION OF KNOWLEDGE

> When a thing becomes important in Germany it is academised, so to speak, and given official standing.
>
> *Arthur Shadwell, 1906*

It was at the level of training, of intermediate education, of professional societies and voluntary associations and of publications that many of the positive linkages between science, technology and industry were forged. The *movement* of knowledge at the higher levels was relatively unproblematic. On the other hand the diffusion of knowledge through the social system and to very large numbers was highly problematic and, in places bore only a loose relationship to the activities of the scientific superstructure. Germany and Great Britain provide somewhat contrasting examples of the relationships between the superstructure of science and its wider infrastructure.

In Germany vocational and professional training in scientific and technical pursuits grew at the rate of over 130 per cent per capita between 1864 and 1911, as against a growth of 100 per cent in general secondary education and 116 per cent for higher education. However, this relatively good growth performance must be set against the small absolute numbers at the beginning of the period, and Peter Lundgreen has estimated that the increase in the stock of human capital in this gross sense was a very minor component of German economic growth.[21] From the early 1820s provincial technical schooling in Germany assumed little educational background and focused on industrial applications. To this extent the early phase of German technical education was similar to that in Britain except that some scholarships were available and training was more closely linked to entry into higher levels of scientific or technical education.[22] It is often stated that German universities or technical institutes were far better integrated into technical education than in the British case and that the very notion of *Wissenschaft* meant something more than merely formal knowledge of science. Merz especially emphasised that German universities were powerful agents of diffusion *because* of their close links with the technical education system, which meant that the laboratory and seminar ethos percolated into the latter.[23] But not every German state was a Wuertenberg with its 450 industrial schools, 523 farming schools, 108 trade schools, 76 industrial academies, its agricultural and building colleges and its polytechnical university at Stuttgart, together serving a population two-thirds the size of London.[24] In addition it has been claimed that induction into science and technology in Germany became *less* socially open just as the industrial economy developed.[25]

If we follow Arthur Shadwell's distinctions of 1906 we may break down the German technical training system into three layers. At the level of the artisan or operative (*handwerker*) were the special trade schools (*Fachschulen*), which in coordination with evening continuation schools were designed to bolster the formal training of industrial apprentices. Shadwell believed that these served the small-industry, craft section of the economy 'and had very little bearing on the large manufacturing industries, with the possible exception of weaving'.[26] This must be qualified. Of the nearly 800 000 apprentices in Germany at the turn of the century, many exposed to the *Fachschulen* system, 70 per cent were employed in the smaller firms, but many of these at a later stage in their careers transferred into the modern, large-scale sector of industry.[27] That is, the progress of high technology in Germany may have depended significantly on the human capital originally formed outside of the modern educational or industrial sectors. At another level were the intermediate technical schools which emerged from the earlier *Gewerbe-Schulen* system.[28] These were overtly geared to the large manufacturing enterprises, involved mainly daytime attendance over two or three years and provided an engineering diploma at graduation. They were usually highly-specialised by trade or locality. Beyond these were the *Tech-*

nische Hochschulen (Section 4.2 above). However, the lower levels of science–technology induction were not especially open to operatives nor did they enrol the sheer mass of workers produced under the *ad hoc* British system. Following the report of the British Consul at Stuttgart, Shadwell – who was no appologist for the British system – considered that 'such figures as these sink into insignificance compared with those of the English technical schools, at which the evening pupils studying the great industries are numbered in hundreds and sometimes in thousands'.[29] Rather, German superiority lay at the technical institute level, which increasingly enrolled students who had previously graduated in the traditional classical courses of the *gymnasia*.

The associational culture of the German professional engineer was certainly impressive. The combined membership of the three major engineering associations (Engineers 1856, Metallurgical Engineers 1860, Electrical Engineers 1893) had reached over 35 000 by 1914, but entry into the profession was probably becoming more difficult. This may have had something to do with the character of the training provisions and a relative underinvestment in institutions which could tap into the talents of the lower social groups. Around the turn of the century, the ratio of enrolment in industrial and commercial improvement schools to all elementary school pupils in Wuertenberg stood at 8 per cent, but in Lancashire it had reached over 20 per cent.[30]

Table 4.4 shows that in Britain, technical innovation remained very much with the individual, most of whom were trained and employed within engineering and the industrial trades.[31]

Of this sample, 45 per cent were engineers, who by 1860 had thus replaced the earlier dominance of skilled tradesmen (see Chapter 3). The role of 'scientists' as professionals or of university engineers was minor. Many direct transfers of foreign technology into Britain centred on tradesmen or engineers. In Manchester, William Stope Lewis patented information received from a scientist in Philadelphia, and Henry B. Barlow patented an improvement in carding machinery based on information received from H. Reiter of Switzerland, and another for screw-cutting machinery derived from a formal communication of J. P. Haigh of Pittsburgh. The engineers John Dale and Heinrich Caro combined to patent a cotton dyeing technique from information obtained from August Leonhardt of Berlin. Within the cities, the majority of engineers were employed in

TABLE 4.4 184 patentees in five English cities in 1860

Occupational category	*Rochdale*	*Birmingham*	*Manchester*	*Liverpool*	*Leeds*
Engineers	15	14	38	5	5
Artisan/tradesmen	5	8	12	4	6
Manufacturers	5	26	16	5	2
Professionals	0	1	1	2	0
Merchants and gentlemen	1	1	0	6	3
Other	0	0	2	1	0

TABLE 4.5 Britain in 1841 – occupations, skills and skilling

Occupation category	*Male journeymen, masters, apprentices*	*% apprentices*	*% of total occupied population*
A Engineering occupations	33 448	16	0.6
B Metal & machine-skilled	60 298	15	1.2
C Traditional metal skills	140 043	17	2.8
D Clock & watch-makers	14 587	15	0.3
E Non-metal skilled	442 026	13	8.8
F Commercial clerks	56 644	18	1.1
G Professionals	87 765	4	1.7

large machine emporiums. Most of the manufacturer patentees were small-scale, expanded-workshop owners, several of whom were producing and marketing their previously-patented machines and products. Formal higher education and large capital were seemingly not predominant features of the British model of invention, which nevertheless managed to generate over 12 000 patents in the years 1855–60 alone.

Naturally, this technology-innovation complex was some function of the tremendous number of skilled working men already existing in Britain (Chapter 3). Table 4.5, which centres on 1841, shows how the British economy had already generated a vast number of engineering and machine-skilled workers.[32]

For this 15 per cent of the occupied population, apprenticeship was of great significance. At any one time well over 10 per cent and up to 18 per cent of the stock of specialised human capital in industry was engaged in training by apprenticeship. Despite contemporary opinion to the contrary, Charles More has recently shown that apprenticeship did not in fact collapse in Britain as a result of increased costs, deskilling or the failure of the older industries. Although the history of the 1840s onwards is certainly one of university, polytechnic and institute formation, an injection of engineering into higher education which was hastened by the Great Exhibition, the Crimea and the industrial competition of Germany and the USA, in Britain much of the adaptation of the scientific enterprise to new circumstances took place within an informal, massive and highly diffused machine culture. Furthermore, what little we know of the occupational characteristics of members of evening classes, mechanics' institutes and local authority associations for knowledge acquisition suggests a positive relationship between apprenticeship and institutions. Over a decade ago Paul Robertson convincingly demonstrated that a *combination* of apprenticeship and non-firm instruction in workshops and in the technical classes of the Science and Art Department in shipbuilding and marine engineering was the basis for training in one of the most successful and major sectors of the economy.[33]

The Great Exhibition of 1851 has frequently been interpreted as an example of industrial success despite a lack of scientific and technical infrastructure. Yet within a bird's-eye view of the Crystal Palace itself there existed a rich and innovative institutional system for the diffusion of knowledge and the gaining of knowledge-based industrial skills. The most pertinent features of this 'steam intellect' culture were the close relationships evidenced between established scientific *savants* and the voluntary institutions of technical training, the conglomeration of social classes within the variety of forums, and the close links existing between lectures and classes and the multitude of scientifically informed technical and trade magazines, manuals, encyclopedias and books. The 'Civil Department' of Putney College, designed for intending mechanics and engineers, gave detailed courses on chemistry, physics, metallurgy, mineralogy surveying and field engineering, civil engineering and architecture, and machine construction, and was staffed by such eminent scientific figures as Edward Frankland, PhD, FRS, FCS, later Professor of Chemistry at both the Royal School of Mines and the College of Chemistry, or David T. Ansted, MA, FRS, Professor of Geology at King's College London, as well as the patentee–engineers Samuel Clegg and T. W. Binns. At modest terms the privately-run Surveyors and Engineers College in Guilford Street, Russell Square, mounted four distinct streams of study in Railway Engineering, Land and Estate Surveying, Engineering and Machine Drawing. Terms were often set 'for the whole, till perfect'. In addition to lectures and library resources, the Inventors' Aid Association in the Strand, operating with a capital of £50 000, provided expert technical and legal services for the provision and disposal of patent rights to all shareholders with a minimum subscription of 1/- per share. Similarly, William Murray and Company, Civil and Mechanical Engineers at John Street, Adelphi, offered their services to all intending exhibitors at the forthcoming Great Exhibition, including negotiations for the sale of inventions as well as their construction and testing.

A variety of publications were allied to such informal forums. *The Artisan*, run from a small office in Cornhill, provided full engravings and descriptions of new inventions, machinery and public utilities, and included critical accounts of new patents as well as lengthier, detailed essays on their own work by such successful inventor–innovators as Robert Armstrong (steam boilers and high-pressure engines), W. A. Adams (railways), L. P. Sturve (railways, mine ventilation) and Frederick C. Calvert (bleaching processes for cotton and flax). The *Mechanical Gazette*, also issued from Cornhill, published a register of all machinery and plant available for sale as well as new patents, and commissioned special articles on recent inventions as well as covering such matters as tax laws, mutual insurance, metal price lists and other items of economic importance to the innovating groups. Extended discussions of all new, important inventions were to be found throughout this highly competitive market, which included such large-selling publications as the *Mechanics' Magazine* the *Practical Mechanics' Journal*, the *Repertory of Patent Inventors* and the *Patent Journal and Inventors' Magazine*.[34]

From material presented in the Census of population of 1851 we have recently estimated that some 72 per cent of all general, scientific and technical institutions in England and Wales came into the category of institutions for the *diffusion* of scientific and technical knowledge to the working and middling classes.[35] Perhaps nearly 500 000 individuals were paid-up members of such voluntary forums in the year of the Great Exhibition,[36] representing approximately one-sixth of the number of persons actively engaged in manufacturing industry at that time.[37]

It was the continuous transformation of this provincial and *ad hoc* mode of knowledge diffusion which dominated the years from the 1870s to the 1890s. It might be noted that Cuthbertson's speech of 1885 (Section 4.1 above) acknowledged the comparative effectiveness of the British model:

> Notwithstanding all these efforts [Continental] foremen and workmen are not, as a rule, up to the present time, technically instructed. . . . Nothing that they have yet been able to achieve is equal to the instruction that we have in this country through the schemes of the Science and Arts Department and the City and Guilds of London Institute.[38]

The advantages held by Britain were brought out even more forcefully three years later by the Nottingham industrialist Lewis T. Wright:

> Were not some of the Continental schools that we have heard about largely founded to originate classes of skilled labour, till then not existent? . . . The best schools are the workshops of these industries, which are naturally established here because we understand them. . . . If our iron and steel industries were to combine to found a school for teaching the practice and theory of these industries, how long would that school represent the best existing practice. . . . [Britain needs more]. . . illustrations, applications, and demonstrations from actual manufacturing processes, supplemented with lectures in the larger institutions, from men actually engaged in, and conversant with technological pursuits.[39]

Shadwell too was convinced that the British system was superior to that of Germany in combining education with skill formation and in percolating down to a wider social stratum. The British organisation of science and technology generated skills 'mainly from below: they educate boys belonging to the working classes – boys at work in the mill or at the forge – into foremen, overlookers, managers and experts', in contrast to Germany where formal technical schooling 'supplies the large industries mainly from above'. Furthermore, the German system failed to deliver knowledge where it was most needed, witness the lack of facilities at Dusseldorf despite the existence there of over 10 000 men in the metal trades. In contrast, the municipally-stimulated trade and technical schools in Lancashire provided relevant practical specialisms: 'All this is not over-lapping

TABLE 4.6 Institutes of the British engineers, 1818–1912

1818	Institution of Civil Engineers
1847	Institute of Mechanical Engineers
1851	North of England Institute of Mining and Mechanical Engineers
1850	Federated Institute of Mining Engineers
1854	The Society of Engineers
1860	Scottish Institution of Shipbuilders and Engineers
1860	Institution of Naval Architects
1863	Institute of Gas Engineers
1868	The Surveyors Institution
1869	Iron and Steel Institute
1871	Institute of Electrical Engineers
1873	Institute of Municipal Engineers
1889	Institute of Marine Engineers
1892	Institute of Mining and Metallurgy
1895	Institute of Sanitary Engineers
1896	Institute of Water Engineers
1897	Institute of Heating and Ventilating Engineers
1900	Institute of Refrigeration
1908	Institute of Metals
1908	Institute of Concrete Engineers
1912	Institute of Foundry Engineers

or over-profusion, but absolutely necessary if working hands are to enjoy the advantage of technical instruction . . . the best are equal to any and the least are superior to the little hand-loom weaving places which count among the technical schools of Germany'.[40]

The professionalisation of industrial or commercial occupations furthered the growth of institutions for information diffusion. In Britain, as in most nations, the setting up of associations was led by the engineers, as illustrated in Table 4.6.

The impact of this process on particular industrial areas was of importance. Chapters 2 and 3 have shown that the organisation of local science and its diffusion had been fairly fluid and spasmodic, but now the professionalisation of the skilled occupations involved the setting-up of a large number of regional branches of specialised institutions, and these combined the functions of trade associations and scientific and technical knowledge diffusion. Table 4.7 illustrates the rise of the professional technical associations and their impact upon the institutional complex for the diffusion of scientific and technical information in Manchester between 1845 and 1900.[41]

TABLE 4.7 Professional technical associations in the scientific enterprise of Manchester, 1845–1900 (no. of institutions)

Year	*Diffusion of S–T information* (a)	*Science and arts (total)* (b)	*% of (b) mainly scientific* (c)	*Professional associations* (d)	*Total institutions* (e)
1845	4	15	73	12	31
1848	2	13	92	8	23
1858	11	22	68	9	42
1868	3	19	58	12	34
1880	3	32	43	23	58
1890	4	47	42	49	100
1900	5	50	34	48	103

The extent of knowledge diffusion in Britain is indicated quite clearly by data from patent office libraries. Between the years 1854 and 1857 the patent office library in the town of Salford, which held some 23 000 patent specifications, reported the number of separate local references to this literature as 12 419.[42] The last four years of the same decade witnessed no less than 115 000 separate cases of references to patent specifications at the Free Library, Camp Field, Manchester, a figure about equal to the *annual* usage of the library of the London Patent Office by the early 1890s.[43] Such readership figures show that knowledge of new product areas grew very fast.[44] Complete specifications of all British patents existed in some 150 free or official libraries by the early 1860s and this represented a diffusion of inexpensive or free information on a scale not matched in any other nation.[45] Only in France after 1858 was there any real attempt to replicate the British example by the *Société d'Encouragement pour l'Industrie Nationale*, and this was extremely limited.[46] There seems to be no doubt that from the 1860s in Britain (and in nowhere else to the same degree) a vast array of technical and scientific information was distributed to industrial provinces and cities. This was made available to a far wider audience of working men and innovators than in any other nation and at far lesser cost, whether the latter is measured in terms of money costs or opportunity costs.[47] Historians who continue to consider this phenomenon as independent of the nation's scientific enterprise must severely qualify their pronouncements upon any lack of input from the scientific enterprise into British industry in the years prior to 1914.

4.4 PRIVATE ENTERPRISE

The employment of scientists or the purchase of consultancy skills represents a demand for knowledge, and in the nineteenth century this was supplied mainly by voluntary or state institutions working external to the firm. Industrial patronage and enterprise research and development represent the creation (supply) of knowledge within the culture of private enterprise itself. In the later nineteenth century, the more extensive activity of industrialists in Germany or the US than in Britain or France undoubtedly reflected the greater demand for specific scientific and technical knowledge as an input into the production process, and this rising demand was in turn determined by the very nature of late industrialisation (Chapters 5 and 6 below). During the industrial drive the fast growth of new industrial sectors created financial rewards and incentives which stimulated the voluntary and official supply of knowledge-producing institutions. At a later stage, industry turned from complete reliance upon external sources of knowledge to the actual provision of knowledge. In this section we consider how industrialists were first involved in the supply of science through endowments, moved then towards an intensive demand for scientific knowledge through consultancies and direct employment and, finally, became committed to the provision of scientific facilities and personnel training by direct investment in laboratories, quality control facilities and research divisions.

Nineteenth-century Britain was awash with endowments for science. The physical chemistry laboratory of the Royal Institution was endowed by Ludwig Mond. In 1843 the Rothamsted Agricultural Research Station was founded by J. B. Lewis on the profits from his superphosphate fertiliser manufacturing. On the other hand, as late as 1873 the establishment of the Cavendish Laboratory at Cambridge depended on a bequest of £8450 from the 7th Duke of Devonshire, erstwhile Chairman of a Royal Commission which had identified supposed inadequacies in the British system of technical training. In England, the existence of the very ancient universities together with a sturdy charity industry (3000 charities were registered during the 1830s) somewhat confused the transition from traditional forms of endowment or patronage to more modern industrial support. However, by the later nineteenth century much of the financing of the new provincial universities was by endowment from local industrialists, especially in the Midlands. Industrial endowments were of particular importance for support of research; e.g. John Owens, Thomas Ashton and Charles F. Beyer at Manchester (a spectacular success), John T. Brunner and Henry Tate at Liverpool, Josiah Mason and Joseph Chamberlain at Birmingham. By the turn of the century perhaps up to £200 000 per annum was being fed into the British university system from private endowments, mostly industrial. But even this substantial flow was only one-twentieth of that in the United States.

American industrial philanthropy was more overtly modern than in the British case, and from the 1860s was frequently associated with a dynamic unison of private funding, government stimulation and learned-society involvement. Popular scientific or technical journals and several associations (e.g. those of the mining 1871, mechanical 1880, electrical 1884, and chemical 1908, engineers) were stimulated by early endowments. A major centre like Chicago was receiving over £1 200 000 in endowments by 1906–7. The $10 million foundation endowment for the Carnegie Institution in Washington was raised from US Steel Corporation bonds.

The growth of science and engineering consultancies was a clear measure of an increase in the demand for specialised knowledge. Consultancy remained a very important ingredient in the British chemical industry throughout the nineteenth century. Specialised consultancy firms emerged in all four nations considered here. Several of the chemical, pharmaceutical and electrical firms which developed in Germany in the 1870s and 1880s depended upon consultants, some of whom then established laboratories within the enterprises. At the Merrimack Manufacturing Company, a textile mill at Lowell, Massachusetts, S. L. Dana was hired to perform expert chemical analysis in dyeing and beaching. In 1864 Benjamin Silliman of Yale was paid the princely sum of $50 000 by a group of Californian oil enterprises. From the 1870s there emerged a tendency to hire key scientists and engineers not merely for *ad hoc* research but to assist in the setting-up of more sustained laboratory and research programmes, and US firms were once more in the lead – the Pennsylvanian Railway Company, Carnegie Steel, General Electric.

The hiring of scientific and technical knowledge was accelerated with the trend towards purchasing the property rights of patentees whose activity was external to the firm. In some cases patentees were bought into the firm, at other times patents were issued in collaboration between scientists, engineers and enterprises. In many cases patent games involved internal research (consultancy and employment) in order to negotiate the rights or around the rights of existing innovations.[48] The massive growth in legislation and strategies surrounding property rights in knowledge (see Chapter 5) undoutedly increased technological competition and the need for internally employed scientists and engineers.

British corporate patenting was of growing importance and included several examples of the corporation simply capturing the patent rights of a corporate entrepreneur's own invention.[49] This is true of the activities of such firms as Pillatt and Co. (furnaces), Incandescent Heat Co. (furnace systems), de Forest Wireless Telegraph Syndicate Ltd, Marconi Wireless Telegraph Co., and so on. More commonly, the corporation had bought into the property rights of an *independent* engineer–inventor, e.g. Weldite Ltd (alloys) or the Western Syndicate Ltd (railway signalling), which latter involved four inventor–engineers from Reading, Cardiff and Hanwell. Amongst British firms, where patenting represented some transfer of property rights from the scientist–engineer to the corporation, were those who took over the rights of foreign inventors – British Westinghouse, Imperial Fibres Syndicate Ltd.

Amongst the most vigorous of the American patentees were corporations explicitly established in order to exploit an interrelated series of new or existing inventions, home or foreign in origin. The many patents of the United Shoe Machine Co. were on assignment from engineers throughout America, Britain and Europe. Other corporations within this group included the Westman Process Corp. (iron ore reduction) and the International Steam Pump Co. (in partnership with the German inventor C. H. Jaeger). A second type of company were those specialising in new areas of production, such as sewing machines, typewriters, refrigerators, tunnelling machines, linotype, rotary engines or waterproofing.

Amongst the most active of patenting corporations was the Swedish firm of Aktievolaget Separator. Their patents all involved farm and dairying equipment and originated in partnerships with several independent Swedish engineers. Rheinish-Nassonische Bergweks of Stolberg patented equipment for ore-dressing and processes for the extraction of zinc from its ores. The firm relied on a variety of engineer–inventors, amongst whom was the partnership of the academic technologist Professor Wilhelm Borchers and the metallurgical engineer Arthur Graumann.

Firms used patent rights to establish their markets abroad. Fourteen British firms were operating transnationally prior to 1914, and these included large operators in the high-technology areas, such as Royal Dutch Shell (oil), Vickers (armaments), Courtaulds (synthetic textiles), Dunlop (rubber), all concerned to move across the tariff barriers of high income nations. The activities of such firms as Dunlop, Lever Bros, Vickers and J. and P. Coats in Japan represented the transmigration of superior technology and enterprise property rights as well as superior marketing skills.[50]

This was the ever-changing background to the development of internal and sustained corporate research facilities. From the 1880s both German and US industry were associated with the steep rise in trusts, cartels and syndicates, good bases for the internalisation of the research function, especially so as such new forms of market organisation were aligned with new, high-technology industries, such as chemicals and electricity. To an extent, then, the movement of organised research and development into the firm was a function of demand, merger, size and complexity. High-technology firms assumed complex corporate forms – General Electric, Du Pont, Eastman Kodak are the well-known US examples.[51] Engineering components were first built into the separate divisions of the corporation, and then this gave way to the development of centralised laboratories (e.g. General Electric from 1900) which served all divisions. Routine product development might remain with the division, as the central research facility concentrated on product improvement or new tehcnologies. This was the American system as adopted by Du Pont (1902), American Telephone and Telegraph (1910–12), Eastman Kodak (1912) and Westinghouse (1916). By 1921, 526 corporations claimed to support some form of research laboratory; by 1927 some 1000 industrial research laboratories employed 19 000 people.

Apart from the internal structural adjustment of firms to new forms of competition, there were other institutional factors involved. In Britain, Burroughs Wellcome (1894), and British Westinghouse (1899) set up laboratory research of this form, but the case of United Alkili (1891–2) was not exactly that of the US system. The United Alkili Co. was itself a defensive cartel of some 48 chemical firms which established a central research and analysis laboratory under the chemist Ferdinand Hurter, a former student at both Heidelberg University and Zurich Polytechnic. Hardly surprisingly, this case seems to have followed that of the German inorganic chemical firms, who somewhat earlier combined in a research cartel. The resulting Central-Stelle was founded by enterprise subscriptions of £100 000 and maintained at an annual cost of £12 000. The research of the central laboratories was determined entirely by the individual requests of the firms.[52]

Changes within national patent systems ensured that industrial research within the corporation began to influence the nature of both offensive and defensive business strategies.[53] The very incorporation of Bell Telephone in 1880 was in order to exploit A. G. Bell's 1876 telegraph patent. Almost immediately the firm established a research group in Boston. From the early years the work of the laboratory was primarily geared to creating a patent cloud around the Bell product to ensure the protection of the firm's market position, although this strategy did not entirely exclude general development work. Employment in research at Bell rose from 60 engineers in 1894 to 95 in 1898 and 203 in 1902. Increased size meant that it was relatively easy for Bell to purchase the patent rights of others, who could not as readily enter the field in competition.[54] A ruling of 1896 meant that henceforth in America a corporate patent holder was 'neither bound to use the discovery himself nor permit others to use it'.[55] Drawing on the graduate system of the universities, enterprise laboratories became a real asset for the capture of strategic patented knowledge.

In both the US and in Germany a major reason for increased enterprise participation lay in the qualitative shift in physical science itself, which occurred around the 1890s and early 1900s. As investigation moved away from electromagnetism and heat to the properties of electrons and the composition of matter, research programmes developed which were of special relevance to the electrical and synthetic materials industries. The work of such scientists as J. J. Thomson and Artur Wehnelt was inherently commercial. Once anti-trust and other restrictive legislation was in place, the alternative strategies for enterprises in high-technology industries were curtailed and the time was ripe from the use and production of science in industry on a larger scale. The German chemical industry is perhaps the most quoted example. Table 4.8 illustrates the growing demand for human capital in the German chemical industry.[56]

TABLE 4.8 Employment in the German chemical industry 1882–1907

	Total employment	*% of total who are unskilled/semi-skilled*	*% of total in dyestuffs sector*
1882	57 530	73	19
1895	102 923	77	19
1907	158 776	80	19

TABLE 4.9 Scientific enterprise and strategic patenting in the six largest organic chemical firms in Germany, 1886–1900

<table>
<tr><th></th><th>Chemists</th><th>Engineers/ technologists</th><th>All other workers</th><th>Completed UK patents, coal-tar, 1886–1900</th></tr>
<tr><td>Badische Aniline Works</td><td>148</td><td>75</td><td>6790</td><td>179</td></tr>
<tr><td>Meister, Lucius and Bruning</td><td>120</td><td>36</td><td>3766</td><td>231</td></tr>
<tr><td>Farbenfabriken Bayer Co.</td><td>145</td><td>175</td><td>4700</td><td>306</td></tr>
<tr><td>Berlin Anilne Co.</td><td>55</td><td>31</td><td>1950</td><td>119</td></tr>
<tr><td>Cassella and Co.</td><td colspan="2">60</td><td>1970</td><td>75</td></tr>
<tr><td>Farbwerk, Mühlheim, Leonhardt</td><td>–</td><td>–</td><td>450</td><td>38</td></tr>
</table>

By the late 1890s some 4000 trained chemists were employed in the German chemical industry, of which one-quarter were in the organic (mostly dyestuffs) sector, 700 in agricultural stations and labs, 400 in chemical metallurgy, 300 in sugar-refining, with the rest scattered amongst a great variety of inorganic, agricultural and other ventures.[57] By 1900 the German dyestuffs industry represented 90 per cent of the world's total output and was very heavily engaged in the provision of scientific facilities and personnel.[58] Table 4.9 details the scientific establishment of the six largest colour works in Germany in the year 1900.[59] The strategic nature of the research in the firms is suggested in the indicated patent figures: the 948 coal-tar chemistry patent lodgements of these firms in the *British* system was hardly matched by the 86 patent lodgements of the six largest British chemical firms.[60]

Strategic patenting seems to have been more a function of the size of the research department than of the size of the enterprise as such. Apart from employing leading scientists (Badische had hired Heinrich Caro as its chief chemist in 1868) the employment growth rate for the research divisions was extremely fast: Meister, Lucius and Bruning (Hochst) were employing 307 chemists and 74 engineers by 1912.[61]

When A. G. Green detailed such facts before the members of the British Association for the Advancement of Science in 1901 he also noted that

> There can be no question that the growth in Germany of a highly scientific industry of large and far-reaching proportions has had an enormous effect in encouraging and stimulating scientific culture and scientific research in *all* branches of knowledge. It has reacted with beneficial effect upon the universities, and has tended to promote scientific thought throughout the land. By its demonstration of the practical importance of purely theoretical conceptions it has had a far-reaching effect on the intellectual life of the nation.[62]

4.5 THE STATE

> Science, like virtue, brings its own reward.
>
> *Nature*, 1911

An article in the *Edinburgh Journal of Science* for 1832 emphasised the positive role of the French state in the financing of science and began a debate on the tardiness of the British state which has continued until the present time. The intervention of the State was generally of importance because it could stimulate so much else. The Institut Chimique was opened at Nancy on the basis of two 500 000 franc grants from each of the central government and local municipalities, and acted as a catalyst for further private developments. The re-opening of the large chemical laboratories of the Zurich Polytechnic School (at which both Ferdinand Hurter and H. Auer of the United Alkiki Company had been educated) in 1886 was made possible by an extensive grant of £70 000 awarded in 1883, followed by government cost maintenance. In this case the government had sponsored the inspection of leading chemical laboratories in Germany, Austria, Hungary, France and Britain (Owen's College).[63] The mathematician Felix Klein's success in founding a department of technical physics at Göttingen in 1894 followed upon the combined support of a state subsidy, industrial endowment and the subscriptions of the Göttingen Society for the Promotion of Applied Physics. So, the intervention of government was by no means independent of other sources of support.

At the same time, the routine interference by government in the setting of standards, control of pollution, pests and adulteration, had a significant impact on the scientific enterprise in all industrial nations. This is well illustrated in the activities of the various government bureaus established in the US. The Board of Agriculture (1862) and bureaus of Animal Husbandry (1884) and Plant Industry (1910) were linked to the land-grant colleges and by the early twentieth century employed thousands of demonstrators, researchers and administrators.[64] The original vote by Congress for the establishment of the National Bureau of Standards was $250 000, and the bureau was soon conducting several thousands of tests annually for both government and industry at a time when increased technological sophistication meant that no single company or private group could evolve adequate procedures for measurement or inspection. The work of the US Bureau of Chemistry embraced the examination of tens of thousands of samples for adulteration, copyright and other reasons, innumerable check analyses, legal disputes and floor inspections, all of which involved a large scientific establishment.[65]

In the nineteenth century it was the contrast with Germany which was most frequently noted in Britain, and occurences in the two nations do highlight the different patterns of state interventions in the area of science and technology. By the end of the century the German government was clearly investing far more than the British in the institutions of higher science. Table 4.10 contrasts total income and state funding in five science-orientated university colleges in England and five similar German institutions (three polytechnics and two universities famous for their science provision) for the year 1899.[66]

Government intervention in this relatively lavish manner meant that scientific and technical teaching in Germany was comparatively intensive. Table 4.11 shows staff–student ratios in the chemistry departments of five major German institutions in that year.[67]

The concentration of staff and facilities meant that specialisation

TABLE 4.10 Funding of ten leading institutions in England and Germany in 1899

	Total income in £ sterling	*% of which is government grant*
English university colleges		
King's (London)	42369	5
University (London)	35456	8
Owen's (Manchester)	47494	7
University (Liverpool)	23792	13
Mason (Birmingham)	17864	15
German institutions		
Berlin U.	143555	83
Bonn U.	63037	80
Berlin Polytechnic	69077	49
Hanover Polytechnic	25240	60
Aachen Polytechnic	22998	50

TABLE 4.11 Staff-student ratios in five major chemistry departments in Germany, c.1899

	Total students	*Staff–student ratio*
Heidelberg University	315	1:13
Strasburg University	48	1:7
Berlin Polytechnic	278	1:6
Stuttgart Polytechnic	88	1:9
Karlsruhe Polytechnic	139	1:9
Total/Average	868	1:10

emerged upon the basis of a broad scientific training in the German institutions, which mixed elements of social science with both theoretical physical science and highly specific applications (eg. machine drawing, electrotechnics, dyeing technology, building construction). Table 4.12 gives the details of a chemistry syllabus of the Stuttgart Polytechnic for 1899.

By the 1890s some 10 per cent of all Prussian university and 13 per cent of polytechnic students were foreign, a measure of the superiority of the German model for formal training.[68] But it has recently been maintained that, on the whole, 'state research' in

TABLE 4.12 Science and technology at Stuttgart Polytechnic

	Hours Per Week		
	1st Year	*2nd Year*	*3rd Year*
Mineralogy/Geology	9	5	
Zoology/Botany	6	9	
Physics	8		
Experimental Chemistry	8		
Theoretical Chemistry	4		
Analytical Chemistry	4		
Organic Chemistry		7	
Technical Chemistry		6	
Dye Industries			3
Building Construction	5		
Elements of Machinery		11	
Electro-Chemistry			1
Political Economy			3
Jurisprudence			6
Microscopic Labs		2	
Chemical Labs	24	24	48
TOTAL	68	64	61

Germany was pursued on a dual basis. The government provided for theoretical pursuits in the university sector, but important 'state research' was undertaken on behalf of government ministries in institutions specifically created for that purpose, especially in the fields of defence, health, veterinary science, crop protection, transport and communications and the establishment of technical standards and controls.[69] Unification in 1871 almost certainly influenced the transition from intervention centred on higher education and intervention through special institutions. Unification brought with it a greater attention to imperial goals – resource-exploitation, testing, standardisation and the development of colonial applied science, particularly in Africa. The first national research institute, the Physikalich Technische Reichenstalt of 1887 was formed from a combination of private pressure (especially that of Werner Siemens who invested half a million marks in the enterprise) with state support.[70]

Private and state interests intersected again in 1911 with the foundation of the Kaiser Wilhelm Gesellschaft, the object of which was to promote the sciences, especially by the 'foundation and support of scientific institutes of research'.[71] Large industrial finance was obtained through the subscription system, which required an initial contribution of £1000 for full membership followed by annual payments of £50.[72] Already by late 1910 the society had £500 000 at its disposal. Such money was to be devoted to the establishment of research institutes acting independently of any teaching function, in which eminent scientists could pursue sustained programmes of research whereby 'the progress of *science*, not necessarily of its industrial application, is contemplated'.[73] But is was from just such combinations of industrial, intellectual and statist interests that the more applied forums emerged. In the same year in which the Kaiser Wilhelm Gesellschaft was proposed so too were the Imperial Textile Bureau and the Imperial Chemical Institute. The imperial functions of the first of these were obvious enough; 'to encourage the use of genuine instead of imitation material (e.g. as regards dyes), to aim at the universal adoption of the metric system in the counting of yarns, to combat the weighting of silk, and to carry out investigations as to technical improvements'.[74] The Chemical Institute was supported by a combination of interests. Property and funds were donated by the Prussian government to the Association for the Protection of the Interests of the German Chemical Industry and large grants were given from both the Kaiser Wilhelm Gesellschaft and the Physikalich Technische Reichenstalt. The central research institute, formed to directly serve the needs of the chemical industry, was organised and financed in such a manner as to prompt close connections between university research, institute research, government and industry. So, whilst government had been important in sanctioning and financing the total set up for research, 'the costs incurred by the creation of this central institute will fall entirely upon the chemical and pharmaceutical manufacturies of the country'.[75]

The role of the state in the support of the scientific enterprise in Britain was probably of no less importance than in Germany, but it was certainly less obvious. In general, the activities of the British government and of the civil service were more complex, but less formalised than in, say, Germany or the USA.[76] As far as the scientific enterprise and the application of useful knowledge to economic ends were concerned, the British tradition of government intervention was very well established by the early nineteenth century. Although the growth of classical political economy during the early part of the century has been seen as the measure of a break with the mercantile interventionism of the eighteenth century – an acknowledgement of the superiority of market forces over bureaucratic interference – in science and in much else, in fact state intervention was transformed rather than reduced.[77]

In the eighteenth and early nineteenth centuries government support for the scientific enterprise was centred on economic applications. The establishment of the Board of Longitude in 1714, the financing of expeditions and surveys and exhibitions all fulfilled obvious military, colonial and commercial needs. These both employed and trained scientists over a wide range from astronomy to oceanography and geology. On the whole, such mercantilist functions required relatively more financial assistance than the laboratory-based university or institute science of the later nineteenth century. Scientists were employed at Kew (botany), Greenwich (astronomy) and at the Assay Office. The Inland Revenue Department sponsored munitions and geology, encouraged the growth of the Museum of Economic Geology and the Mining Records Office and distributed grants to learned societies.[78]

More simply, the State financed direct applications of knowledge to industry and commerce. Excluding the large costs of the above institutions, the British government invested over £206 000 in specific applications of knowledge over the period 1713 to 1815.[79] Such expenditure included rewards to inventors such as Edmond Cartwright or Samuel Crompton, but it also financed the investigative work or large-scale practical experiments that were prerequisites to the development of new types of windmill, dyestuffs, methods of fire prevention, land drainage, transportation and carriage and marine communication. Such grants were in addition to the expenses of intervention in the field of new legislation designed to stimulate technical change, such as the Act 'to encourage the Importations of Pig and Bar iron from the Colonies in America, and to prevent the erection of any Mills for slitting or rolling of iron, or any plating forge to work with a tilt hammer, or any furnace for making steel, in any of the said Colonies'.[80]

In the first half of the next century, government activity shifted to the diffusion and application of existing knowledge. Again, this was nowhere better demonstrated than in the patent system. During the short period 1833–9 seven major bills were brought into the Commons relating to improvements in the patent system[81] and between

1848 and 1851 nine major organisations carried the debate on the manner in which to encourage knowledge-diffusion and application into a very public arena.[82] Undoubtedly, over the 1850s and 1860s the form of government interference was greatly influenced by industrial opinion. During the 1850s dozens of provincial libraries and institutions were receiving technical literature (patent specifications, abridgements, journals, indexes and so on). By the late 1850s the Commissioner of Patents was spending some £100 000 upon the publication and distribution of such technical literature.[83]

Income from the Great Exhibition plus a government grant of £150 000 were used to finance a major departure in the British tradition, the establishment of the Department of Science and Arts in 1853, which took over the existing School of Mines and Museum of Practical Geology and established standards and examinations for technical studies throughout the nation.[84] Between 1853 and 1870 the DSA spent £52 000 on the examination system. From the demonstrated competence of other nations at the Paris Exposition of 1867, through the findings of the Royal Commission on Scientific Instruction and the Advancement of Science in 1870 to the traumas of the Boer War at the turn of the century, public scientists in Britain clamoured for a reform in the system of State support for science and technology. As suggested earlier, much of this was special pleading. Throughout these years the British government was supporting science along a number of lines, most of which have already been noted. In education, the universities received both central funding and in some cases very significant municipal or county support. The university college of Liverpool was instituted and maintained by its municipal administration. By the turn of the century the technical schooling system was receiving over £1 million from public funds.[85] Science within the civil service grew with imperialism and welfarism.[86] Learned societies benefited from grants, especially those for special expeditions, observatories and research stations. Most of such government aid was for specific tasks rather than as recurrent aid to research, although the Royal Society grant after 1850 financed a large number of small projects.[87] In 1876 Richard Proctor estimated that some £300 000 was expended annually across such categories, whilst Roy MacLeod's more recent calculation is that during the peak of 1889–90 over £650 000 or 4 per cent of the national budget was expended on this wide-ranging support structure.[88]

The opening of the National Physical Laboratory by the Prince of Wales in March 1902 represented the arrival of the German model on British soil.[89] The science journalists saw the NPL as Britain's *Reichenstalt*.[90] The original government grant of £4000 was thereafter increased, and by 1912 the NPL received an income of £32 000. But the NPL also depended very much upon private resources, and as with the new universities the British state provided plenty of buildings and land but stopped short of the liberal equippage of laboratories.

4.6 BEYOND THE PALE: THE ORGANISATION OF SCIENCE IN JAPAN

Japan ceases to exist as a separate nation and becomes a button-hook manufacturing appanage of America.

Rudyard Kipling, 1889

The foundation of the Japanese Industrial Laboratory just post-dated that of the NPL in Britain.[91] By the early twentieth century the Meiji government had created an institutional complex for the transfer of scientific and technical knowledge which was far removed from the Tokugawa's Office for the Study of Barbarian Writings of 1855.[92] Kipling's denigration of Japanese sychophancy during the 1880s gave way to the idea that Japan might join the commonwealth of scientific nations as a junior partner. Commenting upon the precision and instrumental ingenuity of Japanese scientists in 1894, F. A. Bather went on to state that 'in the exhaustive and microscopic questionings that the needs of modern science bid us put without cessation to over-elusive nature, the new Japanese recruits will doubtless prove our most valuable allies'.[93] By 1903, with a successful war behind it, Japan had turned the tables and became a model for emulation. In his highly-publicised speech of September 1903 on 'The Influence of Brain Power in History' the president of the BAAS asked his fellow-scientists, 'shall we follow Japan and throughly prepare by "intellectual" effort for the industrial struggle that lies before us?'[94] By June 1905 with a war against a European nation fully waged, *Nature* more soberly proclaimed that the 'lesson which our educationalists and statemen have to learn from Japan is that the life of a modern nation requires to be organised on scientific lines in all its departments . . . it must be consciously used for the promotion of national welfare'.[95]

From the very beginning the Meji government had ranged across the whole field of institutions covered in this chapter in its efforts to install the essence of the Western scientific and technical enterprise into industrialising Japan. The place to start was with foreign experts, and a conclusion must be that the Japanese government went to some lengths to hire foreigners whose expertise was relevant, although it is also true that several quite effective engineers (e.g. E. G. Holtham, resident engineer to the Kyoto–Kobe railway, which was opened in 1877) seemingly arrived 'on spec'.[96] Thus, the appointment of Henry Dyer as Principal of the Engineering College (Tokyo) was made as a result of the extended investigation of the Japanese ambassadors in Britain, who sought advice from such men as Professor Rankine of Glasgow University and H. M. Matheson of the Public Works Department of Japan. Dyer's plans for the college were only approved after consideration by the PWD's Vice-President, a Japanese who had formerly been a student at Anderson's Institute while 'he was learning the practice of shipbuilding in Napier's yard'.[97]

The Westerners appear to have made reasonable efforts to train the Japanese in their skills. At the School of Engineering, Dyer and his colleagues Edward Divers, D. H. Marshall and others encouraged their Japanese assistants to take over teaching and demonstration at as early a date as was feasible.[98] At the other extreme the crude and *ad hoc* railway engineering programme of E.G. Holtham was imbibed by his Japanese 'students' on the job and over a considerable period.[99] During the construction and planning of the railways, supervised by H. N. Lay and several English engineers (Tokyo–Osaka, Tokyo–Yokohama, Osaka–Tsuruga along Biwa Lake) several attempts were made to teach the Japanese involved basic engineering and surveying skills.[100] A sign of success in another field was the Osaka arsenal, which by the late 1870s was being run on modern lines and maintained in excellent condition without the guidance or even assistance of Europeans or Americans.[101]

The formal teaching of Japanese students at the theoretical and practical levels appears to have been substantial. Venues and formats varied enormously. At the School of Engineering in the 1870s a six-year training course was instituted, with the first two general years followed by special study in such Departments as civil engineering, mechanical engineering, telegraphy, architecture, practical chemistry, mining, metallurgy and naval architecture. In the third and fourth years a 'sandwich' course was adopted, with students spending half their time at practical work. In the final two years of study 'mere book work' was considered of secondary importance in an institution financed and organised by the Department of Public Works.[102]

It may be reasonably estimated that such teaching was comparable to contemporary developments in European engineering education. When the mechanical engineer John Perry was Professor of Engineering and Mathematics at the Finsbury Technical College in London in the 1880s he presented a course of studies very similar indeed to that which he had taught at Tokyo, with a particular emphasis on quantitative experiments.[103] In the much more isolated setting of Fukui, W. E. Griffis nevertheless attempted a very ambitious teaching program from 1871. Of the 800 students in the school several had been studying the English and other European languages privately for periods of two to three years at Nagasaki. Nor, seemingly, was there much shortage of funds, for 'they were ready with money and patience, to furnish the new-apparatus and lecture-room'.[104] Notably, the medical department was already stocked with Dutch texts and with French dissection models.

At Hitotsubashi another professor, Alfred Ewing, was teaching 200 students physics and chemistry and incorporating in his lectures the relatively recent findings of Thomson, Joule and Carnot. In quite a different mode were the plans from the 1860s to teach 'the European languages, and the sciences of shipbuilding, gunnery, seamanship and naval maneuvering' to Japanese students at Yokosuka naval arsenal. By 1870 plans were being made for extension of the teaching to include higher 'mathematics and mechanical science' and navigation and marine drawing.[105] Nor was this all wasted effort. There were many Japanese politicians, scientists, and businessmen whose early education owed everything to the pedagogic activities of such foreign experts resident in Japan. Thus, the graduates of the School of Engineering were sent abroad to British educational institutions and returned to become the backbone of Japanese engineering. As the fairly detached Henry Dyer observed at a later date, 'the former students of the Imperial College of Engineering are now to be found not only in all the most important engineering and industrial undertakings in Japan, but a considerable number of them are actively engaged in China and Korea, so that the College has been a most important factor in bringing about the changes in Japan and in influencing conditions in the Far East generally'.[106]

The experts acted as foremost advocates of further and wholesale reform of the Japanese educational system as a vital measure in the diffusion of Western knowledge. Thus Griffis was quick to write from Fukui to the Minister of Public Instruction at Tokyo urging the formation of a polytechnic. The result was a letter from the Mayor of Tokyo inviting him there to discuss plans further. In the same post was a letter from the Ministry of Education inviting him to a chair at the newly formed 'polytechnic' (*Sen Mon Gakkö*). But perhaps the best example of the role of the individual is provided by the American Dr David Murray (1830–1905) who had been Professor of Mathematics at Rutgers prior to his appointment as Superintendent of Schools and Colleges in Japan.[107] Murray's mission in Japan was to institute methods similar to those recently developed in New Jersey. Here the 'Rutgers Scientific Schools' system provided entirely scientific and technical education at a high level for non-classicists and in 1865 the State College was formed under the direction of Rutgers for the 'benefit of agriculture and mechanical arts', with an emphasis on laboratory work.[108] During his six years' stay in Japan (1873–9) he attempted to forge educational institutions along New Jersey lines, as well as establish the basis for universal education. His *Outline History* of 1876 evidenced his educational expertise and understanding of the potential in Japan.[109]

Scientific societies were an immediate forum for the mingling of experts and lay people, the most significant of which were probably the Medical (1877), Chemical (1878), Geological (1879), Engineering (1879), Seismological (1880), Mining (1885), Agricultural (1887) and Electrical (1888).[110] The societies were important agents of knowledge transfer. The seismological would be chaired by either Japanese or Western scientists and it boasted large mixed audiences and produced its own regular journal, *The Seismological Journal of Japan*. Through it, Professor O. Kikuchi of the Imperial University issued circulars to officials of provincial coastal towns and villages, requesting them to submit evidence of any encroachments or recessions on the seabeds. In this manner bradyseismical changes were observed in 'several thousand' instances.[111] The *Transactions* and *Journal* of the society gave evidence of the extent of Japanese involvement in seismology in the 1890s. Similarly, the Engineering Society, possibly the largest in Japan by 1890 (the exception might have been the Geographical), was normally chaired by Japanese academics (e.g. Professor Mano of the University, a graduate of Glasgow), and sponsored lecture courses given to large audiences which, although delivered in the English language, were mainly composed of Japanese.[112] It too published its own *Journal* and this and the society as a whole was gradually taken over into Japanese hands.[113]

In addition to a host of formal institutions, the transplantations of Western science and technology was considerably hastened by learning processes contained in key development projects.[114] The *Japan Herald* approved of the educative effect of railway projects: 'It has, we believe, always been an object with the Engineer-in-Chief to make this line, as it were, a school in which the Japanese should gradually learn something of railway practice, and it is in some measure producing this effect.'[115] The *Kobusho* or Public Works Department was applauded as a 'great centre of administrative action' and the native accuracy in construction (e.g. the bridge at Rokugo) was admitted. Learning and adapting by doing were the orders of the day: '[Japanese workmen] are able to turn out very fine specimens of castings and other iron work, while they were able to alter the gauge of some ordinary navigation trucks from 4 feet to 3 feet 6 inches, turning new axles and completing the work in a thoroughly efficient manner. This speaks a great deal for their capabilities of using foreign machines and tools, and it is no slight praise to say that some castings for screw piles used at Rokugo are equal in every respect to those made by Europeans.'[116] By the mid-1880s surveying, engineering and construction work on new projects was predominantly Japanese, the engines and offices were manned by Japanese, whilst the basic wooden and lighter metal parts were manufactured in Japan (heavy castings and rails were still imported from Britain).[117] Of just as great an effect was the physical opening of the projects: Professor Griffis was present when some 20 000 spectators witnessed the Emperor and his retinue board the railway for the first trip on the Yokohama–Tokyo line in 1872.[118]

It is evident that the process of technology transfer involves much in the way of institutions. But it also requires demonstration, and in the case of Meiji Japan this was eased by the power of government to select and control that which came from the West. That networks of individual became of key importance in the diffusion of knowledge is just as certain. Especially in the strategic and newest industries, Western knowledge was introduced by selected agents of change, and was more certain to be successful if such figures had been associated with the *öyatoi* (hired foreigners) and other Westerners in informal and formal associations.

4.7 CONCLUSIONS: INSTITUTIONS AND INDUSTRIALISATION

> The great new chemical industries owe their existence entirely to a profound and scientific knowledge of chemistry.
>
> *William Whewell at the Great Exhibition, 1851*

> The progress of abstract science is of extreme importance to a nation depending upon its manufactures. . . . The cultivators of abstract science are the horses of the chariot of industry.
>
> *Lyon Playfair at the Great Exhibition, 1851*

The traditional stress upon universities and research programmes has given rise to theories which relate industrialisation to innovative institutions (e.g. Germany or Japan) and deindustrialisation to retardative institutions (e.g. Britain or France). Our broad treatment of the scientific enterprise of major nations suggests that a somewhat more complex series of linkages was involved. The need for a broader approach is of especial importance now that economic historians have thrown very serious doubts upon earlier notions of economic decline or stagnation in Britain and France prior to 1914.[119] The British chemical industry certainly fell behind that of Germany in dyestuffs output, just as every nation was demoted by the tremendous growth in the US steel industry. But in 1870, or even later, neither the German nor the US economy was on a technical par with Britain overall. As David Landes put it for Germany, 'hers was an economy that, for all its capabilities, was well behind Britain in 1870 in assimilating and diffusing the technology of the industrial revolution'.[120]

Just as the fast growth of a chemical or steel industry may bypass shortcomings elsewhere in an economy, so too the rapid growth of

TABLE 4.13 Patterns of national scientific enterprise prior to 1914

	Sources of support				
	Socio-Culture		*Private*	*Government Enterprise*	
Characteristics	*Audience*	*Patronage*	*Enterprise*	*Central*	*Provincial*
Institutionalisation	medium	weak	strong	strong	medium
Specialisation	weak	medium	strong	strong	strong
Professionalisation	medium	weak	medium	strong	medium
Large-scale	weak	medium	strong	strong	medium
Vertical diffusion (social)	strong	weak	weak	medium	strong
Horizontal diffusion (spatial)	strong	medium	medium	medium	strong
ST–industrial links	medium	weak	strong	medium	strong

scientific laboratory research in universities or business enterprises may disguise major inadequacies in the manner in which knowledge is adopted, diffused or applied or in which human capital is deployed. Historians overly-concerned with the origins of 'professionalisation' in science or with 'research and development' may mistakenly equate such processes with the economic role of the scientific enterprise. Yet the social advancement that is promised with professionalisation may well lead to an uneconomic oversupply of certain high-level skills, just as enterprise research and development may be devoted to relatively unproductive gamesmanship over intellectual property rights.

Based only on the most general historical impressions, Table 4.13 charts the stylised relationships between the *sources* of support for the scientific enterprise (as introduced in section 4.1) and certain *characteristics* of the scientific enterprise as they appeared in the major industrial or industrialising nations prior to 1914.

If these judgements are at all accurate, then in nations where fast industrialisation was associated with government intervention over a great range of economic and social matters (e.g. Germany or Russia or Japan), but where a wide base of support for modern science was of little importance, there would be great evidence of modernised universities and other formal institutions, professionalism, a large scale of operations and a crowding of knowledge applications in areas of the economy of strategic importance – military industries, transport and communications. In nations like Russia or Japan scientific knowledge would be treated as a public good, the production of which was the task of state rather than private institutions. A nation such as Britain, with its wide 'audience' for science, might actually seem to be falling behind in science if 'progress' in science is measured in terms of such criteria. But in terms of the creation of new, abstract knowledge, in terms of the diffusion of information through the social system and in terms of the sustenance of routine 'ordinary inventiveness' (see Chapter 3) throughout the industrial system, Britain may well have been significantly ahead of most nations at this time. Writing within a strongly comparative framework, Ben-David depicts the support structure for scientific enterprise in Germany in the later nineteenth century as guild-like and restrictive.[121] As Chapters 5–7 will demonstrate, industrialisation under conditions of relative economic and institutional backwardness involved government interference over a wide range, encompassing the promotion of military industries, the erection of mechanisms of social control and the provision of rewards to the strategic elites. The scientific enterprise could be made to serve all these functions.

5. Technology, Economic Backwardness and Industrialisation – General Schema

5.1 CONTEXT: THE DEMONSTRATION OF FORWARDNESS

As we move across the face of Europe in any given period, we find many differences of culture, religious belief, social attitudes, and so on, and they change over time. They will always affect the way in which economic or technological changes were received, and they did so undoubtedly in our period. However, they were increasingly put in the shade by the progress of industrialisation itself which came to dominate the reaction of a society as it came to dominate so much else.

Sidney Pollard, 1982

AN Anglocentric view of Europe in the mid-nineteenth century draws a picture of the democratic nations, led by Britain, fearful of the autocratic, armaments-orientated colossus in the East. Yet far more real, and certainly more immediate and prophetic, was Russia's fear of the *industrialising* nations of Europe. By 1851 the process of industrialisation dominated European politics.

Militiaman Denis Davidov vividly depicted the effect of Napoleon's invasion of Russia in 1812.

> 'All our Asiatic attacks made not the least impression on the compactly serried European formation. The columns moved one after another, driving us away with their riflefire and making mock of our futile equestrian antics but the Guard with Napoleon in their midst passed through the throngs of our Cossacks like a hundred-gunned frigate through a swarm of fishing rowboats'.[1]

The impact of the Western invasion on the economically backward East was not confined to dramatic military losses, nor to 1812. The Smolensk peasantry complained to the occupying French authorities of how their local landlord master had been guilty of betraying the invading forces! The landowner was shot. Although conquered, the French were, in a sense, victorious in Russia. French ideas, images and fashions multiplied in Russian society after the military defeat of Napoleon.[2] Such occidental influence did not sit comfortably with Tsarist control. After 1820 Alexander made several attempts to suppress Frenchified 'natural philosophy' as deistic, as a demonstration of the dissolute 'nonconformity of various philosophical systems'. Nicholas I (1825–55) supported a similar programme of suppression of Western European influence. But although the reign of the gendarme of Europe produced S. S. Uvarov's (1786–1835) memorandum on 'Orthodoxy, Autocracy and Nationality', and established a pervasive secret police network, these years were perhaps better represented by the public debate between Slavophiles and Occidentalists, populists and modernists, which on the pro-European side produced such traumatic works as V. Belinsky's *Letters to N. V. Gogol* of 1847 and Alexander Herzen's *From the Other Shore* of 1849. By the 1840s the demonstration of forwardness had been effectively accomplished. Even prior to the Crimean War, the machinery of the Russian state was finding it increasingly difficult to move into reverse gear.

Although the case of Russia is a model example of the demonstration of relative forwardness (and, therefore, of relative backwardness), similar conservative reactions to the presence of industrial modernisation elsewhere may be found in the patriotic societies of contemporary Prussia or Italy. Although utilised whenever possible by strategic elites whose immediate focus was the retention of power and internal control, populist reactions to industrialisation were fighting miserable battles by the 1850s.

Within Europe, the period from the 1840s to the 1870s witnessed demonstration after demonstration of the hardening links between military strength and the industrialisation process. Big nations lost wars. Although in 1835 Andrew Ure, industrial apologist extraordinaire, was imprudent enough to advance the thesis that economic rivalry had *replaced* military rivalry within Europe, the years following 1848 revealed the lie.[3] Technological progress had already transformed warfare. In 1836, one year after Ure, Baron Henri Jomini (1779–1869) wrote in his *Summary of the Art of War* that 'the new inventions of the last twenty years seem to threaten a great revolution in army organisation, armaments and tactics'.[4]

None of this was lost on states or strategists. From an early date the emphasis placed by Friederich List (1789–1846) on the strategic importance of railroad systems exerted a powerful influence in Prussia. In 1848–50 railways were to be fully utilised in the movement of troops in both the revolutionary wars and in the French involvements in Italy. The proposal for an ideal Europe of merely eleven nation states, advocated by Giuseppe Mazzini (1805–72) in 1857 cut across all traditional claims to nationhood (e.g. those of the Czechs) and was based on the recognition that in a world of complex military technology smallness of size was tantamount to obliteration. Such elements of a 'new view' were encapsulated in the strategic proposals of the Prussian, Count Helmuth von Moltke (1800–91). Formed from his understanding of the new technologies, von Moltke's recommendations included envelopment (engagement at the flanks) rather than frontal attacks, decentralisation of placements and decision-making, fast movements and ready adaptability to rapidly changing circumstances. The very creation of the German Empire, via the brief battles of the Austro-Prussian war of 1866 and the Franco-Prussian war of 1870–71 owed much to superior Prussian technology (rifles to railways), organisation and tactics (massing of artillery, flank assaults, defensive and prone entrenchments). The rewards were considerable: the war indemnity in the latter case amounted to £200 000 000.[5]

From the 1860s the military demonstration of technological forwardness proceeded apace and well beyond Europe itself. If in America the industrial north defeated the agricultural south, the failure of the Taipings in China (see Chapter 9) symbolised the power of industry even more generally. From 1877 the imperial wars of the eight major powers, who thereby added 11 million square miles of territory to their own, depended upon the efficient development of modern rifles, machine guns and quick-firing artillery, as well as the newer traction engines for supply (e.g. in South Africa), wireless telegraphy, Daimler's internal combustion engine, especially put to the test in the wars of 1898–1905. There was nothing semi-permeable or osmotic about the creeping frontier of Europe.

5.2 THE LOCATION OF PRODUCTION

But the demonstration of forwardness did not occur in a world which could be nicely divided into developing industrial nations and backward agrarian hinterlands. As late as one hundred years ago, the major, specialised pockets of industrial and commercial–agricultural production lay scattered throughout the world. Table 5.1 itemises no less than 1174 major towns or trade centres outside of Britain in terms of their 'best route' distances from the commercial centre of London in 1887, and indicates their respective emphases upon industrial or agricultural production.[6]

Although telling us little about efficiency of production, the extent of actual trading or details of industrial structure, the table does serve to emphasise that as late as the 1880s specialised and urbanised production centres *per se* were by no means confined to a small group of 'winner' economies. What is more, 25 per cent of total and 22.2 per cent of industrial production centres lay on routes over 6000 miles from London; 57 per cent of the total and 50 per cent of the industrial production centres lay on routes over 2000 miles from London and predominantly outside Europe. Modernised, efficient industrial production may have been a feature of the European and Atlantic economies, but throughout the world proto-industry was seemingly rife. Table 5.2 gives more detailed production breakdowns for 235 urban centres in France, Germany and Russia categorised in terms of the two major 'productions' of each centre.[7] Here our focus sharpens a little.

France demonstrates a real emphasis on textile production (42 per cent of all its specialised production centres), Germany less so, and Russia far less so. Compared to the other two nations, Russia shows up as a predominantly agrarian economy, but with a significant number of centres of production in shipping and shipping services and a variety of machine, craft and extractive industries. Even Russia was not simply an underdeveloped agrarian nation.

TABLE 5.1 Location of 1174 major production centres worldwide, c. 1887

Area & distance from London ('best route' miles)	*Industrial centres*		*Agrarian centres*		*Total centres*	
	Number	*% (of 765)*	*Number*	*% (of 409)*	*Number*	*% (of 1174)*
Europe						
less than 100 mls	1	0.1	0	0	1	0.1
100–300	47	6.1	8	2.0	55	4.7
300–400	18	2.3	3	0.7	21	1.8
400–600	61	8.0	13	3.2	74	6.3
600–800	68	8.9	24	5.9	92	7.8
800–1000	57	7.5	25	6.1	82	7.0
1000–1500	80	10.4	24	5.9	104	8.9
1500–2000	37	4.8	10	2.4	47	4.0
2000-plus	50	6.5	45	11.0	95	8.1
TOTAL	419	54.8	152	37.2	571	48.6
North America						
2000–3000	8	1.0	2	0.5	10	0.8
3000–4000	32	4.2	3	0.7	35	3.0
4000–6000	16	2.1	5	1.2	21	1.8
6000–12000	0	0	0	0	0	0
12000-plus	3	0.4	2	0.5	5	0.4
TOTAL	59	7.7	12	2.9	71	6.0
Elsewhere						
less than 2000	15	2.0	16	3.9	31	2.6
2000–3000	17	2.2	9	2.2	26	2.2
3000–4000	43	5.6	38	9.3	81	6.9
4000–6000	45	5.9	60	14.7	105	8.9
6000–9000	84	11.0	71	17.4	155	13.2
9000–12000	73	9.5	36	8.8	109	9.3
12000-plus	10	1.3	15	3.7	25	2.1
TOTAL	287	37.5	245	60.0	532	45.3
Total world						
Less than 6000	595	77.7	285	69.7	880	75.0
More than 6000	170	22.2	124	30.3	294	25.0
TOTAL	765	100.0	409	100.0	1174	100.0

Against this admittedly generalised backdrop, questions arise as to how goods were produced, under what auspices and with what degree of economic efficiency. Why did some production centres represent poles of economic growth and development, whilst others

TABLE 5.2 Major production in 235 urban centres in France, Germany and Russia circa 1887 (percentages)

Major production (categories) %	*France*	*Germany*	*Russia*
Total centres (number)	114	75	46
Total prod-Categories (number)	215	138	71
Agriculture & fisheries	13.0	6.5	29.6
Textiles	42.3	36.2	9.9
Mining	0	1.0	4.2
Specialised agrarian processing	0	13.0	0
Sugar	–	5.1	–
Tobacco	–	6.0	–
Other	–	1.9	–
General commerce; transit, colonial	2.3	11.6	5.6
Shipping/dockyards	7.0	4.3	19.7
Iron & steel production	2.8	2.2	2.8
Chemical manufacturers	4.2	7.2	1.4
Metal-working	5.1	3.6	2.8
Cutlery	3.3	1.0	0
Other	1.8	2.6	2.8
Other manufacturing	10.2	6.5	16.9
Handicraft production	9.8	5.1	2.8
Other	3.3	2.9	4.2
TOTAL	100% (215)	100% (138)	99.9% (71)

acted as drains upon their respective agrarian hinterlands? Why did some centres of production act as receptors for advanced technology transfers and diffusion while others remained outside the modernising, industrial economy? Why did some European nations undergo periods of demonstrable industrial revolution when in others the process of proto-industrialisation petered out in 'failure'? The answer to such questions can not be merely that existing capabilities or populations did not exist, for Table 5.1 suggests another story. Any approach to such far-reaching questions must surely address itself to the technological characteristics of an array of diverse production centres and processes. Chapters 6–10 are devoted to precisely this theme, and in combination illustrate that technological progress and associated industrialisations did not *automatically* result from advances in knowledge, transfers of technique or the possession of a necessary set of natural resources and institutions.

Given that so many points of *potential* technique 'reception' lay scattered throughout the world, it might be thought that in a period wherein the demonstration of forwardness/backwardness was fairly complete, the *advantages* of relative economic backwardness would outweigh its obvious disadvantages. Relative backwardness led to a speedy 'catching-up' process for a select group of nations, whose existing proto-industrial points of production and urbanism were rapidly transformed and added-to by the erection of transport and communication systems which permitted the creation of altogether new points of production.

5.3 CATCHING UP

In heroic mood, Paul Bairoch has provided estimates of industrial production for a selection of nations in the long-swing of 1750–1913.[8] Tables 5.3 and 5.4 below are derived from his presentation and represent index estimates based on UK in 1900 = 100.

TABLE 5.3 Volume of industrial production, index, UK in 1900 = 100

Nation	*Total Industrial Production (Years)*						
	1750	*1800*	*1830*	*1860*	*1880*	*1900*	*1913*
Britain	2.4	6.2	17.5	54.0	73.3	100.0	127.2
France	5.0	6.2	9.5	17.9	25.1	36.8	57.3
Germany	3.7	5.2	6.5	11.1	27.4	71.2	137.7
Japan	4.8	5.1	5.2	5.8	7.6	13.0	25.1

TABLE 5.4 Per capita levels of industrialisation, index, UK in 1900 = 100

	Per capita industrial production (years)						
Nation	*1750*	*1800*	*1830*	*1860*	*1880*	*1900*	*1913*
Britain	10	16	25	64	87	100	115
France	9	9	12	20	28	39	59
Germany	8	8	9	15	25	52	85
Japan	7	7	7	7	9	12	20

TABLE 5.5 Production of steel in the four main producers 1865–1912 (Index, UK in 1865 = 100, volume)

Years	*Nations*			
	Britain	*USA*	*Germany*	*France*
1865	100	6	43	18
1875	321	176	165	114
1885	898	773	534	246
1895	1531	2761	1752	400
1900	2650	9046	4474	982
1912	2973	14112	7942	1812

The early years are dominated by proto-industry production, with Britain in a relatively low position, a reflection of location, population size and resource endowment. Britain's industrial revolution is indicated in relative advancement between 1800–1880. The 'catching-up' process for Germany and Japan centred fairly dramatically on the years 1880 to 1913. Table 5.4 illustrates that the industrial performance of these two nations was hardly less startling in per capita terms after 1880.

Prior to 1830 Britain's per capita performance was even more dramatic than simple output performance; as early as 1800 Britain demonstrated a significant lead. After 1880, per capita measurements show up the improved comparative position of France. By focusing upon particular industries whose growth is known to have been closely related to new technological applications (see Section 5.5) we come closer to the real 'advantages' of backwardness. Table 5.5 shows the tremendous growth of steel production in America and Germany compared to the more established powers of Britain and France; the volume of British production of steel in 1865 is set at 100.[9]

The potential advantages of backwardness are well illustrated in the dramatic rise of the German heavy chemical industry. In 1870 the chemical industry of Germany was barely visible in comparison to that of Britain, whose production of soda was dominated by the Le Blanc process. From that date a series of inducements operated in such a fashion as to create a surge in the German heavy chemical industry: (a) the independent rise of a host of chemical-using industries within Germany itself – especially in textiles, glass, soap, paper, metallurgy and fertilisers; (b) German entrepreneurs faced no substantial *scrapping* burdens as, unlike Britain, there existed no significant prior investment in the Le Blanc process; the demand for *sulphuric acid* could therefore be best satisfied by the new Solvay process of *soda* manufacture, i.e. the interrelationships of production gave rise to spinoffs; (c) the operation of (a) and (b) together encouraged a series of new technological innovations and organisational formats in the German chemical industry generally, the development of enterprise-based research groups engaged in aggressive patenting, and the subsequent emergence of the electrochemical industry.[10] In contrast, the relatively 'forward' chemical industry in Britain was, by the 1890s, facing a declining demand for its products from the *British* textile sector. Industrialists already heavily committed to soda production by Le Blanc methods hence resorted to exporting bleaching powder in the face of falling prices, and to attempts at price control in the face of German and American tariff restrictions on soda imports. From the 1880s the protected German industry found its own export markets and moved towards the creation of an electrolytic alkali industry.

Examples such as these suggest that the demonstration of forwardness had, by the later years of the nineteenth century, given rise to a milieu which promoted technological 'catching-up' on a grand scale in those industries susceptible to rapid technological change. We must yet consider the question as to *how* such catching-up took place (i.e. under what historical conditions 'success' was achieved) and, from this, *why* such transformations settled around (1) *industrial manufacturing* (none of the material so far presented tells us anything of the total growth process in 'winner' economies), and (2) a relatively *small group* of nations which, with the important exceptions of America and Japan, lay within Europe.

5.4 MODELLING 'RESPONSE'

Writing very generally on 'lessons of history', R. M. Hartell touches upon the advantages of backwardness:

> First there is the advantage of 'the late start', the advantage of drawing on a great fund of knowledge and experience, of mistakes made and corrected, of techniques perfected, of institutions proved satisfactory. . . . This is very important, especially as the

developed countries do suffer, in varying degree, from technological obsolescence, so that initially the new countries can have many advantages.[11]

Marion J. Levy, presenting, in a quite different context, the problems of modernisation in the twentieth century, expresses very similar thoughts even more succinctly: 'The latecomers . . . live in a world in which the contrast – between their own states and those of peoples who have crossed the dividing line between the modernised and the non-modernised is one of the most important factors for them. . . . The gaps that latecomers must jump guarantee a considerable impact of strangeness.'[12]

Alexander Gerschenkron has been responsible for the formalisation and extension of such general observations.[13] Gerschenkron's leading hypothesis is that, 'very significant interspacial variations in the process of industrialisation are functionally related to the degree of economic backwardness that prevailed in the countries concerned on the eve of their "great-spurts" of industrial growth'.[14]

Concentrating his attention on the process of industrialisation *per se*, (throughout his argument there is no explicit premise as to the relations between industrial and economic development), Gerschenkron argues that in late nineteenth-century Europe differences in the process of industrialisation between nations resulted from 'institutional instruments' or *creative substitutions* called up by the fact and recognition of relative backwardness. There is no circularity in the reasoning here – backwardness may be defined both absolutely and in terms of the nature of substitutions and the patterns of industrial development in the successful, responsive nations.[15] The pre-industrial history of the backward nation is as nothing to the richness and complexity evidenced in the early development of the relatively forward nations (e.g. England, France). Contrariwise, the history of the industrial 'spurt' in the relatively backward nations (e.g. Germany, Italy, Austria, Bulgaria, Russia) is a rich matrix of institutional responses and innovations. The notion of challenge and response is dominant yet subtle; the greater the magnitude of the challenge, i.e. the degree of relative backwardness and the immediacy of its recognition by those in a position or seeking a position to modify it, then the greater not only the *magnitude* of response but the degree of change in the *quality* of response. A measurement of the 'diminution of backwardness' through time is the degree to which innovative, functional substitutions have given way to a more 'open' working of the total political economy. Thus in Germany the 'liberation' of industry from financing by banks is indicative of the process of diminution of backwardness, as is the increasing independence of industrialisation from state guidance in Russia during the years 1907–13.[16] Although it is never made explicit and possibly is not intended, the idea of 'convergence' of systems *within* the limited later nineteenth-century framework does emerge from Gerschenkron's treatment.

One of the enormous strengths of the schema lies in Gerschenkron's emphasis on the dynamics and discontinuity of the 'great spurt' or *industrial drive* period, this itself being closely associated with his holistic, 'political economy' approach to the subject of industrial revolution: 'To break through the barriers of stagnation in a backward country, to ignite the imaginations of men, and to place their energies in the service of economic development, a stronger medicine is needed than the promise of better allocation of resources or even of the lower price of bread.'[17]

Socio-economic structures and prejudices must be transcended and this becomes that more likely in a relatively backward nation the stronger is the 'tension' between its real position and the needs, fears and ambitions of its industrial and political leaders. The nature and the speed of response means that the notion of 'prerequisites' becomes not only redundant but is translated into *corequisites* or even *effects* of industrialisation. Once more we see the emphasis on discontinuity and variability: 'The line between what is a precondition of, and what is a response to industrial development seems to be a rather flexible one.'[18] For Gerschenkron this breakdown in the concept of preconditions or necessary prerequisites embraces a range spanning on the one hand agricultural revolution and on the other the process of capital accumulation.[19] Because of this the hitherto dominant 'English model' is no longer relevant as a replicable examplar and industrialisation appears 'as an orderly system of graduated deviations from that [English] industrialisation'.[20] If there is *a* precondition for the industrialisation of a relatively backward nation then it is, simply and overwhelmingly, *that of the prior and recognised industrialisation of the advanced nations*. Whilst the preconditions of English industrialisation lay internally within a mercantile past, in relatively backward nations there was 'much less continuity with the preceding mercantile development and instead a simultaneous creation of financial institutions and modern industrial enterprises with their heavy needs for capital dispositions'.[21] Thus Gerschenkron abandons the 'arithmetical quirk' of aggregated continuity for the disaggregated fact of 'specific discontinuities in the form of great spurts'.[22]

5.5 IMPASSIVE MISSIONARIES?: RELATIVE BACKWARDNESS AND THE MECHANISMS OF TECHNOLOGY TRANSFER

enterprise, like love, usually finds ways to laugh at locksmiths.

David Landes, 1968

We have now reached the stage where we might invoke the claim that the underpinnings of the Gerschenkron approach are those of the differential emergence, diffusion and application of technology and its similarly differential transfer between nations.

Although technology is not accorded pride of place in the *formal* presentation of the framework, Gerschenkron appears to conceive of transferred technology as the prime instrument whereby industrialisation is actually realised once the relatively backward nation has 'recognised' its situation and has removed fundamental socio-institutional barriers or 'obstacles' (e.g. serfdom).[23]

There are three major arguments as to why any explanation of industrialisation under conditions of relative economic backwardness *must* focus on the mechanisms and process of technology transfer. Firstly, technology and skills are simply lacking in the 'receiver' nation (i.e. are at a premium). As some confusion might result from this point, a rare instance of Gerschenkron's view is worthy of quotation at length:

> illiteracy and low standards of education, and the resulting difficulty in training skilled labour and efficient engineers, can be overcome to some extent by immigration from more advanced countries and to some extent by using the training facilities of those countries. The same is true, even more importantly, of the lack of a store of technical knowledge. It can be imported from abroad. In this sense, however, one can say that in a backward country there exists a 'prerequisite' to industrial development which the advanced country did not have at its disposal.[24]

On such grounds McKay shows the great value of foreign technology in Russian industrialisation from 1885 to 1913.[25]

Secondly, economies of scale and indivisibilities are associated with advanced, large-scale imported technologies. Technology is thereby a determinant of those *overall* characteristics of bigness and speed exhibited by relatively backward nations during their 'great spurts'. Thirdly, technology is ultimately linked with other fundamental elements of the schema. An essential role of the industrial investment banks was that of carriers of technology, the 'imitative' nature of the Italian case being seen as a result of German technological influence operating directly through financial tutelage.[26] Several fundamental units of the Gerschenkron schema – banks, the state, 'tension' and foreign capital – appear to be functionally related to technology and its transfer. One point about which Gerschenkron does not disagree with Hicks is over the great role of technology in industrial development, that which lends the latter its 'essential novelty'.[27]

However, all of this depends upon whether *in fact* foreign technique, and its related elements are indeed transferred effectively to the backward nation poised on the brink of industrialisation. As Ryoshin Minami has expressed it, 'simply having a backlog of technology available is not a sufficient basis for the successful emergence of a modern economy'.[28] Minami then points to three fairly recognisable elements in the successful absorption by Japan of technology from abroad; its labour-absorbing character, its utilisation of existing social overhead capital and its encouragement by a 'well-organised government'. His last remark is important and true, but difficult to further specify; 'Finally, according to the Japanese experience, policies to develop the social capacity to absorb modern technologies are of great importance'.[29] Our treatment of Japan in Chapter 7 considers such points more fully. In summary, transfer of foreign technology is not automatic and, therefore, is not automatically a solution to the problems of relative backwardness.

The social and institutional constraints on transfer qualify the simple form of 'late starter' thesis and need to be analysed and accounted for when examining late industrialisation. It is this perception which informs many of the arguments presented in Chapters 6–9.

With reference to European and American industrialisation in the late nineteenth century, the effective *mechanisms* for the transfer of technology may be divided into two interrelated sets. As we shall summarise at the beginning of Chapter 6, the *effectiveness* of consequent transfers was determined by the operation of three major sets of *boundary conditions*, within which technology transfers led to a more generalised process of indigenous industrialisation. For heuristic purposes, the two sets of *transfer mechanism* may be identified as follows.

Indirect Transfer Mechanisms mostly relate to processes operating within the international economy at that time. In the Gerschenkron schema, technology is purchased at a price, not transferred as a free gift. So such factors as relative prices and the terms of trade, the balance of payments of the 'receiver' nation, and the interest rates available on investment or capital goods to capitalists or agencies in the technology-exporting nations are of relevance. Below, such mechanisms are dealt with under the four subheadings of Commodity trade, Capital movements, Migration and Technological progress and technology transfer.

Direct Transfer Mechanisms incorporates the more normal meaning of the term 'transfer mechanism' – trade journals, translations, knowledge flows, patent agreement, movements of people and machines, as well as *existent* or created *institutional* arrangements for the transfer of knowledge, techniques, training procedures, and so on, as well as for the efficient 'settlement' of the new process.[30] We interpret such institutional arrangements as part and parcel of the 'creative substitutions' of the Gerschenkron schema (above). Below, such a variety of mechanisms is surveyed under the headings of Individual agents; The State, armaments and autonomy; Institutions for Dissemination; Commercial Institutions; and International Patenting.

Indirect Transfer Mechanisms

Commodity Trade In 1850 some 20 per cent of world commodity trade passed through Britain, the most technologically advanced nation in the world; French trade, at half the level of the British, was yet second in rank. In Britain intellectuals such as Mill and Cobden advocated the unrestricted expansion of trade, which would lead to a spread of 'liberalism' (i.e. European ideas and ideologies) as well as of artefacts, predominant amongst which would be the new technologies of the industrial revolution. Although already proclaimed by Frederich List, the alternative nationalist and protectionist ethos was in its infancy and still echoed the mercantilist fears of earlier years.

If world trade expanded at an annual average rate of 4.2 per cent between 1820 and 1870, a more startling measure is the 260 per cent increase in world trade over the period 1850–70.[31] Britain by no means lost her dominance of total trade; in 1870 perhaps 25 per cent of all commodity trade was British. The volume of world trade (with an inflation in prices) was growing faster than that of the *European* economies, i.e. several European nations experienced a substantial rise in their external trade/national income ratios, Belgium from the 1820s, Spain from the 1840s, Germany from the 1850s. Liberal notions, an elastic supply of gold and the expansion of new means for settlement of international payments *permitted* (they did not create) a freedom for international enterprise never witnessed prior to 1850. Although the expansion of world trade decelerated somewhat after 1870, total transactions were increased by an acceleration in the movement of capital between nations. Not all economies could gain relatively; the shares of Russia and Latin America in world trade were falling prior to the 1890s. Obviously, the general bouyancy of trade reflected industrialisation itself, but was further aided by improvements in the mechanisms for trade (payments and insurance systems), transport improvements (which reduced prices), the increased *specialisation* within the industrialisation process, and the acceleration in the movements of both capital and people.

The age of *Gewerbefreiheit* (freedom of market entry) preached the destruction of guilds, the removal of internal and external tariffs, the abolition of usury laws and of official controls over extractive industries. Furthermore, the two largest trading nations prior to the 1880s, were substantial *debtors* on their trading accounts, Britain after 1850, France from 1866, imbalances which were offset by earnings from interest and dividend payments and shipping, insurance and brokerage services.

Until the 1880s and perhaps beyond, the explosion of world trade and the dominance of Britain, created specific mechanisms for technology transfer.

Nations whose export industries boomed gained surpluses which could be used for the purchase of superior technologies. This was particularly true of nations who traded substantially with Britain or France. As early as the 1840s Platts of Oldham exported 50 per cent of their cotton machinery. Between then and the 1870s, the volume export from Britain of rail iron and steel increased over threefold, the export of machinery multiplied ninefold. At least 30 per cent of US imports by value during the 1850s were in the form of iron and steel and machinery.

Also the expansion of foreign demand encouraged the transfer or application of new technologies in order to increase the efficiency of potential export industries. This mechanism was possibly of greatest importance in industries which had already expanded considerably on the basis of increased home demand, for example in the German steel and chemical industries of the late years of the nineteenth century.

In addition, the penetration of traded manufactured commodities (cotton textiles, iron and steel, machinery) into technologically backward nations from Britain, Belgium or France *competed* with *existing* indigenous manufacturing (especially where external transport systems had been improved) and thus induced a search for efficiency improvements amongst indigenous entrepreneurs.

A rising trade *ratio*, even in the context of a negative trade balance, was often associated with a period of *capital* importing by the trading economy. An increase in German trade led to an influx of foreign capital, entrepreneurs *and* engineers into the mining and metallurgy sections of that economy.

Finally, the exploitation of comparative advantage via *increased* specialisation of production meant that trading economies secured positive 'economies of specialisation' in the strategic area of producer goods. The demand for chemicals or machinery or tools from specialised, export-orientated purchasers of final goods *permitted* the producers of intermediate goods to utilise technologies which would otherwise have not been cost-effective.[32]

Capital Movements Until the 1850s foreign investment flows mainly took the form of purchases of foreign government securities, mostly for political purposes (Latin America) or public works (the individual borrowings of US states). The break in this pattern was led by a growth of railway securities and followed by increased international investment in mining, metallurgy and, eventually, manufacturing industries. There were exceptions in earlier years. During the 1830s

British capital had combined with British technique and engineers in thwarted attempts at the development of steam shipping in France. The boom in the German mining sector during the same decade was a result of foreign capital imports into that sector and the employment of French, Belgium and British skills in the Ruhr ironworks. The European revolutions of 1848 encouraged British financiers to seek outlets outside the Continent, particularly in America and the Empire, and her earlier financial assistance was replaced by a surge of French investment throughout European mining, transport and industrial sectors. British capital went into land-abundant, temperate areas for food and raw material exploitation. One reason for the British pattern lay in the nature of the British capital market, which followed a risk-aversion strategy for foreign investment; capital was most likely to search out foreign or colonial government securities, then railways and only then, perhaps foreign private enterprise. Such risk-aversion was less characteristic of the later foreign investments of the US or Germany, whose capital markets were more concentrated and influenced by long-term banking interests. In contrast to the British case, French capital penetrated both transport and manufacturing enterprises within Europe itself, and was led by the railway, mining, metallurgy and shipping interests of the Credit Mobilier.

In addition to these underlying patterns, foreign investment was most likely to move into regions of high population or economic growth. On the demand side this was because the need for capital in such nations was likely at times to outstrip the supply of internal savings (e.g. foreign capital *substituted* for domestic savings in Sweden, Russia, Australia, Argentina). On the supply side (e.g. France) capital was most likely to be exported when savings ran above the given demand for internal investment, or because there was an autonomous *fall* in the demand for domestic investment in manufacture or transport (e.g. Britain).

By 1875 British investment abroad was valued at around £1000 million, whilst French foreign investment had increased nearly tenfold from its 1850 level. In the boom period 1907–13 the British were investing abroad at an average of £176 million per annum; in today's terms this represented an excess of £20 billion per annum! By 1913 the pattern of foreign investment of major nations was fairly firmly established. Of *all* foreign financial assets in 1913, Britain had invested some 70 per cent in government and railway securities and only 4 per cent in manufacturing. Thirty-seven per cent of British investment was held in North America, 6 per cent in Europe, the rest in Australia, Asia and elsewhere. French investment abroad in 1913 represented approximately 50 per cent of the British, but 61 per cent was in Europe (nearly one-half of which was in Russia). In the same year, 53 per cent of all German foreign investment, about one-third of the British total, lay in other European nations.[33]

Recent economic historians have written somewhat sceptically of the *financial* impact of foreign investment in this period.[34] However, our own prime concern is with foreign capital as a mechanism of *technology* transfer more specifically; financial valuations in donors do not necessarily reflect economic impact in receivers.

Foreign capital at times served to satisfy negative imbalances in industrialising nations (for example, in Russia in the 1890s). Borrowing therefore permitted some limited (capital had to be repaid, often with *guaranteed* interest) breathing space within which such economies could revamp their infrastructural or export sectors. It could provide the financial wherewithal for the technologies that might be required in order to do so.

As is already obvious, the increasingly capital-intensive nature of major technologies in transport and manufacturing meant that many 'transfers' were in fact *embodied* in 'capital' movements or were strongly *associated* (via entrepreneurs, engineers and skilled workers) with foreign private or government investment. Thus Cameron has suggested that the greatest contribution of French capital to German economic development in the 1850s and 1860s was not in its *amount* but in the skill transfers and technological *demonstrations* which were associated with it: 'in the timing, rate and direction of German industrialisation foreign contributions, especially French, played a decisive role'.[35] The early employment of French, British and Belgium capital in Ruhr mining was followed by French investment in zinc and other non-ferrous metals from the late 1840s. The exploitation of German lead, iron, silver, lignite, zinc and other resources was only made viable through the work of French engineers and artisans. From the 1850s the *Societé de la Vieille-Montagne* rolled zinc and produced zinc oxide in France, but drew its resources from its mines and smelters in Belgium and Prussia. In an effort to replace material imports from Germany and Russia the company then invested in or took over German smelting and rolling mills, increasingly dominated German copper production, and reorganised the

Silesian zinc industry into a company which became the second largest producer of zinc in Europe.[36]

Migration It was precisely during the 1840s, when US population growth clearly reflected immigration from Europe, that manufacturing efficiency in that economy began to improve, and this lends some credence to the notion that a significant mechanism for technology transfer lay in the migration of human capital. Furthermore, between the 1820s and 1850s the proportion of farmers and mechanics in total migration to the US rose from 27 per cent to 47 per cent.[37]

Europe's population growth ran significantly ahead of world population growth throughout the nineteenth century. Between 1800 and 1850 Europe's population grew by 36 per cent, world population by 29 per cent; between 1850 and 1900 Europe's population grew by 42 per cent, the world's by 30 per cent.[38] One estimate of emigration from 17 European states between 1846 and 1932 offers a figure of 51 million people, of whom 18 million originated in Britain, 10 million in Italy, 5 million in Germany.[39] This great European migration represented 97 per cent of all world migration in the period. By far the greatest receiver was USA (58 per cent of the world), followed variably by Argentina, Canada, Brazil, Australia, South Africa and the West Indies.

United States economic history, and to a lesser extent the histories of Australia and Argentina, illustrate how such flows of people *could*[40] represent a mechanism for technology transfer. The greater number of *consumers*, with tastes which drew from European patterns of production, created a structure of demand (not merely a *level* of demand) conducive to technology transfer and diffusion into indigenous industries. Furthermore, high levels of immigration were associated with high levels of capital importation. In addition, migrants carried with them a general *knowledge* of the manufacturing system, a *culture* already adapted to industrial organisation and specific *skills* in the operation of more advanced techniques. Many migrants transferred their *savings* with them. Though this was often followed by substantial remittances from subsequent income, initial savings stimulated technological development via the general demand for goods – housing, furnishing, services, etc. In a minority of cases such savings could be directly invested into newer manufacturing or extractive sectors.

A variety of literature suggests that in any period of modern world history, migrant groups exhibit a pattern of innovative behaviour and a level of energy which is not necessarily related directly to the environment from which they came, but is rather a response to the challenge of a new, foreign environ.[41]

Technological Progress and Technology Transfers In a number of ways, technological progress in some fields itself acted as a mechanism for technology transfer and diffusion in others. Thus inventions which allowed the economical exploitation of hitherto dormant raw materials (e.g. phosphoric ores, non-ferrous metals, gold tailings) could be directly transferred to regions of abundance. The larger and low cost supply of such materials induced still further transfers in order to exploit their potential in hitherto unexplored areas of production. Again, the activities of a host of heroic explorers, mineral prospectors, surveyors and railwaymen opened up areas to international trade which had literally not existed before. The exposure of the rich resources of the Ukraine by the railway linking the Donetz basin to Krovoi Rog in 1886 heralded an enormous flow of advanced technology into that region. In a multitude of ways, technological change extended the market for manufactured products and machinery.

Perhaps the strongest arguments relate to transport and communication systems. British shipping rose from a position of superiority in the early nineteenth century to outright predominance in the later nineteenth century. By 1880 British vessels represented 50 per cent of the world's steamer tonnage. Harbours, roads and railways combined with such spectacular projects as the Suez Canal to extend the international market for technology; that is, they served to reduce the price of technologies at their point of entry, and were therefore approximately equivalent to a reduction in their capital cost. Between 1840 and 1880 railway mileage expanded from around 5000 miles to some 230 000 miles, 50 per cent of which lay outside Europe. By 1875 the technological diffusion represented by some 62 000 locomotives scattered throughout the world was mirrored in enormous feats of engineering in construction, cuttings and tunnels. Within developing regions railways helped in the creation of national markets, increased specialisation of production. More goods at lower cost were more available for export. The earlier improvements in sailing ships cut the costs on long-distance routes as steam ships did the same for shorter trading routes. Improvements in steam shipping[42] were bolstered by the effects of the opening of the Suez Canal in 1869, which halved the rates from Liverpool to Bombay. By 1885 the immigrant passage from Hamburg to New York had reduced to $7.

In the field of communications scientific research was of more importance.[44] The electric telegraph was quickly adopted within Europe from 1848 and was followed by the construction of submarine cables throughout the world. The International Telegraph Union was formed in 1865, the Universal Postal Union in 1875, the International Meteorological Organisation in 1878. At the same time linguistic barriers were assaulted by an industry of literary, commercial and technical translation. Esperanto was a product of the 1880s.

Faster and more reliable forms of communication permitted a quickening and increased certainty of commercial decision-making, which in turn enriched the flow of goods and capital. Such technical improvements opened up temperate-zone agriculture throughout the world: for example, the long-distance wheat trade and refrigeration. The importance of such cost-reducing tendencies must be set against the backdrop of increased tariff protection in the later years of the century. Indigenous entrepreneurs exploiting natural resources benefited from lower prices at the point of entry at a time when they were themselves increasingly protected by tariff and quota systems.

From the 1820s to the 1850s (and perhaps later) and in the years prior to 1900 *technological progress* in Britain reduced the total cost of production for such major export products as cotton, iron and steel. From the 1840s to 1870 improved technique reduced the price of British coal exported. Sidney Pollard has constructed an insightful schema around the notion that although total costs of export-goods production in Britain were falling, the price of British labour was not. Increased productivity and demand combined to increase *wages* in Britain relative to other European nations. The rise in wages, although due to technical change in innovatory industries, was passed on in a variety of less efficient British manufactures also. This meant that in Britain the less innovatory sectors of production suffered from 'artificially' high production costs, especially in more labour-intensive sectors. In such product lines European producers could benefit by exporting woven fabrics or ironware to Britain, whilst obtaining cotton or machinery at relatively low cost. Thus, the impact of technological progress on the general level of wages in Britain induced a particular pattern and high level of trading, financial surpluses in other nations and a consequent increase in the ability to procure superior British technologies. According to Pollard's careful conclusion, 'the technology gap became large enough to become the major component in comparative cost differences'.[44]

Direct Transfer Mechanisms

Individual Agents If Hobsbawm's 'impassive missionaries'[45] dominated the process of technology transfer after 1850, prior to that time individual change agents were still of considerable importance. In the early nineteenth century private rewards promoted movements of technicians and artisans in much the same way as they had done in the eighteenth century (Chapter 2). It is true that after 1815 foreign innovative *guidance* may have become of less importance than more pervasive artisanal immigration, which provided the key personnel both for particular processes (e.g. puddling) and supervision of workshop operations. But, whatever their status, individuals remained of some importance. Furthermore, it was at this level that British influence on the Continent continued to operate despite the movement of British capital away from Europe.[46]

The form which was taken by such individualised transfers varied considerably. The military engineer William Müller, an instructor at Gottingen, carried the eighteenth-century tradition into the 1800s by travelling throughout France, Holland and Austria, inspecting the installations of advanced technology. When the French seized Hanover in 1807 Müller settled in Britain, where he lodged a series of patents into the 1820s. The emergent industrial chemists of Germany became employees of such British firms as Roberts, Dale and Co. or Williams Bros, a tutelage which was prelude to the creation of an indigenous training fabric. The agricultural chemist Justus Liebig singlehandedly transferred his findings in the field of agricultural chemistry to Britain by means of an all-but-'royal' tour of the nation.[47] The graduates of French technical and military academies poured into the railway systems of the Mediterranean, the near East, Russia, Austria-Hungary, Switzerland, Belgium and elsewhere.

But the major form of individual enterprise continued to be the movement of skilled individuals into the original setting-up of production in other nations or the introduction of new technologies into existing foreign operations. In Italy, French engineers drained harbours, built ports (Genoa, Leghorn), canals, bridges and railways. Amongst the early French change-agents were the Saint-Simonians, often graduates of the Ecole Polytechnique, who included amongst their number the Péreire brothers, founders of the Credit Mobilier. Thus, French capital and French expertise travelled together to the same destinations. From France came Charles Lambert, invited to Naples to introduce French machine techniques into the manufacture of woollens. Similarly, Philippe de Gerard was invited to Russia in the 1830s to instal machinery for linen production. At an earlier date, Gerard's improvements in flax-spinning, neglected in France itself, were transferred through personal networks to John Marshall at Leeds and became the basis for a new industry in the 1820s.[48] British engineers were even more widespread abroad. From the 1820s British machine-builders and ironmasters were vital agents in the transfer of improved technique into the Czech areas of Austria-Hungary. Bohemia became 'little England'. The partnership of John Hughes and A. N. Pol was influential in the establishment of a Ukrainian iron industry. In 1869 Hughes founded his New Russia Company for the establishment of iron manufacturing at Donetz. Stimulated by grants of land and money, the firm was operating blast furnaces by 1872. British skills were utilised in key operations; by 1893 the company employed 6000 men.

Such personal transfers also took place between nations and, more unusually, very specific technique flowed into advanced areas from relatively backward nations. British engineers such as Aaron Manby and D. Wilson set up demonstration plants around Paris, and later contracted with French firms in the introduction of British machines, designs and artisans. Humphrey Edwards acted as an employment agent in France for the strategic placement of British artisans. A number of creative engineers, often the owners of patent rights (e.g. James Cockerill, James Jackson, Joseph Dixon, Richard Roberts) were employed to install particular techniques into existing French enterprises in the fields of foundry and ironwork, machinery construction (e.g. the Woolf high pressure engine), metalworking (the Huntsman process), cotton spinning, locomotives and lace manufacture. France and Britain by no means monopolised the transfer process. Swiss entrepreneurs and mechanics were important to the establishment of modernised cotton plants around Naples, and followed in the wake of the Müllers (from Germany), who had first introduced the spinning jenny into the Italian cotton industry. Another Swiss engineer, J. C. Fischer, the inventor of improvements in the manufacture of crucible cast steel and steel alloys, set up plant and workshops under specific orders from French industrialists. In 1843 William Siemens arrived in England (not long after Liebig's successful tour) in order to exploit a new method of gilding by electro-deposition. The purchase of his invention by Elkington's of Birmingham for £1600 provided him with the funds to support his researches into the regenerative furnace (designed to conserve the waste heat of furnaces) and other techniques. Siemens and Halske, with William as partner, set up a London office in 1858.[49]

The most that can be said of such varied activities is that they *tended* to focus on very immediate, new and specific machine improvements, most of which were patented or at least purchaseable. Increasingly, the actual *commercial* working of such technologies relied upon some positive conjunction of the 'impassive missionaries' itemised above. When they were absent or insufficient, superior knowledge often led nowhere at all. By the later years of the century this was even more evidently true, and the effectiveness of individual agents depended upon either their own abilities as organisers of production, or upon a firm, purposively erected, support structure. A former pupil of Pasteur, the Count de Chardonnet, perfected a method of dissolving natural cellulose using alcohol and ether as solvents. His first enterprise for the manufacture of 'artificial silk' was founded in France in 1891, but the transfer of the technology into Belgium, Germany, Hungary and England depended entirely upon his personal supervision of the nitro-rayon plants.[50] Ludwig Knoop's success in Russia depended upon a complex support structure; (1) as a representative of a group of Manchester merchants he reduced his 'entrepreneurial' risk; (2) his machine spinning was installed first into an existing, innovative and expanding firm (C. W. Morozov); (3) his subsequent erection of numerous cotton factories elsewhere in Russia drew upon the resources of Manchester and Oldham for raw materials, steam engines, machinery, managers and skilled workers.[51]

The State, Armaments and Autonomy To this point our review of eighteenth- and nineteenth-century technology transfers has illustrated how state support could be given to the activities of individual transfer agents, indigenous or foreign. By means of invitations, salaries and the award of prestige, by the setting up of government-aided or -owned 'model' enterprises, by the payment of subsidies, the granting of land or land revenues or tax concessions, and the purchase of manufactured goods, and by the granting of tariff or other protection to infant industries, the State directly intruded on the pace and direction of technology transfer. Obviously, any State encouragement of foreign trade, migration, capital inflow or transport and communication facilities may be seen as potential indirect means of encouragement through their positive impact on indirect technology transfer mechanisms (above). Finally, Chapters 2 and 4 have illustrated how officialdom at times supported the development of educational, technical and research institutions in the fields of both science and industry. These acted as agents of technology transfer *insofar* as they stimulated the increased *diffusion* of existing knowledge within the nation, promoted innovative breakthroughs or direct transfer of knowledge via translations, training, and so on.

As Section 5.1 showed, military motives *continued* to dominate State *policies* even in the years and places where *laissez-faire philosophies* were dominant: between 1851 and 1854 Britain spent an annual average of £20 million on military development; between 1855 and 1857 the average rose to £34 million, or 57 per cent of net capital formation in these years. But it was particularly in the more economically backward of nations that modernised military–industrial complexes emerged. The development project of La Sociedad Española de la Construcción Naval was designed to cement links between the reconstruction of dockyards and arsenals and the development of the metallurgical and machine-building sectors of the Spanish economy. Utilising a coherent package of British armaments and advice, British capital, management and skills, machines and material, by the 1900s the complex represented Spain's capture of a high-technology 'sector'. Similarly, the Newcastle firm of Armstrong supplied and supervised the naval works at Pozzuoli, near Naples, which from the 1880s emerged as a major nexus of machine shops and industrial enterprises. The integration of this complex within other parts of the economy was accomplished in a series of contracts with domestic suppliers of railway stock, iron and steel. As ruler of Egypt, Mohammed Ali introduced Frenchmen into his civil service, medical, surveying, irrigation, canal and cotton enterprises.

The military-minded rulers of Russia established the support structures for the activities of J. J. Hughes in metallurgy, the Nobels in the oil of Baku. The Russian state maintained the 'communities' of foreign artisans in metallurgy, coalmining, chemicals and electrical engineering. Although recent historical research has questioned the *financial* effects of State activity in a number of nations, there has yet been no effective critique of the notion that State activity in the field of technology transfer – at whatever the cost – was instrumental in the modernisation of major industries or regions.[52]

Institutions for Dissemination Chapter 4 dealt in some detail with the scientific and technical institutions and policies of major nations in the nineteenth century. The scientific communities increasingly devoted their energies to the dissemination of information within and between nations; the more the former was accomplished, the more the latter was feasible. In this at least, science was often encouraged by the State. Statesmen in backward economies were well aware of the transforming power of knowledge. Camille de Cavour, who as a railway administrator was quick to appoint French expertise, was also closely influenced by foreign advances in agricultural chemistry. Through William de la Rive, Cavour received very specific knowledge from England which he used in establishing fertiliser manufacture at Turin.[53]

But, more normally, transfers of knowledge were accomplished through the constant flow of information in the scientific and technical media and by international patenting. In this respect France remained the metropolis of science for some time. The German scientific enterprise was boosted by French training (Alexander Humboldt, Justus Liebig), textbooks, exhibitions and research institutions. The common use of the French language meant that the nation's 'applied', technical journals became major vehicles of knowledge dissemination – the *Journal de Commerce*, the *Journal des Chemins de fer*, the *Semaine financier*. To 1804 one-quarter of the students of the Ecole Central were foreign, as too were many students at the Polytechnique and the Ecole des Ponts et Chausées. French-trained Germans exerted a significant impact upon the German techno-structure; Liebig returned to Giessen to set up the first experimental chemical laboratory in Germany; encouraged by the State, Heinrich Magnus returned to become a leading teacher in the field of industrial sciences. The direct importation of dynamos to

Britain followed from the publication of Z. T. Gramme's findings in the *Engineer* in August 1871, which had first been presented in papers to the French Academy. The purchase and commercial use of the dynamo in the US followed upon the disseminating activity of American university academics.

In relatively backward areas the settlement or refinement of transferred knowledge was particularly affected by a series of institutional innovations encouraged or instigated by the states. Following the production of public goods and incentives and subsidies to industry, State policy focused on the diffusion of best-techniques and advanced scientific knowledge. The transfer and control of complex technologies, from lighthouses to chemical installations, demanded the active presence of commercial clerks, artisans, foremen and a host of skilled lower managers and technicians. Patents, licences and trademarks had to be enforced and monitored, and this on a scale never before witnessed. The rush from backwardness resulted in tremendous misallocations and overcompensation in the entire field of training for industry, from continuation schools to the theoretical science of research institutes. At a time when Imperial Russia was desperately sending students to Western Europe for training in law, commerce and industry, that nation's leading scientists were creating breakthroughs in such new fields as probability theory, crystallography, bacteriology, soil science, immunology, embryology, and the new chemistry.[54] Similarly, Hungary may have witnessed a shortage of foremen and managers with technical skills, but the invention and development of such high technologies as the alternating current transformer, the induction Watt-meter and the three-phase electric locomotive all belonged to the engineering genius of the Ganz Electrical Works prior to the turn of the century.[55] Although the vanguard of development was the new industrial sector, the maintenance of traditional skills and crafts boosted by government training schemes were crucial for the securing of both employment and consumption levels.

In late industrialisers, social disruption and the concentrated nature of industrial advance meant that government universally saw the need to maintain existing skills in older areas and to stimulate the original formation of skill in the strategic, yet small, but fast-developing sectors. In many cases apprenticeship could play little part, and was at any rate difficult to monitor.[56] Japan and Germany invested heavily in both control and training levels, and this may be extended to other nations. In Russia the State concentrated on university training in an effort to create a new functional, technical elite. The new meritocracy was further strengthened by training abroad – by 1912 some 8000 Russian students were studying in Western Europe, approximately half of whom were in Germany. By the early 1900s state provision for practical industrial training had also increased. Of 649 middle and lower technical and trade schools, many provided purely industrial skills, others offered rural artisan training. Even here, however, institutions were organised by a voluntary combination of firm, *zemstva* and municipalities.[57] But in the slightly earlier period of the 1890s, when literacy was low and industry fast growing, an even more ancillary provision was apparent. Indeed, Olga Crisp has argued that the seemingly high literacy levels of adults engaged in industry was possibly a function of adult educational facilities.[58]

It should be emphasised here that, even in the case of successful industrialisers, there existed definite boundary conditions on institutional innovations in the fields of science, technology and education. Ideally, policy determined that entry into scientific or technical training establishments should not disturb entrenched elites. The meritocracy must not swamp the existing powers. Following this, the transfer of technologies and of abstract knowledge was to be divorced from the transfer of advanced 'liberal' or 'radical' ideologies. State-promoted education served directive, control or technical functions. The exception to this rule were the elite universities.

Finally innovative formations must not destroy the *existing* base of skills. As Yoichi Yano has shown for Japan, and as may be extended to the German states, technical education was aimed at preservation and mild adaptation rather than wholesale transformation.[59] Creative innovation would ideally take place in a context of skill preservation, else much of previous human capital formation would be lost.

In addition, bureaucrats had to *justify* the transfer of *public* funds to training institutions, for this entailed the use of tax revenue to increase the potential profitability of individual firms. For this reason amongst others, State-financed training tended to focus on the provision of general scientific or industrial knowledge (i.e. of 'processes' and 'concepts'), rather that on the introduction of specific worker skills.[60]

Commercial Institutions Any institutional innovations which, by reducing risks or concentrating capital, acted so as to increase the level of foreign investment from advanced nations, might be seen as indirect agents of technology transfer. The most spectacular instance of this was the investment bank. But even the emergence of 'mutual funds' from the 1860s may be seen in this light. Mutual funds allowed smaller investors to contribute to large investment portfolios and thus spread risks over a number of different stocks.[61] In an era when *The Economist* was referring to overseas governments as 'barbarous' or 'semi-civilised', such devices reduced the seeming risks of foreign government bonds. In this form of investment, however, there lay little suggestion of active management in the affairs of foreign industry. The transnational corporations and investment banks were quite another matter.

Industrial corporations in advanced economies began to set up branch plants and foreign subsidiaries for various reasons. On the 'push' side, large, advanced technology firms in steel, chemical or textiles were gaining monopoly profits at a time when their *home* markets were becoming saturated. The reasons why they did not simply continue to export abroad relate to such 'pull' factors as tariffs (tariffs could be surmounted by manufacturing or assembling *behind* tariff barriers), the need to protect existing markets in periods of heightened competition, or the wish to exploit *patent* rights successfully lodged abroad (see below). When cheap labour, stable government or tax concessions on foreign enterprise were also available, such firms were most likely to transform their export activity into direct production abroad.[62] On the basis of the ownership of specific knowledge, Solvay established branches in a number of nations. In Italy, the electrical industry became dominated by a series of German, Swiss and American firms, all of which owned or exploited patent rights. Much of American interest abroad by 1914 was commanded by some 30 to 40 large US companies, mostly investing in Europe. Generally, transfers of technique by transnational corporate activity tended to be in the advanced fields of chemicals, food processing, machinery and electrical engineering. From Britain, the Vickers Company built and operated armaments factories in Italy, Russia, Spain and Japan. Kenwood and Lougheed mention some 17 European transnationals by 1914, eight of which were primarily concerned with chemical industry.[63]

To the producing company must be added the international

investment bankers. The investment bank originated in Europe (not Britain) as a result of the need to mobilise and concentrate scarce *internal* resources for railway, industrial and public works programmes. Those such as the Bank of France or the Deutsche Bank were direct offshoots of the State, others, such as Rothchilds, were private merchant banks. The French Credit Mobilier 'accelerated the movement of French capital abroad and at the same time, by financing ventures in other countries, it facilitated the transfer of French banking techniques and French technology and expertise to those other nations'.[64]

Banks which had hitherto serviced the growth of indigenous, advanced technique industries in their own nations, began to transfer funding to a replication of such technologies in relatively backward nations. The electro-chemical industry of Austria-Hungary, oil exploitation in Rumania, railways in the Balkans and a variety of colonial ventures all benefited from foreign bank funding. The Anglo-French banking house of Laffitte and Blount financed a company in 1840 to build the Paris–Rouen line using British capital, and engineers such as John Locke and contractors such as Thomas Brassey and William Mackenzie. The French Northern Railway (Paris–Lille) was started in 1846 and co-funded by James Rothschild. The Paris–Orléans Railway Company was contracted out to Mackenzie. By 1847, 50 per cent of French railway shares were in British hands. The later transfer of both capital and expertise from France to Russia was simply an epiphenomenon of the extention of French investment banking.

International Patenting Perhaps because patent lodgements are not a universally accurate measure of *technological progress*,[65] the transferal effects of the nineteenth-century patent system have been grossly neglected by modern economic historians. Yet the character of patenting within and between nations in our period is of interest to historians of industrial development. The laws and institutions of national patent systems reflected the nature of the penetration of bureaucratic institutions into the economic process more generally. The mode in which governments assisted in the protection of property rights over specific knowledge loosely accorded with their general development strategies.[66] Patent lodgements by residents of *other* nations measured the degree to which the host nation was open to new processes and products. A sole foreign lodgement might result from random factors, but a surge of chemical patents from one nation or group into another measured both the viability of the former and the receptivity of the latter: the high immediate and transaction costs involved in foreign lodgements of patents means that such patents were indeed likely to represent the transfer of potentially lucrative (therefore potentially productive) techniques. As will be seen below, the general pattern of patent application serves as a rough indicator of relative economic backwardness/forwardness. Variations in patent laws undoubtedly affected the transfer of technologies in *newer* industries. For instance, until 1907 the British patent system required no obligation on the part of foreign patentees to actually work their process in Britain. Thereby, and by one day only, Caro, Graebe and Liebermann lodged an English patent which forestalled Perkin, who perforce had to licence the process from his German competitors.[67] In contrast, the complete absence of a patent system in Switzerland allowed Swiss entrepreneurs to exploit the Perkin analine dye patent without restriction. By 1859 (i.e. within a two-year span) manufacture was established at Basle and newly-founded firms speedily diffused the technique throughout Switzerland.

Renowned contemporary innovators continuously invested in their national patent systems: Henry Bessemer was said to have spent some £10 000 in patent office fees alone! It was through *international* patenting that inventors and businessmen protected their rights over specific knowledge-improvements and were thereby able to exploit such breakthroughs by direct manufacturing under protection, licensing, purchase or the hire of their operating skills. More importantly, the securing of patent rights was very often a prelude to a fervour of technology transfer by active change agents. When John Platt of Salford secured the patent rights to John Collier's (of Paris) machine for combing short wool he laid the foundations for a machine-building empire which earned much of its revenue from technology transfer.[68] In America, General Electric, previously the Edison Company, diffused advanced technology by a combination of aggressive patenting, patent purchase and mergers. Similarly, Westinghouse's contemporaneous rise was based on patent ownership. Hardly surprisingly, in 1896 the two large US concerns signed a 15-year patent-pooling agreement, strengthening their near-monopoly position in both the economic and research areas. In almost identical vain, in Germany the AEG was founded in order to exploit the Edison patents and by takeover and stock acquirement set out to control all further patenting in the electrical area. In Germany, such concentration of knowledge-ownership and market shares was associated with a spurt in the total output of electrical equipment, followed by a recession which resulted in the survival of two giant firms (AEG and Siemens), who together exerted technological dominance over the whole of Europe. The massive invasion of Britain by these four major world electrical-goods producers meant that there was very little subsequent development of a *British* electrical industry, the exception being the cable-making industry, it alone remaining free of the knowledge monopoly commanded by branches of foreign firms. In a world of international patenting, even the absence of a national patent system could induce transfers. In the early 1880s emerged the Swiss Chemische Industrie (CIBA), strongly based on purloined patents, imitation and *foreign* patent lodgements. Similarly, the development of the Swiss electrochemical industry, especially in the generator field was based on German and other patented inventions (e.g. Gramme, Siemens), and was further developed by improvements within Switzerland and subsequent lodgement of Swiss patents in other nations.[69] But in the case of the transfer of the viscose process for the manufacture of rayon, the ownership of the major patents by the Viscose Spinning Syndicate Ltd (secured by purchase of all previous patents) allowed it to dominate foreign usage through licensing to foreign (e.g. French) firms. Such agreements and relationships suggest in combination that the nineteenth-century patent system would bear some further investigation as both a measure and mechanism of technology transfer.

Table 5.6 shows the dominance of France, Britain and America amongst 15 nations patenting in the years 1842–61.[70] Of 155 000 patents 44 per cent were French, 22 per cent were American, 16 per cent were British.[71] The industrialisation of Belgium is nicely reflected in the figures, as is the relative backwardness of Russia. Within Germany, the rise of the Prussian economy is not yet evident; 1320 patents represented only 25 per cent of the German total, 2 per cent of the French. On the other hand, Austria (not included in the table) in the short period 1855–60 issued 3785 patents or 630 per year.

In the later years of the century Anglo-American dominance over the industrial patent system emerged strongly and the united Ger-

TABLE 5.6 Patents granted in 15 nations, 1842–61

	Britain	*France*	*USA*	*7 German States*	*Sweden and Norway*	*Belgium*	*Italy*	*Russia*
1842–44	1 241	722	1 550	398	127			5
1845–49	2 460	11 497	3 429	1257	297	7 497		32
1850–54	6 413	16 641	5 744	1193	374			41
1855–1860	12 226	33 959	20 503	1973	402		1136	155
1861	2 047	5 941	3 340	372	99	11 911	224	26
1842–61	24 387	68 760	34 566	5193	1299	19 403	–	259
Average per Annum	1219	3438	1728	260	65	606	194	13

TABLE 5.7 Patent applications in Britain, US and Germany 1877–94

	Britain		*USA*		*Germany*	
	No. applic.	*% granted*	*No applic.*	*% granted*	*No. applic.*	*% granted*
1877–82	33 139	67.5	83 417	60.3	37 449	56.7
1883–88	93 524	46.0	125 094	67.1	55 900	45.0
1889–94	139 881	47.0	135 070	58.4	78 801	42.2
1877–94	266 544	49.2	343 581	61.6	172 150	46.3

many replaced France as a main contender in terms of patents granted after inspection. Table 5.7 gives information for the years 1877–94.[72]

The lower proportion of patents granted in Britain and Germany suggests that, compared to the US, their patent systems were more closely geared to sifting and selection of patent applications in these years.[73]

As the nineteenth century progressed, successful patent-lodgements increasingly reflected the pattern of innovation in newer, manufacturing technologies. Thus, of all patents lodged in Germany between 1877 and 1904, over 12 000 related specifically to the chemical industry, with 3447 devoted to bleaching and dyeing processes, 3733 to the preparation of colours more generally. Table 5.8 below reviews patents actually granted in Germany between 1877 and

TABLE 5.8 The German patent system, 1877–94; categories of patents granted

Patent groups	*No. of sub-groups*	*Patents granted 1877–94 No.*	*%*	*Patents granted 1893–94 No.*	*%*	*% of grants to Applications 1893–94*
Mining & metallurgy	11	6 852	8.6	1 045	8.2	53.1
Chemical & electrical	10	9 128	11.5	1 809	14.2	44.5
Textiles	10	6 996	8.8	1 077	8.5	45.1
Agricultural	1	2 902	3.6	463	3.6	44.3
Transport	4	4 762	5.9	751	5.9	44.3
Construction	1	1 099	1.4	131	1.0	21.3
Motive Power	8	8 792	11.0	1 025	8.1	34.3
Food & drink	8	5 863	7.4	835	6.6	38.0
Misc. Mnftrs	17	20 067	25.2	3 524	27.7	49.1
Other	19	13 158	16.5	2 050	16.1	40.2
Total	89	79 620	100	12 710	100	43.5

1894 in terms of the 89 official categories used by patent authorities.[74]

Innovations in mining, metallurgy, chemicals, transport, motive power and food processing represented some 44 per cent of all registered invention in Germany during these years, and the trend was rising. Advanced techniques were also spread around miscellaneous manufactures and other categories.[75] In addition, 28 per cent of all applications were by nationals of other countries, which suggests that the German patent system was acting as a mechanism of technology transfer into Germany. In extreme contrast, outright discrimination in the American patent system meant that of over 7000 patents lodged there between 1836 and 1849 only 3 per cent were foreign in origin; even during 1870–1900, when an average of nearly 19 000 patents were lodged in the US annually, only 9 per cent were foreign.[76] As Chapter 10 suggests, US technological capability was added to primarily by direct inflow of 'human capital' and by selective *purchase* of European technique, or informal transfers.[77] In addition, this all occurred at an earlier period.

More generally, foreign lodgements are some indication of the extent of technology transfer between major nations. Table 5.9 gives information on patent lodgements in Britain originating from 39

TABLE 5.9 Patent applications in Britain originating from 50 major cities, 1867–9

Cities	*No. patents*	*No. cities*	*% of All UK patenting*	*Largest city in group*		*Largest 3 cities % of group*
British	4333	11	38	London	(2480)	79
European	1215	25	11	Paris	(966)	69
USA	844	14	7	New York	(393)	85
All UK Applications	11500	–	100			

foreign cities and 11 British cities in the three years 1867–69.[78] Together the 50 cities represented 56 per cent of all applications in these years. The last column of the table suggests that urban patenting was fairly concentrated within large, major inventive centres.

Finally, we may indicate how technology transfer through patenting reflected national technological capabilities and degrees of industrialisation. Foreign lodgements are a more accurate indication in that measures of gross invention as registered were affected by the vagaries of patent law. Small-population, high per capita income developing nations are useful focuses as foreign innovation loomed particularly large. Australian patent applications illustrate the emergence of American influence and provide a good ranking of industrialisation across a number of nations. Table 5.10 gives full details of nearly 17 000 foreign patents in Australia between 1886 and 1918 and ranks nations in terms of number of applications.[79]

On the whole, high-ranked nations had undergone periods of industrial revolution prior to this period. Those ranked below 10 witnessed contemporaneous industrial revolutions, e.g. Russia or Italy. The lowest-ranked nations had either late industrial revolutions (e.g. Japan) or 'failed' to pass from sporadic industrial drives (see below) into more general industrial revolution.[80] There can be little doubt that the evidence assembled even so briefly suggests that patented technologies and specific information about them passed across borders fairly readily in this period, and that such transfers indicated approximately the stages of industrialisation reached by the respective patenting nations.

TABLE 5.10 Foreign activity in the Australian patent system, 1886–1918

Nation	*Applications in NSW* 1886–88	1889–91	1892–94	1895–97	1898–1900	1901–03	*Applications in Australian Commonwealth* 1904–06	1907–09	1910–12	1913–15	1916–18	*Rank 1886–1903*	*Rank 1904–1918*	*Rank 1886–1918*
Australasia	1044	1701	1509	1775	1701	2028	5566	6558	8 225	8 111	7513	1	1	1
Britain	270	413	289	398	418	406	810	846	1 068	910	825	2	3	2
British Dependencies	10	12	12	33	39	68	166	197	145	152	79	5	5	5
USA	96	162	160	234	332	408	894	904	1 121	1 024	921	3	2	3
Germany	26	21	32	38	39	48	138	205	292	179	17	4	4	4
France	20	17	9	32	45	28	97	102	124	77	55	6	6	6
Sweden	8	6	5	10	7	29	57	58	56	78	96	7	7	7
Denmark	1	1	2	5	23	8	21	34	24	13	20	8	9	9
Austria	2	10	3	3	6	8	26	20	19	12	0	9	14	12
Belgium	6	4	1	8	9	3	14	33	27	24	3	10	11	10
Russia	3	2	2	2	4	3	4	17	15	3	0	11	16	16
Italy	0	0	1	6	2	5	26	25	15	23	22	12	10	11
Switzerland	2	0	1	1	1	7	20	21	20	8	13	13	13	13
Norway	0	0	1	1	1	4	8	6	14	12	24	14	15	15
Latin & C. America	2	2	0	3	0	0	0	3	0	10	11	14	17	17
Netherlands	0	0	0	0	5	1	4	18	6	24	33	16	12	14
Spain & Portugal	0	2	2	1	0	1	2	2	4	3	0	16	19	19
Japan	0	0	1	0	0	0	0	2	3	6	11	19	18	18
Elsewhere	0	3	0	0	1	0	13	21	27	69	55	18	8	8
Hungary	0	0	0	0	0	0	0	0	2	9	0	20	20	20
Total	1490	2356	2030	2550	2633	3058	7866	9072	11 207	10 747	9702	–	–	–
% Foreign	30	28	26	30	35	34	29	28	27	25	23	–	–	–
% UK & Dependencies	19	18	15	17	17	15	12	11	9	11	9	–	–	–
% USA	6	7	8	9	13	13	11	10	10	10	9	–	–	–

6. Industrialisation: Winners and Losers

6.1 INTRODUCTION

> Some kinds of economy (mostly backward) and some stages of growth (mostly early) may *require* state assistance; others may render it almost entirely marginal.
>
> *Clive Trebilcock, 1981*

THE economic success of technology transfer into *particular* nations obviously depended upon the degree to which they were positively affected by the available direct and indirect transfer mechanisms outlined in Chapter 5. But even given some positive coalescence of such transfer mechanisms, successful *diffusion* and usage of superior technologies depended upon the satisfaction of three sets of *boundary conditions*.

The first set of conditions relate to the *technological imperatives* of the introduced techniques. Capital and skill intensities, scale of production and the environmental requisites of technologies might or might not be appropriate to or *adaptable* to the economic environment of the receiver nation. Related to this is a simple but often neglected point: the later the industrial drive, the more likely it was that (a) *indirect* transfer mechanisms would be less effective, e.g. later nineteenth-century protectionism was probably less conducive to transfer through trade than the earlier phase of 'free' trade, and (b) the more likely it was that the technological imperatives of imported techniques were inappropriate to the existing economic and social conditions in the receiver nation, i.e. introduced steel and chemical techniques of the 1890s *on the whole* required more capital, skill and organisation than the best-techniques of the 1830s.

Secondly, given the role of the State in both transfer and diffusion processes, the ability of governments and bureaucracies to react to the demonstration of forwardness without political breakdown represented a condition of success. Apart from the advancement of technology transfer, the reformist yet stable State could construct an economic environment for innovative business behaviour, i.e. followership.

Already touched upon in Chapters 4 and 5, the last set of boundary conditions were those economic and other institutions deliberately created by the State or other authorities which were designed to settle or diffuse new technologies, as well as those which functioned so as to *adapt* new technologies to existing internal conditions. Ineffective or irrelevant institutional responses could be of less utility (because they were costly) than no response at all.

Even a brief outline of late industrialisation prompts the conclusion that the satisfaction of such boundary conditions was of critical importance in the process of successful industrialisation during these years. Admittedly, it is difficult to measure the extent to which such conditions were 'satisfied' in financial terms, and thus to quantify the response process. Successful industrialisation took place in three phases, approximately equivalent to the distinctions between proto-industrialisation, industrial drive and industrial revolution. Transition from industrial drive (Gerschenkron's 'industrial spurt') to industrial revolution was crucial, for it marked the effectiveness of the institutional and ideological responses to backwardness which were characteristic of the second phase.

6.2 THE PHASEOLOGY OF LATE INDUSTRIALISATION

Nations which industrialised in the approximate period 1860–1914 included Germany, Japan, Austria, Hungary, Italy, Sweden, Russia and the somewhat special case of the USA. Excepting the last case, we may generalise upon the nature of industrialisation in this group of late starters. From the early nineteenth century in the German states, the late 1840s in Austria, 1861 in Russia and 1868–72 in Japan, economic growth was associated with movements to emancipate serfs or their equivalents, together with related institutional reforms in the areas of noble privileges and access to land. Especially after the Crimean War, the industrial strength of Britain, France and the Low Countries evoked both fear and hope elsewhere. Fear of the territorial advance of powerful industrial nations, hope that their economic and political success might be emulated by adoption of institutional reforms, and transfers of personnel, capital and technology. Economic growth was also desired as a means of keeping at bay the social tension which had arisen within such nations, often itself a product of military and economic failure on a large scale. Thus the Austro-Hungarian Compromise of 1867 or the Meiji Restoration of the next year.

The earliest structural phenomenon characterising the industrial drive was the complex of foreign capital, joint stock companies and railway boom. At this point, circa 1850s–1870s, the State encouraged rather than entered into such development. With the possible exception of food processing, the military and foreign elements in the mobilisation process tended to ensure an emphasis upon heavy industry as the most modern sector, although of course not the largest. At this stage governments tended to encourage speed and direction through the granting of strategic concessions to favoured groups, just as they withheld them from others. The expansion of government-guaranteed franchises, licenses, minimum rates of interest on select projects, and so on, meant that the second stage of industrial development in late starters was hardly the result of classic free competition. Indeed, it was seldom associated with the working of sophisticated markets for capital, land and labour. But during this interim stage, inefficient allocation made under conditions of fear and reform was crucial to the process of economic mobilisation (that is, the process whereby more resources are brought into the system), and eventually led to the involvement of the State in the creation of transfer institutions. At this stage even most backward nations could record impressive achievements. After the Compromise of 1867 Hungary was building some 600 kilometres of railway per annum.[1]

Industrial revolution involved continued direct government intervention and the rapid rise of new industries. Until this phase, industry was not only a minor sector, but still based on guild and craft, upon traditional skills rather than the factory, deskilling and reskilling.

As a national market emerged, the transfer of technology increased, transforming agriculture, (ploughs, threshing machines, milling, fertilisers), and the older industries (roller-milling, sugar-beet processing, food processing, coalmining), and gave rise to the very new (Bessemer and Martin steelmaking, Solvay soda, munitions, railway and machine production, electrical industry, power and modern communications).

In this third phase governments stepped in to provide public goods, often the crucial sector for technology transfer, or to substitute for the lack of capital or risk-taking behaviour in the private sector, as in railway development in Japan and in Hungary, or in the large-scale textile manufacturing sectors of most of these nations. With tax exemptions and incentives, subsidies and guaranteed loans, government intervention boosted the industrial growth rate.

The influence of all this upon political behaviour must remain debatable. It might be that liberal parliamentarianism was irrelevant and possibly inimical to the functional requirements of a society facing late industrialisation prior to 1914.[2] Gradual economic development based on a rise of demand, and consumer sovereignty was replaced by producer goods, munitions, and government orders. This limited choice and consumption, and curtailed individual freedoms. At the lower levels, the opportunity for *individual* advancement was severely limited. This provided fertile ground for the virulent political movements which followed late industrialisation, both prior to and after the First World War.

6.3 ASPECTS OF PHASE TRANSITION

As might be expected, the *proto-industrial phase* harboured contradictions and inconsistencies. Podkolzin has referred to nineteenth-century Russian industry as composed of 'a medley of various modes of production'.[3] Prior to the industrial drive of the 1890s, established textile and woodmaking *kustarni* industries existed side by side with modernised ore-dressing and metallurgy (e.g. the replacement of 'finery' by puddling, the use of Bessemer converters). Between 1867 and 1876 output of the Russian engineering industry multiplied threefold. Although the peasant disturbances which resulted from the Crimean War and poverty had at last unmuffled the 'alarum for change'[4] and a limited peasant emancipation has been accomplished (1861), Russian economic policy remained *ad hoc* and partial. The ministry of M. K. Reutern (1872–8) aimed at railroad-building and the encouragement of foreign enterprise, that of N. K. Bunge (1882–6), a former Professor of Economics at Kiev, continued railway development against a background of reform in the tax base. The international aspects of the development process were better addressed by I. A. Vyschnegradskii (1887–92), who aimed at amassing gold resources, rouble stabilisation, revenue increase, and grain marketing and export. But only under the authority of Sergei Witte in the 1890s was a more general, industrial development policy implemented.[5] So, although the output of Russian industry grew at an average of around 6 per cent per annum during 1861–85, this introduced tremendous employment dislocations and exacerbated agrarian retardation. In 1850 handloom weavers outnumbered power looms in the ratio 49.1, but by 1880 domestic weaving contributed only one-fourth to total textile output, and henceforth all but disappeared. Industrial development remained highly enclavist and, therefore, partial.

The fervour of State activity in Prussia up to the 1840s was geared primarily to military technologies which did not generate efficient resource usage. Commercial policy was aimed at revenue collection in a traditional mercantilist manner. In Germany more generally, the formation of the *Zollverein* undoubtedly induced some industrial growth, but the real surge of industrialisation associated with the development of a national market awaited the 1860s. Any limited transfer of funds to industry prior to that time was dominated by local markets for capital. Railroads absorbed the bulk of foreign technology transfers.

The *industrial drive* was another matter. In Germany after 1860 railway development was added to by a quick growth of machine-building and the emergence of a banking system which concentrated scarce capital and directed it into a wider field of industrial investment. But even in 1870 German textiles were still only partially mechanised, and wood remained a prime resource in power production. As late as 1882 well over 40 per cent of the employed population worked in rural areas, a measure of the continuing importance of proto-industrial organisation. In this second phase, the agencies of the State moved from mandatory controls to the encouragement of a stable business environment. The earlier 'Prussian' style of direct intervention gave way to demonstrations of best-techniques and awards of subsidies to selected and efficient domestic producers. The rise of the entrepreneur in this stage was associated with business integration and a high level of capital formation and technology transfer, e.g. the steel industry. Capital formation was significantly higher than in Britain, France or the USA, representing 14 per cent of GNP in 1851–70, 24 per cent in 1896–1913. The latter period most clearly indicated a phase of industrial revolution, entrepreneurial zeal and a decline in economic and technological dualism. In Germany, entrepreneurs responded to the buoyant environment which had resulted from the earlier infrastructural investments of the State.

In Russia during the 1890s, Witte's achievements were considerable in terms of technology transfer and industrial growth.[6] *Industry* grew at an average annual rate of some 8 per cent, railway mileage (already considerable) grew twofold, iron and steel production fourfold. The less sturdy growth in the proto-industrial phase meant that the industrial drive was not a replica of the German model: State direction and ownership continued in railways, telegraphs, mines, oil-wells and fisheries. Here was a tsarist version of Lenin's economy. The years 1893–8 witnessed a growth of foreign capital importation of 120 per cent.

Far more partial were the industrial drives in such nations as Italy, Austria-Hungary and Spain. Whereas Germany exhibited definite industrial and economic dualism prior to the mid-1890s and Russian industry remained enclavist until at least 1907, in this group of nations the industrial drive did not break and possibly increased localism, structural and geographical dualism and agrarian poverty.[7] In terms of agricultural productivity or per capita income, this group almost certainly outreached Russia in, say, 1800. But a weak proto-industrial background led to an inability to benefit from the mechanisms of technology transfer at an early date. Consequently, the industrial drive was held back into the 1890s and beyond. When it came, and with the possible exception of the railway sector, industrialisation here was characterised by the simple exploitation of available, recognised natural resources, rather than by an extension of the resource base. Although we may refer to localised industrial spurts in the Basque region, in Bohemia or in northern Italy, these were normally associated with specific advanced technology transfers, with the exploitation of iron, lead, coal, zinc, aluminium and petroleum deposits and with the continued underdevelopment of agrarian hinterlands. Thus industrialisation impinged hardly at all upon the general process of economic growth. Existing, strong regional inequalities were transformed into economic dualism and industry did not let loose the 'spread' effects witnessed elsewhere.[8] In other nations dualism within *industry* was some function of technology transfer and State policy; in Italy or Spain general *economic* dualism was associated with the continued domination of a backward agrarian sector employing a majority of the working population. At its worst, a combination of population pressure and revenue extraction for industry led to underconsumption, political turmoil in the localities, and a dynamic process of economic underdevelopment. Foreign capital tended to concentrate around resource exploitation which, in the longer term, was asset-reducing. At the same time, many industrialised nations were by now well-supplied with inexpensive raw materials, from ore to grain, silk and fruits. The traumatic combination of falling export prices and negative savings (debt) in agriculture resulted in the truncation of the industrial drive. None of this precluded the transfer of 'high' technologies to enclaves within such nations. By 1900, Budapest in Hungary boasted 400 factories, including advanced machine shops and electrical engineering. Austria developed an automobile industry. Italian technology ranged from electric power and lighting to relatively sophisticated tramway systems.

The third, or *industrial revolution* phase of development in relatively backward economies was measured by a rise of non-government activity and a spread of technologies from particular regions and industries into agriculture, new regions and new industries. From this, industrial revolution generated a spread of high incomes, i.e. a rise in consumption per head. The activity of the German State after circa 1890 was even more indirect than before. Tariffs and company legislation replaced direct inducements. The State and big business still flourished, but now so too did the middling business organisations. If 50 per cent of the industrial workforce was engaged in enterprises of above 500 persons, nevertheless 95 per cent of all firms in manufacturing and mining employed less than 10 workers. Utilising superior technologies, the competition within this newer group of firms resulted in lower prices for manufactured consumer goods. The pressure groups of the new bourgeoisie impacted upon the economic policies of the State. The growth rate of the economy, which had faltered in the 1870s and 1880s, accelerated during

1896–1913. Similarly, the economic recession in Russia from 1901 to 1905–6 was followed by a significant increase in industrial growth between 1907 and 1913 (around 6 per cent per annum), a growth now dictated by business groupings rather than the State.[9] The recession could not destroy the infrastructure of transport and heavy industry laid down in the Witte years, and technologies transferred into the application of oil products (with Baku technology overtaking foreign tutelage), chemicals and electrical production. The resumption of a military programme did not detract from the emergence of strong market forces, the rise of a bourgeois class, gains in the agricultural sector and an increased production of manufactured consumer goods. In this sense at least, Witte's 'gamble' had finally paid out dividends.[10]

6.4 TECHNOLOGICAL PROGRESS

Although steam-power remained a herald of technology transfer (world steam power increased fivefold between 1850 and 1870), after 1850 technological progress centred on metallurgy, chemistry and machine tools. From the early 1870s the focus of technological progress shifted to new power sources (turbines, electricity, oil, the internal combustion engine), new machines based on new materials (steel, alloys, non-ferrous metals) and science-based industries such as organic chemicals. Many *developments* were the result of specific research in scientific societies, technical institutions, and business organisations, although original *breakthroughs* were often the product of very personal research programmes. It was the scientific success of C. Wheatstone in London and William Thomson in Glasgow which gave rise to the telegraph. The development of artificial dyestuffs, explosives, photography and various improvements in steel were the result of similar, directed research efforts. A cause of the German economic boom of the 1890s was the national research programme in such areas as electrical engineering, dyestuffs, optics, and precision engineering.

The importance of specific research programmes in the generation of significant technologies is well-illustrated in the steel industry. The scientific metallurgy of the years 1857–79 was directed at removing the 'brittle' quality of Bessemer steel.[11] Metallurgists spent a great deal of effort in attempting to compound or volatilise the phosphorus in the production process, or to purify the ore magnetically or chemically. Eventually they turned to the puddling process itself and from then technological advance depended on high-quality research. The problem was to reduce the heat in puddling to a level which would remove the phosphorus without allowing the iron to become unmanageable. A series of patents in Germany, France and Britain was followed by the discovery of the 'basic' process in 1879 by Sidney Thomas and George Snelus.[12] This allowed the use of phosphoric ores in steel production at a time when hermiatite ores were rising in price.

Industrialisation created its own resource base. American oil was not a 'resource' until advanced chemical processing allowed its utilisation in production. From the 1850s the Bessemer converter (1856) and the Siemens–Martin open-hearth furnace (1864) permitted charcoal to be replaced by low-grade coal in smelting, and throughout industrialising regions cheap steel ousted wrought iron during the 1870s.[13] In turn, new manufacturing processes and products induced a *search* for new raw materials – the tin of Malaya, the manganese of India and Chile, the nickel of New Caledonia – and transport technologies allowed the hitherto untapped sources of fruit and wheat in faraway places to become resources in Europe. Wastes became resources. In 1864 Pierre Martin had shown that steel could be produced by adding a large proportion of scrap metal to the pig iron of the open-hearth furnace. This not only refined scrap as a resource, but meant that a greater variety of ores could be utilised. The cyanide process of gold production converted waste gold tailings into assets.[14] In the early 1900s the Haber–Bosch process, extracting nitrogen from the air, replaced the fixed nitrates of chemical fertilisers.[15] Just as an abundance of recognised resources did not guarantee their efficient exploitation, so too a scarcity of resources was first defined and then solved by the very process of industrialisation. The extension of the natural resource base was one of the major economic functions of technological progress in the years prior to 1914.

6.5 THE RESPONSE OF THE STATE

In our period nations did not respond to the challenge of relative forwardness. Governments and elites did. No European state (nor even the US) operated on the basis of a truly democratic suffrage in the 1850s, and, with the exceptions of France and Switzerland, few did thereafter. In mid-century the British electorate comprised perhaps 4 per cent of the population. In Belgium, the government was elected by 60 000 people out of a population of nearly 5 million. Where representative government did emerge, nevertheless, the estates and the upper chambers continued to exert great power. Thus, even bureaucracies well-aware of development elsewhere, at times protected vested interests (e.g. Prussia) or failed to satisfy the needs of industry (e.g. Italy). From 1878 Italian tariff policies simply reduced the likelihood of technological innovation in metalworking and machine industries. German state policy after 1870 was obviously pro-Junker in terms of both levels of expenditure and tariff policy. Nevertheless, historians who are sceptical of the industrialisation role of the State might well take note of three obvious points.

Firstly, the protection of vested interests may at times have retarded development in particular industries or technologies, but the retention of some degree of political stability created a general environment for the emergence of innovative business activities; revolutions produce shortsightedness in the market-place.

Secondly financial failures were by no means equivalent to industrial failure *per se* (see Chapter 7). The laying-down of an infrastructure of transport and other services, at whatever cost, served as a base for more efficient, profit-orientated production in a later period (e.g. Russia in the 1890s and 1907–13 respectively).

Finally critics of the State are in danger of implying that the *existing* alternative to State activity was a well-structured, finely-tuned market system, which would distribute factors and generate prices in some optimal fashion. In fact this would represent an ahistorical implicit counterfactual hypothesis, and ought to be challenged as such. The institutions of the State may have directed investment away from alternative industrial requirements, but they also generated investment.

6.6 DEVELOPMENT PROJECTS: RAILROADS

David Landes has judged that from the 1850s railroads 'had displaced textiles as the drummer of industrial activity, setting the beat for short cycles and long trends alike'.[16] When examining the development process, historians often think in terms of whole economies, specific sectors or industries, individual firms, or, more unusually, geographical–resource regions.[17] However, late industrialisation was more often associated with a development process which emerged from the combined positive effects of a series of *development projects* which together harboured advanced technologies. It was the sometimes tremendous success of these which created technological and industrial dualism in relatively backward economies. It was often on the basis of projects that new forms of industrial organisation emerged as appropriate responses to dominant imperatives (e.g. in Germany or the US). Projects possessed a life of their own, beyond the macro-policies and political machinations of state bureaucracies. Initial assumptions or estimates altered, not as a result of changing belief, but because the *working* of a project forced upon its organisers the need to adapt and to modify technologies, organisations, resource bases, and so on.

Railways epitomise the nineteenth-century development project in that they demanded purposefulness of decision-making, they were located in specific geographical settings, and they were mounted on the assumption that 'a sequence of further development moves [would] be set in motion'.[18] In addition, large railroad projects satisfied the requirement of being 'units or aggregates of public investment that, however small, still evoke direct involvement by high, normally the highest, political authorities.'[19] Railways were privileged particles of the industrial development process.[20] Railway development represented one of the earliest and most obvious examples of advanced technology transfer. Railway investment illustrated general economic tendencies. Prior to 1850 British capital was of importance throughout Europe, but henceforth shifted to Egypt, India and the USA. Between 1850 and 1870, of 50 000 miles of new rail laid down in Europe, the French built nearly 20 per cent, the Germans less than 10 per cent. As early as the 1850s France was placing more capital in foreign railroad projects than in those at home.

In cases where railways developed within an environment of generalised industrial revolution, the measurement of 'social benefits' (freight rates on railways versus those on alternative transport systems e.g. canals) illustrates the greater part of their contribution to economic growth. In cases where railways represented original 'modern' technologies in proto-industrial environments, their wider ramifications on the process of economic development must be considered. Admittedly, such ramifications are easier to postulate than to measure.

During the 1840s the skill-flows from such nations as France or Belgium were more congregated around railways than in any manufacturing sector, enabling widespread technology transfer. Railroadisation involved not only innovations in iron production or locomotive-building, but also in such fields as safety, speed, tractive power and points and signals. Railways demanded improvements in steam excavation, stone-crushing, track-laying and haulage and lifting. The first successful use of Bessemer steel for rails at Crewe in 1862 was quickly transferred to America (1867). Internally, railways exploited high-level skills. Nearly 30 per cent of the graduates of the Ecole Centrale between 1829 and 1885 were employed in railroad projects, as compared to 9 per cent in other public works. Between 1880 and 1905 nearly 20 per cent of the graduates from the technical university at Stockholm sought employment in railroads.[21]

Hardly surprisingly, new railway projects involved the introduction of new techniques into a variety of related fields. The locomotive engineer William Pattison moved from Newcastle to a position as foreman in the French railways and in 1851 helped in the formation of a large mechanical engineering firm in Italy. Thomas Brassey (1805–70) at one time employed 80 000 men over five continents in an array of railway projects. The flow of locomotives from Britain and Belgium into Prussia, was replaced during the 1840s by German suppliers. Taylor and Prandi established a machine plant in order to service the needs of the state-instituted Genova–Torino line, the first railway project in Italy. In 1851 the plant was transferred to the State and by 1853, when it employed 500 workers, constructed not only boilers but complete locomotive engines. More interestingly, this servicing complex developed alongside a series of foreign, independent machine shops in the Genoa area. Most of the ancillary, independent plants were successful, and all but one were dominated by English workers.[22]

The ancillary technological requirements of railway projects acted as a textbook 'inducement' mechanism for diffusion and adaptation of technology *within* the host nation. America best illustrated the process during the 1830s and 1840s. An initial dependence upon Britain gave way to a search for appropriate techniques. Large imports of rail iron stimulated a rejuvenation in America's iron industry. During construction wood substituted for iron; specifications regarding gauges and tunnelling were altered in order to reduce capital expenditure; to the 1860s wood was utilised as a cheap fuel source.[23] The reduction of freight costs and the extension of the eastern railroads expanded the market for *specialised* machine goods. The 1840s saw the emergence of a large number of specialised machine producers. Not only did machine-shops act as centres of diffusion, the subsequent railway developments demanded ever-changing technique solutions. The substitution of steel for iron, the production of telegraphs, air brakes, increases in locomotive force and speed and the reduction of fuel intake stimulated incremental technological progress.[24]

In France and Germany, the development of a railway sector in the 1850s encouraged the transition to coke-smelting and the adoption of the Bessemer and Martin processes for cheap steel-production. Finally, the improved and extended transport system of railroadisation increased the supply of raw materials and encouraged a further usage of exploitative technique within the host nation. The actual process of 'frontier' construction led to fortuitous geological discoveries, e.g. the mineral fields of Canada. New routes prompted transfers into new area in order to exploit available raw materials (e.g. Australia, Russia).

In addition they allowed large-scale mobilisation of capital. By the early 1850s railway requirements accounted for 12 per cent of all net investment in Germany. This soon rose to 20 per cent, and by the 1870s remained at over 20 per cent. Between the 1850s and 1870s the share of railway investment in net capital formation swung between 15 and 25 per cent, and in the 1870s moved countercyclically to the general pattern of industrial investment. At its peak in the late 1860s German railway investment was 60 to 70 per cent of investment in mining, manufacture and finance combined.[25] In contrast, the share of railway investment in total investment in France during the second half of the century was small (around 10 per cent) and matched by investment in road and waterway systems and in merchant shipping.

But railway projects took over one-third of all new French foreign investment between 1852 and 1881. During this time, French investment went hand-in-hand with French *polytechniciens* and incorporated a degree of French control over ancillary mining and metallurgy enterprise (e.g. in Austria) within a system of government-guaranteed returns to investors (e.g. in Austria, Hungary, Bohemia, Russia). As a result, French entrepreneurs could be found throughout Europe, not only in railway projects but in ironmaking, steel production and engineering.

Several of the direct production linkages between railroadisation and other sectors have already been indicated *en passant*. In Germany over 30 per cent of the domestic production of pig iron in the years 1840–59 was destined for rails, a figure which excludes iron in locomotives and bridge construction. Even during the 1860s rail demand absorbed over 25 per cent of pig-iron production. Undoubtedly, it was this sturdy level of demand which encouraged the German transition from charcoal-using furnaces to coke utilisation. By 1875 in Prussia, 75 per cent of all steam-engine horsepower was satisfying railroad needs. At the same time, throughout Europe, railways generated increased demands for wood, glass, leather, stones and a vast array of machinery and tools. In metallurgy and engineering, Schmookler has made much of the positive relationship between railroadisation and the increased demand for and induced supply of patented techniques. Indeed, it was his findings for the railway sector which stimulated Schmookler's general observation that variations in the sale of equipment induce the variations in inventive effort.[26]

Finally, one of the greatest contributions of railroad development was as an inducement to regular and sustained government intervention. Here the military and development goals of governments most nicely coincided. Railways served clear military needs, and by stimulating heavy industry generally, could generate developments in other, military-orientated sections of the economy, e.g. construction or shipping. Prussian policy in the 1840s and 1850s created a close relationship between railway investment and general public sector expenditure. The railway sector was the strongest stimulation to government involvement in Russia after 1882. The first stage was the official purchase of bankrupt private lines, followed by the large-scale direct construction of lines as well as continued purchases. Railways were a particularly attractive proposition for new governments seeking legitimisation (Russia, Japan, even Belgium between 1834 and 1843), and competition between the German states undoubtedly hastened the development of railways. Contrariwise, violent changes of government inhibited railroad construction (e.g. Spain). The scale of railroad projects, together with their strategic importance, forced upon governments the need to construct relatively sophisticated measures for encouragement and regulation. From 1842 the Prussian government instituted a complex system of state guarantees. In Russia during the 1890s officialdom developed a system of concessionary freight rates for industrial producers and exporters. Such monitorship tended to be most significant in nations where canal and road systems were not already well-developed or where for topographical reasons they were not appropriate.

6.7 WINNERS AND LOSERS

During the Great Exhibition in London in 1851, 14 000 firms mounted exhibits. In the Paris Exposition of 1855, 24 000 firms were represented, in 1862 in London, 29 000 firms exhibited. The largest technological and industrial exhibition of the century, the Philadelphia Centennial Exhibition of 1876, was visited by 10 million people. Yet the demonstration of forwardness, in peace or in war, did not serve to industrialise all of Europe, and left untouched or badly mauled most of the rest of the world. Although historians may be tempted to address this profound shortfall in terms of an absence of 'prerequisites' in 'loser' nations (e.g. low levels of savings, primitive market and agrarian institutions, a lack of entrepreneurial talent due to atavistic value systems), we have argued that this approach is of little use in generating a consistent explanation. Proto-industrial production existed at many points on the world's map, technique was transferable and values and institutions could be transcended. The most 'forward' of Europe's economies of the eighteenth century were not all amongst its successful industrialisers in the later nineteenth century. Russia, a most backward economy on almost every count in 1800 still managed to mount an industrial drive at the end of the century. As we shall see, Japan, with a value system and social structure and knowledge base which were vastly different from those of Europe in 1850, and which had been amazingly stable over a long period, rejuvenated its economic structure under the ironic banner of restoration. China and India joined the 'losers' not because of a lack of necessary prerequisites in 1750 or 1800, but because the nature of their interaction with the international economy of the nineteenth century – and thus the character of their exposure to transfer mechanisms – was conditioned by colonialism and economic imperialism.

Alternatively, we may view success and failure in terms of transfer mechanisms and the boundary conditions of response and positive action. We would, therefore, suggest that industrial revolutions (true economic watersheds) followed from successful, fairly short-period industrial drives, rather than from the distant past of malleable economies. Furthermore, successful industrial drives occurred when national economies 'responded' to a coalescence of transfer mechanisms, which were made available through the expansion of the international economy after circa 1850. The only major exception to this outside Europe were the patterns of staple exploitation followed by areas of recent settlement.

Expanding trade links with relatively advanced nations were often the prelude to technology transfers. But it must be emphasised that until the 1870s, although European trade expanded at a faster rate than European economies, the bulk of trade was *between* European nations (and between Britain and America), and coincided with a decline in the actual share of world trade for the rest of the world. The Singer–Prebisch thesis argues that industrialising nations eventually exhibit a *fall* in their propensity to import food and raw materials. Engel's law effects (the propensity to consume 'basics' falls as incomes per head rise) combined with raw material saving-or-replacing technological progress (see above), will create a fall in the demand for primary products relative to manufacturing products. Falling transport prices are not sufficient explanation of the lower prices of raw materials. Although the thesis has been attacked it may be sustained for the years after 1875.[27] Furthermore, the available evidence suggests that the fall in price for primary products was most significant in peripheral, low-income nations. Thus, between 1872 and 1938 the value of Europe's *combined* exports and imports of primary products fell by 22 per cent but the price of Europe's *imports* of primary products fell by 38 per cent.[28]

Chapters 8 and 9 consider the failure of technological progress as an agent of economic transformation in China and India. In these two nations economic sovereignty over whole fields of production passed into the hands of 'colonial' speculators. At this point we may simply generalise that colonial connections between large industrial economies and relatively backward economies (excluding areas of recent settlement) yielded frequent transfers of technology (from

Turkestan and the Russians to Algiers and the French) but few examples of technological transformations; and that most of the densely populated areas of the world were more or less colonised or externally controlled by 1914.

Rather than claiming a failure of national will-power or an absence of 'prerequisites', we might argue that *any* force or combination of forces which inhibited a nation from benefiting from technology transfers is a sufficient explanation of industrial 'failure' in the years prior to 1914. Looking elsewhere yields a clouded view of the nineteenth century. Finally, we suggest that the major forces inhibiting industrialisation related to the imperatives of advanced technologies and thus the degree to which they were appropriate to the social *and* economic needs of receptor nations, to the strength of transfer mechanisms and to the *ability* of the relatively backward nation to maintain a degree of political stability and economic sovereignty during the period of the industrial drive.

7. Technology, Economic Backwardness and Industrialisation – the Case of Japan

7.1 JAPAN, TECHNOLOGY AND LATE INDUSTRIALISATION

> The Japanese are very ingenious in most handicraft trades, and excell the Chinese in several manufactures, particularly in the beauty, goodness, and variety of their silks, cottons and other stuffs, and in their Japan and porcelain wares. The women are subjected to a most wretched state of degradation. A husband may put his wives to a more or less severe death, if they give the least cause of jealousy.
>
> *The Christians' Penny Magazine*, 9 June 1832

THE image of the Japanese as at once peculiar and ingenious lies at the heart of much nineteenth- and twentieth-century commentary and historiography. Relatively backward, isolated and village-based in the mid-nineteenth century, they were at the same time literate, organised and ambitious. The Japanese were lacking any interest in 'metaphysical, psychological and ethical controversy of all kinds', and were 'inclined to be satisfied with a ready-made knowledge', but 'their genius leads them in the direction of accurate, detailed investigation'.[1]

War and territorial advancement did not alter such images, rather it highlighted them. The Japanese were peculiar but successful. In his presidential address to the British Association for the Advancement of Science in September 1903, Sir Norman Lockyer chose the subject of 'The Influence of Brain-Power in History'. He suggested that in such matters as skill-provision and application to industrial processes, the under-utilised workshop of the world might learn from the full-capacity workshop of Asia: 'shall we follow Japan, and thoroughly prepare by "intellectual" efforts for the industrial struggle that lies before us'?[2] In June 1905, with a war fully waged between Japan and Russia, *Nature* isolated the basic lesson of Japanese experience:

> The lesson which our educationalists and statemen have to learn from Japan is that the life of a modern nation requires to be organised on scientific lines in all its departments, and that it must not be directed chiefly to personal ends, the attainment of which may, to a large extent, intensify many of our problems, but that it be consciously used for the promotion of national welfare.[3]

The most authoritative modern interpretations of Japan's economic success since the Meiji Restoration of 1868 all agree with those who witnessed it: productivity-gains in agriculture *and* industry arising from the transfer of advanced technologies into Japan, were essential determinants of the achieved high rates of industrial and economic growth.[4] The outstanding monograph of Thomas C. Smith on the role of government and political change in Japanese industrial development is a testament to the positive function of foreign technicians and their technologies, whilst for these and later years Ohkawa and Rosovsky 'assume that autonomous investment based on borrowed technology is the major driving force of Japanese economic growth'.[5] The latter writers go on to link technology transfer with the workings of *creative institutional substitutions* of the type stressed in Chapter 5 above: 'Japanese economic history shows both the significance of a rapid absorption of imported technology via the repeated investment spurts and also the development of specific institutions that facilitated the entire process.'[6] It is for this reason that students of nineteenth-century Japanese development are confronted with the itemisation of the modernised shipyards of Nagasaki, the Kagoshima Spinning Factory, the Osaka Spinning Company, the Yokosuka Iron Foundary and Arsenal, the Osaka–

Kobe railroad, the silk-manufacturing of Tomioka, and the several examples of foreign-built or -inspired canals, river embankments, harbours, telegraphs, lighthouses and so on.[7] But description does not serve in place of analysis, nor does concentration upon selective best-techniques uncover the basic improvements in agriculture and traditional industries, which probably contributed most to the increasing industrial growth rate during the 1870s and to the acceleration of overall economic growth during the 1880s.[8]

It would seem that as far as technology transfer is concerned, Japan falls into that group of relatively backward nations which succeeded in capturing the advantage of backwardness rather than succumbed to its obvious politico-economic disadvantages. In terms of our presentation in Chapter 5, Japanese industrialisation may be divided into three phases as follows:

Proto-industrial (to 1868) The major features of the Tokugawa economy (1603–1868) included a firmly rising proto-industrial sector,[9] a large but steady population at around the 33 million mark, an intensive use of land on small-size plots, relative isolation (see below), a highly developed road transport system between very large cities,[10] a rising but dependent merchant class, an increasingly relatively deprived *samurai* and *daimyo* noble class,[11] a large, urban-based bureaucracy of communication and control, and a reasonably strong sense of nationhood.[12] *The* outstanding feature of the Tokugawa economy was the relatively small space in which a large, increasingly urban population was grouped.

Industrial Drive (1868–81)[13] As might be expected from Chapters 5 and 6, the major features of this period were associated with a congregation of institutional substitutions designed to remove yet satisfy feudal privilege, create central government revenue (e.g. the land-tax reforms of 1872–3) and establish property rights and commercial security,[14] with the speedy introduction of a limited range of infrastructural, military and industrial technologies from Europe and America yielding an increase in the *industrial* growth rate, with a high level of government *direct* interference in the economy, and in areas of social reform and control,[15] with a high level of inflation and exploitation of all wage-earners, and with wholesale confusion.[16] *The* outstanding feature of this phase was the retention of central authority in the face of cultural conflict, economic exploitation (especially of female urban workers) and emergent, contending political ideologies, most of which entered from the West.[17]

Industrial Revolution (1881–1912 and beyond)[18] The major features of this period include the acceleration in the rate of growth of the *overall* economy, associated with a greater spread of efficiency improvements in industry (e.g. cotton), infrastructure (e.g. railways) and agriculture (e.g. fertiliser inputs),[19] greater control and government financing (measured by a reduction of inflation and more efficient use of official resources), a shift to *indirect* government activity and *indirect* foreign tutelage (e.g. with patents and licencing arrangements replacing foreign technicians and Japanese missions abroad),[20] and the growth bonuses derived from the impact of earlier, expensive institutional innovations and earlier, expensive infrastructural investment. *The* outstanding feature of the period was the spread of superior technologies (both introduced and *existing* best-practice) into a far greater variety of 'leading sectors': Rostow nominates cotton manufacture, railroads, pig iron production and steel as dynamic leading sectors in these years.[21]

The rest of this brief chapter will suggest reasons for Japan's industrial and technological success during 1868–1912, at the beginning of which the nation was opened up to a host of foreign economic and political influences and on the very brink of dynamic underdevelopment.

7.2 TRANSFER MECHANISMS

Indirect

The movement of capital and of people into Japan were of far less importance than trading relationships in explaining the process of technology transfer.[22] Williamson and de Bever argue that the decline in foreign raw material prices (cotton, wool, etc.) during the later nineteenth century was a cause of Japan's foreign exchange viability, and hence her high rate of accumulation, technology importing and industrialisation.[23] Richard Huber estimates that between the 1850s and the mid-1870s, Japan's commodity terms of trade, determined greatly by prices received for her staple silk product exports, improved by over 300 per cent and boosted her real national income growth rate by some 65 per cent.[24]

But if we use as our benchmark-years those of the industrial drive itself (1868–81) it would seem that the period of technology transfer was associated with a continuous deficit on the balance of trade. A combination of loss of tariff control, mounting demands in Japan for iron, machinery and cotton imports, and internal inflation, led to a huge outflow of gold and silver, with the bullion stock diminishing by 60 per cent in the years 1869–72 alone.[25] It would appear that technology importing was being financed in this crucial period by expropriated feudal accumulations, which permitted a postponement of deflationary policy (which might well have dampened technology imports), until 1881–5.[26] Secondly, the precarious trade balance forced government intervention in the form of a coherent import substitution policy which, as we shall see further, itself had implications for a speeding-up of strategic technology importation. We might conclude that the net financial gains from trade were of little importance prior to 1881 in explaining the pattern of technological transfer. The ability of the authorities to react to financial pressure seems of more importance. Trade forced structural changes upon the economy, and these involved further technology transfers. Thus, the early windfall gains arising from silk disease in Europe and the opening of the Suez Canal, served to place silk at the head of the vital export industries. The rise of the import bill and the bullion rundown stimulated technical change in cocoon-raising and silk-reeling (Section 7.4 below) in order to sustain a surge in export output at low prices.[27]

During the industrial revolution phase (1881–1912) deflation, import substitution and an exogenous increase in demand within Europe and America for traditional, crafted Japanese products forced an export surge. During this second phase of industrial modernisation there is little doubt that the trading sector did help to act as a financial mechanism for technology transfer.[28] In the years 1890–1913 the annual average increase in volume of exports from Japan was 8.6 per cent, for Germany the figure was 5.1 per cent, and for the world as a whole, 3.5 per cent. Poised on the brink of financial disaster around 1868, Japan had escaped a debt trap.

Direct

Compared to the Meiji years, mechanisms of technology transfer prior to 1868 were rudimentary. In the so-called *Nanbangaku* (period of the Southern Barbarians) years from 1543 to 1639, interaction with the West was dominated by Spain and Portugal. Artifacts such as muskets, globes, navigation charts and printing presses were introduced on a limited scale. In mining there is some evidence of Western influence, with galleries replacing open-cut operations and the adoption of cupellation in refining during the second half of the sixteenth century. It is also known that the introduction of the matchlock gun by the Portugese set off a process of technology replication throughout Osaka and Shiga and elsewhere, based on the precise skills of indigenous swordmasters.[29] Lastly, there is some evidence of a wider intellectual transfer with the publication around the end of the period of *Kenkon bensetsu* (Heaven and Earth), a compilation of Western thought on astronomy and meteorology.[30]

The machinations of the Jesuits created internal strains which led to the period of *Sakoku* (Seclusion) between 1639 and 1720, although internal hunts for Christians took place in 1790, 1841 and as late as 1856. Seclusion was effective but not absolute. In 1641 the Dutch took over the previous Portugese factory at Dejima at the head of Nagasaki Bay. Strong censorship and limits on internal passage confined the transfers of knowledge to the small group of *rōjū*, or government inner councillors, who were the recipients of Japanese translations of official reports on Western developments (*fusetsugaki*) delivered to the Shogun during compulsory annual visits to Edo.[31]

There is little doubt that it was during the years 1720 to 1868, the period of *Rangaku* (Dutch learning) that effective Western tutelage began. In an attempt to solve administrative and industrial problems, the Tokugawa regime relaxed the exclusion rules and allowed a firm inflow of knowledge of Western medicine, science (Newtonian dynamics) and technologies. With an increased emphasis on *jitsugaku* (the study of real things) a flow of discrete information took place not only through Dutch settlement but through European publications directly imported from China.[32] In 1796 the first Dutch–Japanese dictionary was published, and this was followed by a series of academy and school foundations in which Western learning was included. An important element of *Rangaku* which developed from then was *Bussangaku* or 'knowledge of production', which emphasised practical rather than abstract themes in Western culture. Under this influence was instigated the first artisan exhibition in 1857: in the following decade thirty provinces sent exhibits to metropolitan exhibitions.[33]

The effect of earlier *Rangaku* learning was to make Japanese elites *aware* of their relative economic backwardness. The effect of later developments was to *introduce* specific technologies.[34] There was a speeding-up of transferences. P. van der Burg's *Erste Grondbeginselen der Naturkunde*, a survey of new industrial technologies, was first published in 1844 and translated in Japan in 1854.[35] The military needs of the clans dominated the immediate pre-Meiji years.[36] The Saga fief established the *Nirayama Ransho Honyuko kata* (Nirayama Dutch book translation office), designed to bring together scholars of Western learning and workmen engaged in the casting of cannon. Dutch scholars were vital to the military ambitions of Saga, Satsuma, Higo and Nanbu clans in the working of iron and the construction of reverberatory and blast furnaces, boring machinery, mortars, drydocks and steamships.[37]

In summary, pre-Meiji technology transfers were based on *small networks* of men and books, and were closely controlled by elite groupings, although there is some limited evidence of a more middle-class Westernised culture emerging in the large cities.[38]

The enormous range of transfer mechanisms established during the Meiji period may be briefly outlined under three headings; those designed to establish a *selection environment* (including original searching and screening processes), an *innovation environment* (measures promoting diffusion and adoption of techniques) and an associated *service environment* ('institutional substitutions' which *indirectly* induced diffusion, adaptation and adoption of advanced techniques).[39]

The Selection Environment Between 1868 and 1895, 2500 students and officials were sent to Europe and America for study and investigation of best-techniques.[40] Between 1873 and 1885 Japanese officials participated in over twenty international industrial and commercial expositions. To the 1873 Vienna Exposition were sent seventy Japanese delegates, including twenty-four technical workers from the private sector whose task was to search and buy appropriate techniques. From Austria the group brought back the latest textile looms, which were then tested and leased to various provincial entrepreneurs after regional public display.[41] More spectacularly, the government established its own model factories and utilities, which were designed to not only demonstrate existing best-technique, but to establish the *appropriateness* of alternative Western modes of production.[42] The government's Akabane Engineering Office, established in 1874, was designed to construct selected and tested machinery for purchase by private Japanese enterprise.

The Innovation Environment The bulk of government activity was aimed at diffusion and adoption.[43] This was especially so during the years 1879–85 when the authorities sold up model enterprises on advantageous terms to private interests. At a lower level the *kyoshin-kai* (cooperative-competitive exhibitions) were established from 1879 for the spread of *existing* best-techniques and were normally product- or industry-specific. During their heyday in the 1880s successful exhibitions would each last from 30 to 100 days. In 1887 alone, 317 exhibitions were opened for a combined total of 2410 days, involving 180 000 producers displaying 430 000 articles before an estimated audience of two million people.[44]

Through a vast range of regulatory and promotive devices, the central authorities ensured that interaction between Europeans and Japanese was maintained throughout the industrialisation process. Between 1870 and 1885 an average of 42 per cent of the total expenditure of the Ministry of Engineering was spent on official foreign experts (*ōyatoi*), the bill peaking at 66 per cent in 1877. The hiring policy of the *Kobusho* (Ministry of Industry) was in fact directed to hitching foreign skills not to manufacturing but to the provision of strategic infrastructure: in 1872, 80 per cent of its 800 foreign employees worked on railway, telegraph, lighthouse and mining projects.[45] In microcosm, the foreign communities settled in Yokohama, Kobe, Osaka and elsewhere served as living demonstrations of Western technical achievement and also as a focus for the much more important and pervasive intellectual and social interaction between Europeans and key Japanese agents of change.[46]

Such stimulative influences were possibly of greater importance than the more spectacular direct investments of government. Similarly, the standardisation movement of 1875–91 cost little in the way of financial resources, but metrication, standardisation in medicines and pharmaceuticals and clearer specifications in government contracts for supply of raw materials hastened the process of technology diffusion. The establishment of hundreds of trade associations under official auspices from 1884 were primarily designed to control the quality of a variety of products, but soon developed as regional technical extension centres. The establishment of a patent system paved the way for a series of licensing agreements and joint ventures, especially in the field of general and electrical machine production after 1904. Thereby Mitsubishi gained ownership of the Parsons turbine and Shibaura obtained entry to General Electric's manufacture of heavy electrical equipment.[47]

The Service Environment It should be noted that the central administration of Japan was controlled by a Cabinet system (rather than the Diet established in the 1890s). The Cabinet exerted monopoly power over the appointment of prefectural governorships and was at most minimally responsible to public opinion.[48] However costly, this arrangement permitted a relatively coherent programme of institutional formation. The task of the industrial drive was to remove Tokugawa restrictions and noble privileges, and to alter the system of land tenure. In itself this tended to increase the likelihood of innovative behaviour.[49] The next task of government at the wider level was to pass legislation which at least protected innovative behaviour amongst both bureaucrats and industrial producers by reducing the risks and increasing the *potential* reward of such behaviour. The establishment of channels for social mobility would seem to have encouraged creative adaptation amongst many in the middle and upper echelons of both central and provincial administration.[50] In addition, government educational expenditure as a percentage of total government expenditure reached 12 per cent in 1885, remained around 10 per cent in 1895 (a year of war expense) and had moved to 15 per cent in the 1920s.[51] More significantly, between 1880 and 1940 there was a fortyfold increase in the number of students enrolled in higher scientific and technical education and an eighty-fold increase in the number of students in further technical and commercial education.[52] This not only expanded the training in modern technology but provided an extra avenue of social mobility in a society previously dominated by systems of ascribed status.[53]

The introduction of Western systems of commercial law and practice, of patents, copyrights and utility models, hastened the emergence of individual entrepreneurship.[54] At the same time began the system of specific legislation promoted by research councils organised by relevant ministries. Such legislation was often backed by government subsidies to firms operating under the new rules. Apart from such strategic legislation as the Promotion of Navigation and Shipbuilding Law or the later Electrical Power Supply Enterprise Law, there were also a series of important enactments encouraging innovative behaviour amongst smaller enterprises in the more traditional industries.

Sections 7.3 and 7.4 suggest the reasons for the economic success of this clutch of deliberately-created transfer mechanisms.

7.3 BOUNDARY CONDITIONS

> Knowledge shall be sought throughout the world so as to strengthen the foundation of imperial rule.
>
> *5th Article of Meiji Charter Oath, 1868.*

Much recent Japanese writing revolves around the idea that the nation was successful in technology transfer because of its *already* high level of handicraft and industrial production.[55] Certainly, prior to 1868 individual fiefs had financed and encouraged handicrafts in the more advanced provinces of Japan. By-employment regions emerged in metalworking, agricultural implements, mining, production of cast-iron goods, copper, bronze and tin wares, ceramics, paper, cotton and textile manufacture, watermill construction, all set against the evolution of a wide variety of skills in water-control technology, i.e. flood and hydraulic techniques. Of 420 families surveyed in the village of Iwakura (proximate to Nagoya castle town) in 1835, 42 per cent claimed to be wholly engaged in trade, 42 per cent partially so. Trades included cotton dealers and retailers, dyers, indigo dealers, weavers, rice dealers, traders of sardines (for fertiliser), millers, oil pressers, cracker-makers, tea dealers and shopkeepers. T. C. Smith's work on Kaminoseki (a Choshu fief at the extreme south of Honshu) shows that by the 1840s 18 per cent of its population produced 55 per cent of its total income through non-agricultural occupations, although 82 per cent of workers regarded themselves as 'farmers'.[56]

However, given that such by-employment and specialised regional markets were hardly absent elsewhere in Asia (see Chapters 8 and 9), we would argue that their pertinence to Meiji transfer processes lay in certain demographic, spatial and political factors *not* shared by other nations in Asia and elsewhere. By the 1860s there were 33 million people settled on the Japanese islands, mostly living around the alluvial plains of the largest, central island. By historical and contemporary standards the population was very large.[57] Furthermore, topography ensured that the settlements were densely packed, encouraging a movement of men and ideas at a speed and frequency not easily atttained in China and India, then or now. Also, the level of urbanisation was extraordinary. Edo (Tokyo) was one of the largest cities in the world, with a population of at least one million, mostly composed of administrative and aristocratic *consumers* of goods. Edo was surrounded by a network of other large administrative, commercial and cultural centres of over 60 000 population, such as Osaka, Kyoto, Kobe, Nagasaki and Nagoya. According to Toshio Furushima's early calculations, by the *mid-eighteenth* century the urban population represented 22 per cent of the total, far higher than that of another successful island, Great Britain, prior to its industrial modernisation.[58] The large urban centres generated a consumer demand of great impact and increasing sophistication which ensured that by early Meiji the modern manufacturing techniques from the West secured bouyant levels of demand.[59] Secondly, urbanism meant that Japan benefited from so-called 'neighbourhood effects'; within an economic neighbourhood pressure on laggards builds up as the number of innovative adopters grows.[60] Lastly, the city phenomenon engendered some homogeneity of culture and institutions, a side-effect which might well have assisted the Meiji government in its efforts towards economic modernisation.[61]

Geographical and spatial characteristics may also have gone some way in determining Japan's political position within the economy of nations during a phase of remarkable Western aggression, i.e. may explain the retention of economic sovereignty.[62] Continued economic sovereignty requires that destabilising political forces from *both* outside and inside the nation must be either minimised or somehow contained by the State and its strategic elites at a cost which is not so high as to detract from the industrial programme itself.[63] Undoubtedly, location had ramifications on the *manner* in which Japan was penetrated by the West, particularly in contrast to China.[64] But internal space had similar ramifications on the *ability* of the State to retain social control during the crucial years of the industrial drive.

Political and cultural historians have explored features of Japanese society and administration which served to produce political continuity amidst economic change.[65] In their typology of industrialising elites Kerr, Dunlop, et al. label Meiji Japan as 'dynastic' and held together by a common alegiance to the established cultural order, orientated towards tradition and its preservation, divided, if at all, between 'realists' and 'traditionalists':

> The authority of superiors stemming from the restoration of the emperor, was maintained into modern Japan . . . the traditional culture favoured the dynastic elite as the prime movers towards industrialisation, and this elite in turn preserved as much of that culture as possible.[66]

Michio Morishima has argued that a crucial ingredient of this 'traditional culture' was the distinction between Chinese and Japanese forms of Confucianism, with the latter's historical emphasis upon hierarchy and loyalty.[67] Silberman has proposed that the efforts of the lower samurai to carve out paths of 'status maintenance' resulted in innovative, constructive behaviour in both industry and the bureaucracy rather than denial, revolt or disengagement.[68] From his study of a sample of 253 upper civil servants during the early Meiji, Silberman found that 35.5 per cent of the total were formerly of the lower samurai and that the bureaucratic structure thus offered a degree of upward mobility, 28 per cent were of the *Kuge* (court aristocracy) and some 20 per cent were of the upper samurai.[69] In 1880, according to a contemporary estimate, 91 per cent of senior government officials were of samurai origin (the ex-samurai representing at most 6 per cent of the total population).[70] In conclusion, there was a significant motivation for the lower samurai to engage in 'innovational behaviour' as a result of their inaccessibility to older forms of achievement, and this carried forward as a facet of the Meiji upper bureaucracy, promoting inner cohesion and a Western orientation amongst a strategically placed, upwardly mobile group of decision-makers.

However, such factors do not adequately explain the undoubted political *turmoil* of the 1870s and early 1880s, which produced political movements engaged in the formulation of economic and industrial policies at odds with those of the State itself.[71] At the same time, many of the 'traditional' cultural features enlisted as ingredients of continuity and control were in fact reconstructed by the State in a deliberate effort to produce a 'usable past'.[72]

Another approach is to argue that mechanisms of control were forged *during* the Meiji years as part and parcel of the 'institutional substitutions' of the industrial drive and beyond, and that their effectiveness derived at least in part from the spatial factors noted above.[73]

The very weight of administrative formality circumscribed political behaviour; 4555 major directive notifications were issued by government departments between 1873 and 1877.[74] Provincial discontents were met by bending to the wind. As early as May 1878 at least nominal representation in local matters was increased through the introduction of a general assembly of provincial governors. When in early 1880 the administrative power of the local assemblies was furthered, many of the popular societies which continued to press for democratic institutions were pushed further and further into an outside position. The forced removal of the means of diffusion and sustenance of radicalism (the press, public meetings) shattered the opposition groups.[75] The selling-up of government enterprises and introduction of subsidies during the late 1870s, which provided occupational avenues for the *shizoku* (ex-samurai) and others, weakened support for radicalism. Finally, technology borrowing and the process of its application in the economy affected political development in a multitude of ways. The comparatively simple levels of technique permitted merchants to move fairly comfortably from their traditional commercial functions to a 'management' function.[76] Bureaucrats returned from the West could comprehend and control technological innovations quite readily. The nature of technology was such as to minimise its 'displacement' effects on employment, skills and regional balance. Indeed, technology planning in Hokkaido was seen as a means of relieving employment and political problems as well as creating growth.[77] The fact that both agriculture and most by-employments thrived under the new technological regime allowed for a continuity of organisation in older sectors and a retention of non-urban commercial and social institutions.[78] The political appropriatness of Meiji technological advance is illustrated further in Section 7.4.

7.4 THE APPROPRIATION OF TECHNOLOGY: CASE STUDIES

According to the econometric results of Kelley and Williamson, 'the role of total factor productivity growth in agriculture was unusually high and this impressive rate may be related to the underutilised technical potential bequeathed as a Tokugawa legacy.'[79] In explaining technological change in the rural sector, the relations between existing and introduced techniques seem crucial.

The industrial drive began with enthusiasm for the best from the West. From 1872 were established a series of agricultural implement stations, agricultural schools and translation services.[80] A leading theme was acclimatisation of Western staple lines – thus the establishment of the Mita Botanical Experiment Yard (1874), Shimozawa Sheep Farm (1875), Kobe Olive Farm (1879), Harima Grape Farm (1880). With the possible exception of the Shinjuku experimental station from 1873, which by 1876 had experimented with 313 strains of foreign wheat and 398 foreign trees and grasses, the early dominance of British and American ideas and personnel was inappropriate and wasteful of resources.[81] The Komaba and Sapporo agricultural schools emphasised large-scale mechanical farming unadaptable to the plot-sizes, scale and climate of Japan. Particularly in Hokkaido, the American Agricultural Secretary, Horace Capron, was a leading advocate of new cereals in preference to rice, a policy from which the island took some time to recover.[82]

Tokugawa agriculture had incorporated Dutch, Chinese and indigenous techniques in irrigation and canal construction (dikes, sluices) and raised the level of fertiliser usage and vastly increased the crop acreage. Nevertheless, topography, demography and the nature of rice cultivation together ensured that farming remained based on small (1 hectare), family-owned and labour-intensive cultivation, assisted by periodic bouts of village and inter-village cooperation in sericulture and water control.[83] From the mid-eighteenth century the major irrigation technology of rice-paddy areas such as the Saga Plain was the raising of water from man-made channels into the fields by man-powered wooden waterwheels, the major form of equipment investment for small households. This tied up labour in the crucial season when it was desperately needed for planting and weeding. The principal solution to this problem in the Meiji years was found *within* established technology adaptation, e.g. with the introduction of special ploughing to reduce the rate of water-leakage combined with the staggering of planting times. In areas where the labour problem had been so solved there was a resistance to Westernised irrigation technology *until* a combination of increased demand and labour migration, both stemming from industrial expansion, gave rise to the need to produce more rice at rising labour costs. (The Saga Plain labour force in agriculture fell by 50 per cent between 1890 and 1920). Market forces then induced a new stage of technological change in such rice areas. Only then did the institutions established by government (see below) begin to take effect, evidenced in rising rice-yields and mechanised pumping. A second lesson from Japan's experience in irrigation improvement is that even at such an early stage, techniques are highly interdependent; only when mechanical pumping became community financed, electrically powered and dependent upon machine manufacture in Japan for Japanese conditions, did mechanisation spread rapidly amongst smallholders.[84]

However, in several instances of agricultural technology change, transfers and the role of government in supplying discrete information at low cost were of far greater importance. In encouraging the development and diffusion of new seed varieties, cooperation in technology purchase and knowledge of agricultural chemistry (especially German), the Department of Agriculture and Commerce (founded in 1874), through its regular Bulletins, was of fundamental importance, as it increasingly concentrated upon information dispersal at the prefectural level. The prefectures responded. Despite small budgets, prefectural governments ran lecture courses on new Western and Japanese varieties and technologies and reported local innovations to the Shinjuku experimental station.[85] They also subsidised testing and seed improvement programmes, as well as local agricultural colleges. The prefectures were essential to the success of the *rono* or veteran farmer system (1880–95) which focused on the interaction between new methods and established skills.[86] From this milieu emerged the practice of *ensuisen*, a method of selecting rice seeds using salt water.[87] Epitomising this 'indigenisation' of technique was the work of the Experimental Farm for Staple Cereals and Vegetables at Nishigahara, which favoured simple field experiments comparing the viability of the existing range of varieties. At the local level developed the *hinshukokankai* (seed exchange societies) and *nodankai* (agricultural discussion societies). Apart from seed selection by salt water, such societies diffused new knowledge on the preparation of nursery beds, checkrow planting and selection of non-rice varieties. Investigations became increasingly location-specific.[88] New varieties encouraged the use of better fertilisers (herring meals, soybean and later industrial products) and in turn these encouraged further Japanese experimentation to maximise the fertiliser responsiveness of varieties.[89] The 1890s were dominated by the development of fertiliser-responsive high-yielding rice varieties.[90] Government encouraged competition. Apart from the major Promotion of Industries Exhibitions after 1877, a nationwide system of prize shows was instituted, experimental stations and colleges were instructed to display exhibits and a Museum of Agriculture was maintained in Tokyo after 1874.[91] By 1896, 46 provincial agricultural schools were established, and after the turn of the century some 60–80 per cent of all government agricultural expenditure was prefectural.

Apart from double cropping, the introduction of hybrids in sericulture, dry paddy cultivation and the introduction of farm animals, the activity of government resulted in a diffusion of pedal-operated threshers and a thirtyfold increase in fertiliser-input between 1878 and 1913. As a result, from the 1880s although land and labour costs were rising in agriculture, the prices of fertiliser and machinery inputs were falling.[92] Inputs from the industrial sector were land-saving and labour-improving, and technological change permitted a reduction of underemployment in many rural areas.[93] Kelley and Williamson argue that a growth rate of 2 per cent to 2.2 per cent per annum was the direct result of increase in labour productivity, approximately half of which is to be explained in terms of the application of purchased machinery and chemical inputs from the industrial sector.[94]

Technological progress in Japanese agriculture may be summarised as involving three phases; a relatively unsuccessful, mechanism-orientated 'Western mania' phase, associated with the industrial drive and important in setting up an institutional complex which was to prove productive later; a phase of 'Japanisation' associated with the beginning of the Industrial Revolution and stimulating real indigenous breakthroughs;[95] and a successful, growth-sustaining period of Western-orientated technical change, primarily characterised by biological rather than mechanical improvement, and therefore at once less capital-intensive and land-extensive than the form of technological progress often associated with the 'Western' model.

Silk was an industry which emerged as strategic both in linking the agricultural and manufacturing sectors and in generating export revenue. Silk products contributed to over 40 per cent of all commodity export values prior to 1919, and even during the interwar years retained a 30 per cent footing.[96] Sericulture (mulberry-bush growing and silkworm weaving) and silk manufacture (reeling, weaving and finishing) spanned the range of Japanese technology, whilst the growth of the latter during the 1870s has been seen as a foundation for the emergence of the cotton manufacturing industry as a leading sector during the 1880s and beyond.[97]

The international price of Japanese raw silk products increased fivefold in the years 1859–68.[98] Full capacity with existing techniques led to a rapid decline in cocoon and raw silk quality in the immediate post-restoration years.[99] For this reason a range of Western techniques were introduced in an effort to increase quantity, quality and standardisation. Introduced foreign technologies in the form of thermometers and hygrometers, egg-selection and mulberry-tree production were of crucial importance during the industrial drive. Sericulture emerged as the most profitable product of agriculture due to the increase in quality production arising from Western methods of egg-selection and temperature-maintenance in silkworm breeding, both stimulated by the activities of official research institutes. From 1874 a series of official missions to France and Italy culminated in the foundation of the Silkworm Research Institute. The development of improved culture (by postponing hatching by storing eggs at cool temperatures) made practical the most significant innovation of the era, the development of summer–autumn rearing technology.[100] This replaced the traditional spring period of cocoon culture and allowed the use of off-season labour from the rice economy. By the end of the Meiji period cocoon production by the new methods was contributing to nearly 50 per cent of all raw silk production.

In reeling, traditional methods survived for some time by carving out specialised home markets.[101] Through demonstration plants the government quickly introduced waterpower winding and steam boiling, and established Italian and French systems of filatures.[102] The Tomioka Mill was French-built and -designed, staffed by European technicians, steam-driven, provided a formal system for indigenous training and maintained low costs by employing 400 or so female operators. Soon under Japanese management, this one mill had trained over 1000 reelers in modern techniques by 1892. The second mill, built in Tokyo in 1873 utilised both French and Italian equipment and was leased to private interests in 1879. Using a technique just developed in Italy, but serviced in Japan by German and Swiss technicians, the government also established a scrap-reeling mill at Shimmachi. As early as 1878, 20 per cent of all Japanese silk was reeled by machine; during the 1880s machine filature overtook hand-reeling. The inappropriateness of this blueprint technology to the cocoon quality of Japan, and its high setting-up costs (not necessarily *production* costs) led to government instigation of a series of *adaptations*: iron was replaced by wood in construction, steam engines by manpower and waterpower, and overall scales were reduced. Such adaptation led to a definite speeding-up of the diffusion process and to further innovations designed to increase labour input at the expense of capital during the *gestation* period of the machines, e.g. longer or double shifts, increased repair facilities. This phase of *adaptive diffusion* pushed hand reeling out of the general silk markets. Diffusion was further stimulated by government search and information activity in foreign markets, centred on Japanese consulates abroad, and was soon followed by the formation of foreign trading companies which gained expert information – e.g. Mitsui Bussan and Hara Gomei.[103]

Cotton was the industrial revolution sector which most stimulated the emergence of Japanese entrepreneurship in the private sector. In this case, the adoption of Western technology was held back by its initial costs and by previous *Japanese* responses to changing *international* economic conditions.

Costly early experiments at Westernisation were set against a background of the emergence of the *gara-bō* ('rattling' spindle), an indigenous improvement induced by the surge of manufactured cotton imports.[104] Whilst a machine of 50 spindles of the new type yielded a labour productivity of six to eight times that of existing hand spinning, could be constructed of tinplate, wood and waterpower and purchased at the cost of one month's salary for a skilled worker, multiple machines harnessed to larger waterwheels generated improvement of productivity of up to twentyfold.[105] Simplicity, low costs and use of plentiful raw materials meant that the system spread rapidly in the 1870s and stemmed the tide of Westernisation in this field.

Most accounts date the foundation of the Osaka Spinning Company in 1882 as the first Western-style spinning mill of any significance.[106] By 1900, 70 *private* companies operated 80 plants, 1 135 000 spindles and 3000 power looms, importing raw cotton from China, India and later, America.[107] Why did the introduced technology spread so fast after 1883 and why was it replicated so closely? There are several contenders as explanation. First, the mimicking process in fact followed some adaptation away from the blueprint technology of the British through introduction of double shifts and replacement of the self-acting mule with ring spinning. Second, the dominant position of Mitsui Bussan in the importation of manufacturing equipment (to perhaps 80 per cent of the total) and their sole agency agreement with the British machine-builders, Platt Brothers of Oldham, meant that original choice of technique was curtailed. Eighty-seven per cent of the nearly 2 million spindles in Japan in 1909 had been supplied by Platt Brothers.[108] Thirdly, Saxonhouse has argued that the information dispersal function of the All Japan Cotton Spinning Association (the *Bören*) created a communal technological culture in the cotton industry.[109] The last and simplest possibility is that the adapted blueprint technology was exactly appropriate to Japan's economic conditions in terms of labour skills (heightened by previous silk industry experience), labour relations, raw material qualities and capital availability. Certainly, when in 1900, 150 Northrop automatic looms were introduced by the government into the weaving sector of the industry, they failed because available labour skills could not be accommodated and because Japanese cotton yarn was not strong enough to withstand the process. As a result, the already adapted, wooden, narrow-framed Japan-style *power* looms thrived in their stead.[110]

7.5 CONCLUSIONS

> Discontinuity occurs when an item or set of items is borrowed from outside a culture, and when that borrowing alters the whole style of the relevant activity in the recipient culture.
>
> Lynn T. White, Jr 1963

In Figure 7.1 we have attempted to sketch in the contrasting processes at work in the industrialisation of relatively forward and relatively backward economies. The standard example for the left-

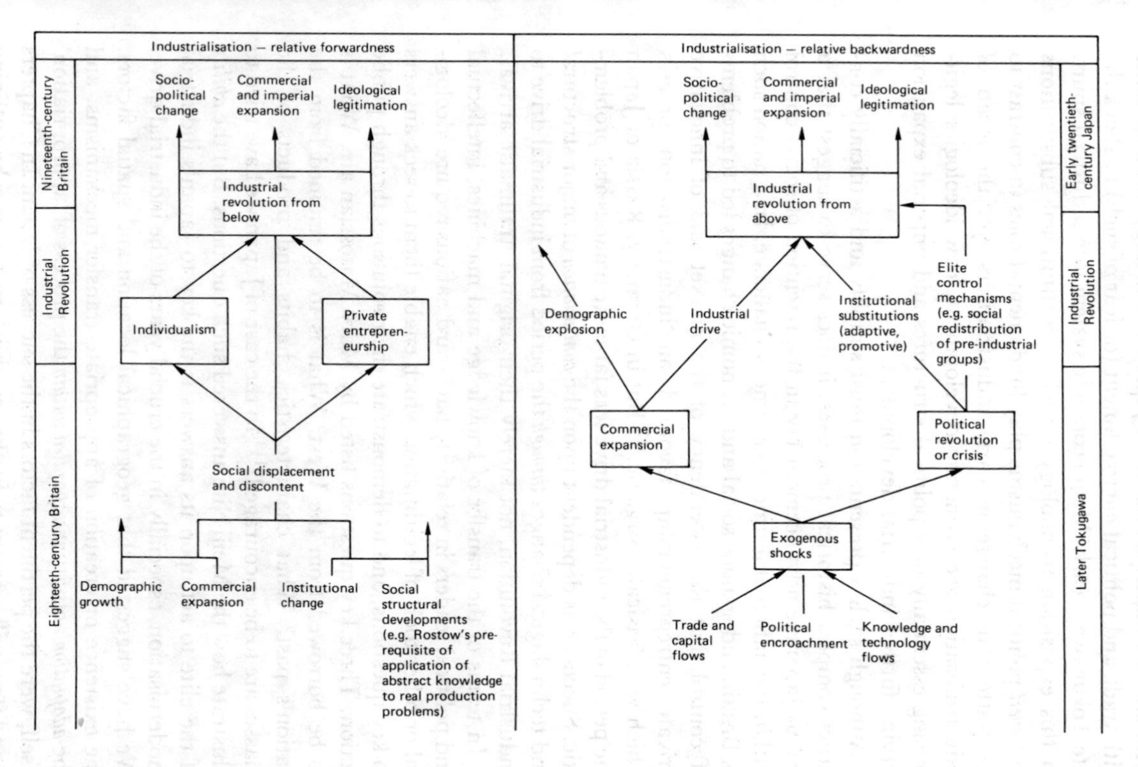

FIGURE 7.1 Patterns of industrialisation

hand side of the chart is Britain, the right-hand side embraces the 'winners' of Chapter 6 and Meiji Japan. Technology-flows combine with trade and political encroachment (or fear of such) to shock the late Tokugawa – early Meiji economic system into industrial spurt. In this exposition technology transfers, institutional substitutions and *overt* political mechanisms play fundamental roles in contrast to the pattern of change in early industrialisers, yet the *results* of industrialisation are economic development, new ideologies legitimising essentially new political structures and imperial expansion serving further industrial development.[111]

Although such a presentation must simplify and artificially construct complex historical processes, it does serve to suggest that a key to Japanese advancement lay in the retention of social control within the nation at a time of enormous challenge. In a nation such as Britain endogenous social and economic changes led to problems of control, but the ascendency of the market and of innovative private entrepreneurship favoured an industrialisation process which was, basically, organic.[112] But in Germany, Russia or Japan the period of the industrial drive was far more tenuous and problematic. Success was dependent upon the *continuation* of major structural and technological changes *through* the period from industrial drive to Industrial Revolution, not merely their original, traumatic arrival.

In terms of the transfer of knowledge and machines, intellectual and political leaders in relatively backward nations require ideological or other tools of legitimation which enable them to seek answers to key questions, and to demonstrate their solutions through public action. Three key questions listed by Mary Matossian are, What is to be borrowed from the West? What is to be retained from the nation's past? What characteristics, habits and products of the masses are to be encouraged?[113] In the case of Japan we have tried to illustrate how the Meiji elite answered such questions. But the *ability* of the elite to act upon its answers is the key to Japan's industrial modernisation, especially in the crucial years of the industrial drive. We have suggested that geographical location and spatial factors, the existence or creation of appropriate transfer mechanisms, and the *adaptation* activity *incorporated within* the process of 'appropriation' itself, were the pertinent factors in the success of Japan. In Chapters 8 and 9 we will attempt to further establish such points by outlining the economic histories of China and India during the nineteenth century, cases of dramatic 'failure'.

8. Science, Technology and Imperialism: (1) India

8.1 IMPERIALISM

The structure of the economic elements of society remains untouched by the storm clouds of the political sky.

Karl Marx, *Capital*, I

UNTIL relatively recent years the diverse writings upon the causes of the wave of colonialism heralded by the 'Scramble for Africa' in the 1880s accepted the phenomenon itself at face value. The economic history of an A. K. Cairncross or the diplomatic history of an A. J. P. Taylor were brought to bear on the classic economic arguments of J. A. Hobson, V. I. Lenin and R. Hilferding. Although centring upon British colonialism, these important writers had explained the New Imperialism of the years prior to 1914 as a natural outcome of the changing character of industrial capitalism in Europe. Imperialism represented a 'vent for surplus', a source of strategic raw materials and foodstuffs, a market for the cotton manufactures of Europe.[1] From the early 1950s the work of J. Gallagher and R. Robinson served to alter the basis of discussion. These authors emphasised the virulence of pre-1880 territorial encroachments; introduced more explicitly the notion of 'informal imperialism' – where an industrial nation brought its political power to bear upon nominally independent states; and sought for an explanation of 'the Scramble' not in Europe but in Africa itself.[2]

The subsequent revitalised debate has given rise to a confused terminology (new imperialism, colonialism, social imperialism, informal imperialism) and has opened up major questions concerning the origins, functions and impact of nineteenth-century empire. The collapse of the watershed of the 1880s and the evidence for the importance of indigenous uprisings and disturbances in Africa and Asia (e.g. British annexation of the badly mismanaged state of Oudh) do not serve to entirely devalue the broad economic explanations of even formal imperialism. If British intervention in Egypt resulted from her need to protect the route to India, an acknowledgement of India's substantial economic importance to specific interests in Britain serves to place her initial aggression back into the economic sphere of explanation.[3] We may admit that the further penetration of Britain into Africa was then conditioned by a series of events internal to Africa, together with increased French, British and German diplomatic and military competition. But the changed power balance within Europe, and particularly the rise of Prussia, was itself a function of industrial changes there, as shown in Chapters 5 and 6 above. In addition new military, transport and communications technologies provided the *means* for renewed expansive activity, but were a result of industrial development within Europe and America.[4] How an account of all this concluded would surely depend upon how its author visualised the causal chain. Here, perhaps, E. H. Carr and Alexander Gerschenkron would have found some common ground, for both have emphasised the complexities involved in the historian's analysis of causal regress.[5]

Nor does finding that the social returns of empire to the British economy were minimal dispel the older economic arguments.[6] As we have suggested, 'imperialism' may include more than 'colonialism', with all the costs of administration to the home economy entailed in the latter term, and in Britain as elsewhere *expectations* of large returns from colonial acquisition may well have prompted activity which did not eventually bring that reward. Finally, concentration upon the social returns of colonial adventure obscures the economic importance of colonies for a wide range of powerful private interest groupings, from colonial settlers, civil servants and military and naval forces, to individual traders, investors, insurers, agents and bankers. It has yet to be established that the combined interests and *private* returns of such groups were not sufficient to ensure a national urge to expansionism.

But it is the theme of the socio-economic *impact* of imperialism on the imperialised nations which is the one most open to debate. It is the theme which is least competently researched.[7] It is a theme which throws wide open our definition of imperialism. More importantly, it is a theme which embraces one of the most political questions which may be asked of modern history: Did imperialism prevent successful capitalist development in a range of nations which might otherwise have either (a) developed independently into industrial and commercial economies prior to 1914 or 1929 or 1948, or (b) developed successfully through major internal structural changes instigated by the free and sovereign participation of such nations in the fast-growing international economy of the nineteenth century?

For brevity if not for beauty we will define as 'imperialist' any force acting upon a relatively undeveloped nation of the nineteenth century (India or China), which did not directly arise from the workings of the international economy *per se*. This formulation induces some exploration of a fairly wide range of historical events and processes, and represents for the examination of 'impact' a perspective similar to that of Edelstein, itself utilised for his examination of 'function'; 'the gains to Britain from ruling India (for example) can be measured as the amount by which Britain benefited from its economic relationship with India minus the benefits she would have gained if Britain had not ruled India but merely traded with India as she did with Germany, France or the U.S.'[8]

With this perspective in mind, our principal concern in this chapter is with the impact of imperialism upon India in the years prior to 1914. Obviously, 1880 no longer serves as a watershed and our treatment of impact or intervention may not be confined to trade, investment or technology. Hopefully, our presentation will permit us to draw some systematic comparisons between nations of relative economic backwardness who retained their sovereignty after circa 1850 (Chapters 5 to 7) and those whose incorporation into the international economy coincided with their subjugation to external political pressures. We will address two related questions. First, did colonialism or other forms of political and military aggression remove national sovereignty to such an extent that the ability of nations to 'respond' to the Western impact was severely curtailed? In other words, was the so-called failure of such nations as India or China a result of lack of effective sovereignty and choice rather than a necessary consequence of cultural or social 'barriers' to industrialisation inherent within (and supposedly limited to) such nations? Second, did the imperialist regimes directly or indirectly inhibit the successful modernisation of economies? The answer to this question requires some examination of both indigenous attempts at industrial development and the activities of imperialist regimes in the areas of institutional and industrial reform. We have seen that the successful transfer of technologies depended upon much more than just immediate transfer or diffusion mechanisms. In addition, such mechanisms were affected by a wide range of social and economic factors. India and China represent good test cases, for they were by far the largest and in some ways the most likely candidates for incorporation into the brave new world of Western industrial technology.

After admitting to the supremacy of the Dutch in South East Asia, the British established bases in India at Madras (1639), Bombay (1661) and Calcutta (1690). The decline of the Mughal empire after the death of Aurangzeb and the defeat of the French in 1763 provided opportunities for further Britain territorial expansion and direct rule over the large province of Bengal. This was a period of adventure, with the private enterprise of the East India Company serving as the motor of rule. The Bengal base gave Britain the ability to protect her commercial and strategic interests elsewhere in Asia.[9]

An acceleration of Britain's creeping frontier in the sub-continent took place under the administration of Lord Dalhousie (1848–56) and was based on the Doctrine of Lapse. By this, a state was taken over whenever a problem of succession arose. Such a method led to the annexation of Satara, Nagpur, Jaitpur, Sambulpur, Baght, Udaipur and Jhansi. Annexation and its associated policy of land tenure reform caused widespread unrest, culminating in the Indian Mutiny of 1857, the best-known of several uprisings against British rule in the nineteenth century.[10] One result of the Mutiny was the final exit of the East India Company, when the government of India was assumed by the British Crown in 1858. From thence a large administration and efficient police force removed effective sovereignty from the Indian people, and the white man's burden became a collection of files. Throughout the rest of the century India was not only of great economic importance in itself but also of strategic importance to further expansionary activity in Asia.

8.2 ROOM FOR DEBATE

The causes of Indian poverty are to be sought, not in overtaxation and the 'drain' but in the inherent social structure of Indian society.

H. G. Rowlinson, *The British Achievement in India*, 1948.

The most explicit and embracive account of Britain's failure in India is provided in an article by E. N. Komarov, published in 1962, which surveyed no less than 72 of the major Russian-language works on this theme during the 1950s.[11] After rehearsing Komarov's seven-claim condemnation we will go on to summarise the recent more empirical work which serves to bolster his position.

Komarov's first claim is that prior to British incursion the Indian economy 'showed signs of transition to late feudalism'. Although the self-sufficiency of the village community, the institutions of caste and remnants of slavery, the predominance of state property in land, and the preservation of tribal relations hindered economic development, pre-colonial India evidenced a growth of towns as centres of handicraft and trade, some development of merchant capital and regional markets, a differentiation of the peasantry (so emphasised by Marxist writers on nineteenth-century Russia) and the emergence of private land ownership. Secondly, the establishment of East India Company rule entailed a loss of economic sovereignty in the form of the trading monopolies secured by the Company, and these directly caused 'a decline of Indian merchant capital', i.e. immediately inhibited the natural growth of indigenous capitalist relations. Thirdly, British reforms served to intensify 'the feudal exploitation of the Indian peasantry'. The British altered the fundamental nature of tenure by 'abolishing the former traditional limitations of the amount of feudal rent as well as by investing the feudal landlord, be it a *zamindar* [native landlord] or the colonial government itself [in the *ryotwari* areas], with greater powers over the person and property of the peasant.' Fourthly, the British industrial revolution inevitably demanded increased foreign trade and investment, so to revenue-exploitation was now added 'exploitation through non-equivalent exchange between Britain and the colony'. Precisely when India was being drawn into the world's commerce and exposed to the potentials of Western techniques, so too 'the economic and political domination of British capital in India necessarily retarded the development of new [indigenous] productive forces and the replacement of feudal relations by capitalist relations'. Fifthly, the later nineteenth century witnessed the further destruction of indigenous capitalism through the 'extraction of super-colonial profits on capital investments in India by ruthless exploitation of her industrial workers'.[12] Plantations produced raw materials for export; railways transported them. Sixthly, in agriculture further reform of the tenure and revenue systems now served to break open village society and differentiate the peasant class. In both *ryotwari* (tax returns to the Raj) and *zamindari* (rent) areas the occupancy rights of the peasantry were now more or less secured, and the resulting land market meant that peasant holdings began to pass into the hands of traders, moneylenders, landlords 'and some of the prosperous peasants who combined farming with trade and manufacturing', creating a class of landless labourers and unemployed. The labour surplus combined with a lack of employment opportunity in both industry and commerce to force down wages and thereby retard the emergence of labour-saving technological innovations in agriculture. Seventh and finally, the Indian industrial sector was not even a pale reflection of the Western model, wherein industrial states had historically encouraged industry by financial aid, protective tariffs or monetary incentives. In contrast, the domestic industries of India were faced with a flood of colonial imports which competed with manufactures of both consumer (e.g. textile) and producer (e.g. iron) goods, and were burdened with excise duties and other taxation which 'averaged 40 per cent of the country's industrial profits', the revenue from which was spent on administrative facilities which bore little relationship to the needs of the indigenous industrial sector. Where the *kustarni* industries of Russia confronted the international economy, the native industries of India faced sudden death. In summary, 'the industrial revolution in India was really held up at its early stage'.

If true, such basic claims serve to define the British presence as reducive of India's society and economy. Sovereignty is removed, commodity surpluses are exchanged by force and unequally, surpluses of funds are reduced, and the general environment for the successful transfer of modern technologies is destroyed. Komarov has seemingly answered all three of the major questions of Section 8.1 in the affirmative.[13]

The fundamental work of Irfan Habib clearly depicts the Indian economy of the Mughal period as one of growth and change.

Differentiation of the peasantry was significant and based upon an agriculture which produced an enormous range of crops in both spring (*robi*) and autumn (*kharif*) harvests. Over large areas, perhaps some forty crops might be produced. New crops such as tobacco or maize were diffusing generally and in the case of sericulture the Hindu traditions which forbade the taking of life did not prevent the inevitable death of the silkworm.[14] The diffusion of simple but superior techniques, ranging from the bamboo seed drill to pin drum gearing (transferred from Europe), the Persian wheel for lifting water and the parallel worm for the cotton gin and roller-crushing in sugar-mills was not hindered by any features of the traditional value systems. The barriers to the generalised use of new techniques resulted from the operation of objective economic forces, such as the lack of capital and skills required to operate a system of geared rollers. Indian occupations were composed of a range of agricultural workers, village artisans (blacksmiths, carpenters, potters, weavers, warehousemen, watercarriers), many others engaged in the handicraft industries of cotton textiles, silks, jewellery and weaponry, as well as the small traders in saltpetre, Indigo, sugar, opium and ginger.[15] Seventeenth-century Indian craftsmanship in the working of alloys, soldering, lacquerwork, oil distillation, the use of saltpetre to cool water, rivetting and sophisticated haulage techniques were as advanced as anywhere. Habib sees a major problem in the relatively limited application of such advancements, but explains this more as the result of low labour costs and the specificity of skills than of inhibitions resulting from the caste system. As with Morris D. Morris, Habib argues that under the force of economic incentives castes could crumble.[16] It is true that, perhaps apart from mining, many merchants limited their activity to financing and organisation rather than innovative production, but so too was this the case in such nations as Tokugawa Japan.[17] Nor did a lack of merchant interest prevent artisans from gathering together in *Karkhanas*, whilst a relative lack of financial control might well have encouraged an independent spirit amongst artisan groups. This position is not unlike that claimed for pre-British India by Komarov. In a recent paper Habib has concluded that 'there was no inbuilt resistance in the economic system to technological changes'.[18]

Similarly, the trade and revenue effects of British dominion are difficult to dismiss.[19] If new tenure arrangements as such had not quite the disruptive effects postulated by Komarov, the extent of the financial demands imposed by the Raj did indisputably result in land sales, peasant indebtedness, subletting, the rise of a relatively small, rich peasant and *comprador* class, and the consequent unsettling of the traditional village community.[20] The resources of a peasantry facing greater burdens on the land were now further culled by trade policy. Between 1700 and 1825 imports of silk manufactures from India to Britain were prohibited by law and heavy duties on cotton textiles were established during 1797–1825, precisely when Samuel Crompton's invention of 1779 allowed machine spinning of fine yarn for muslins which had previously been imported from the hand-spinners of the subcontinent.[21] The Benthamite radicalism of East India House (chief executive officer of which for 1831–6 was James Mill), the subsequent victory of the Lancashire free trade school, combined with the absence of any developmental monetary, banking and financial policy, dictated no trade protection except that which distance and ignorance might afford.

Amongst scholars of India analysis of such relationships has collapsed into the two major debates over the 'drain' and 'deindustrialisation'. The argument that a net drain of resources took place under British rule originated in numerous contemporary writings.[22] Nevertheless, the notion remains difficult to define and measure. In conformity with our initial definition of imperialism, 'drain' must exclude profits paid to foreign investors if we assume that such outward transfers would have occurred even within an environment of independent international trading. After all, from Argentina to Russia, foreign investors were repatriating profits from their investments.[23] Drain involved all 'homecharges', interest paid to Britain on the Indian public debt, all military charges (in the years 1870–1900 defence spending represented some 30 per cent of total British expenditure in India), the cost of purchasing 'stores' in Britain and the 'civil charges' of the British administration in India.[24] Those who emphasise the negative financial effects of the drain argue that such imperialist charges on India's capital account more then offset any gains from trade which might have resulted from British rule, and therefore significantly reduced the ability of the Indian economy to import developmental technologies and capital equipment.[25] This flow held down net investment to 3 per cent of national product and meant a loss of foreign exchange and the removal of the possibility of successful technology transfer by direct import of goods. India was in the highly unfortunate position of having a large debt but a relatively small level of foreign investment.[26]

In contrast, other writers point to the increased revenue from trade as a result of a fivefold increase in export values between 1870 and 1914, and to the less measurable gains to the economy resulting from efficient administration, internal stability and external defence.[27] Further research may well increase our understanding of the net financial effects of Raj dominance. However, what must remain of crucial importance is the effect of the British trading and investment connection upon the overall structure of Indian *production*. Myrdal has emphasised this as the vital clue to both Raj policy and impact: 'In pressing for the conversion of India into a market for British manufactured goods, the British inhibited India's own manufacturing industries and gradually converted India into an agricultural hinterland of Great Britain.'[28] Here we have the widest definition of 'de-industrialisation'. Export revenue may well have increased, but from around 1815 such gains depended more and more upon the exports of raw cotton, opium (twice the value of cotton) and sugar, and less and less upon exports of silk products, cotton piece goods and indigo. Thus, revenue gain was associated with a process of agrarian extraction rather than production in industry.[29]

We may summarise our own sceptical view of the general economic role of the British Raj in the form of a series of simple but central analytical points.

First, given that a reasonable expectancy would be that Indian handicraft industries would gain through trade, the finding that during the period 1880–1930, when industrial growth accelerated, the proportion of the labour force employed in industry did not alter, is certainly of significance.[30] Whereas in Prussia and Japan, developing handicraft industries provided a basis for manufactured exports, income (demand) rises and capital accumulation, that was not to be the case in India.[31]

Second, financial gains from trade for a nation are inflationary or illusory if the resources gained are not shifted into productive avenues at home or abroad. Even more generally, although Marx claimed of the Indian that he 'merely hoards in an aesthetic form', there is much evidence to dispute this.[32] Resources gained from production or trade will only be moved into productive channels if there exists an ability to do so. In India, the absence of essential services and tariff protection *combined* to discourage any rational movement of resources into either handicraft-based or entirely modern industries. The chief accumulator of resources, the Raj itself, invested the great bulk of these into administrative, policing and military activities, and into bolstering the position of those very conservative elements in Indian society who were least likely to turn to industry.[33]

Third, however successful railways or even canals were in generating high private returns to investors, the loud voice of the British 'free-trade imperialists' from mid-century ensured that such infrastructural investment was advanced for short-term revenue, rather than long-term developmental purposes.[34] Tapan Raychaudhuri has argued cogently that such projects were *enclavist* rather than linked into the indigenous economic structure. Not only did the Raj refrain from large investments in the Indian private industrial sector, the guarantees provided in its infrastructural projects served also to draw scarce indigenous capital into themselves and away from Indian industries, and in all probability reduced the cost-effectiveness of such resource usage.[35] John Thavaraj similarly shows that for the period 1898–1938 fluctuations in general public investment in India were significantly related to the British policy of 'balanced budgets'. Public investment was neither developmental nor contra-cyclical.[36] This created a severe imbalance in the modern sector, a lack of mechanical power, of chemical and electrical enterprise, and a continuity of the regionalism (Bombay, Calcutta and the Bengal–Bihar coal area) encouraged by the Raj from its earliest days.[37] In essence, structural change took the negative form of a transfer of Indian revenue into the enclavist, British-orientated sector of the economy.[38] This process was certainly aided by the institution under the Raj of the 'managing agency' system, which placed the power of decision-making over joint-stock companies in the hands of professional managers rather than Indian shareholders, and which enterprises tended to employ European staff, capital and expertise, therefore reducing the learning effects of new technologies which we have noted as so important elsewhere.[39]

Fourth, the dictation of external trade by British policy was unquestionably of negative impact upon structural change within the Indian economy. Given that the financial gains from trade are now disputed, then in the Indian case we might also question the structural gains resulting from increased trade. Structural change led by the trading sector depends upon economic sovereignty in the form of (a) specialisation in areas of greatest advantage in terms of factor and raw material endowments, (b) a free (market-led) movement of internal resources into the 'import substitution' area, perhaps instigated by government activity (Japan 1881–96), coincidental with or immediately followed by (c) tariff or other protection of infant industries (Japan 1896–1914).[40] Under the dominance of Raj trading policy, India was in fact unable to instigate any of the requirements of (a)–(c).

Such points serve to rehearse our previous distinction (Chapters 1 and 3) between economic growth and economic development. It is often the failure to make this distinction which has confused many of the major issues of Indian history. Growth in particular projects, industries or regions, or amongst certain social groups, may occur just as development is retarded. A clarification of this might be gained through a very brief consideration of one large area, the Punjab.

At the heart of Raj policy in the Punjab after 1892 was its scheme for the development of the Canal Colonies. With this project the government departed from the earlier Company emphasis upon improvement or restoration of existing cultivation (e.g. the Ganges and Agra Canal systems, the construction of the Khadakwasala dam in the Bombay Deccan), and aimed to introduce cultivation into the sparsely populated *doabs* or wastelands.[41] Gadgil has summarised that here the returns to investment were high and productive settlement was achieved. Furthermore, the Punjab has been depicted more generally as a developing area, for which might be cited the introduction of power machinery into cotton ginning, pressing, cleansing and spinning, and the successful acclimatisation of foreign silk worms under the auspices of Lister and Co. and Deputy Commissioner Coldstream.[42] However, a carefully-wrought thesis by Imran Ali has questioned the development effect despite the undoubted acreage, output and trade impacts of the Canal projects. Policy in the region was dictated by political, military and financial goals. As to the first, Dr Ali concludes that 'land distribution made it possible to avoid or overcome tensions in agrarian society'.[43] Land grants to ex-soldiers were hardly pursued on economic grounds and resulted in the rise of rentiers, absenteeism and sub-tenancy. Much land was granted for the breeding of horses for military purposes, which dominated the land used in the colonies of Jhelum and the Lower Bari Doab. The substantial net profits (reaching 40 per cent of capital outlay) were spent upon administrative and military purposes rather than relocated in the regions. The room for bureaucratic corruption was left large and with its door wide open. No infrastructural servicing was introduced to complete the project, i.e. the provision of seed, cattle-farming, land reclamation and tubewell irrigation. The 'project' had merely created revenue, left the peasant groups landless, and strengthened the traditional power and class structures.

The remaining agnostics would perforce turn to India itself in order to uncover a set of *explanans* for such a dismal *explanandum*! Admitting to only a 'positive' or 'neutral' role for the Raj, several authorities have pointed to supposed inherent, retarding features of Indian life and social structure. We have space only to note a few of the less objectionable claims, and to preface these with the remark that 'obstacles' to development may be found in any pre-industrial society.

We may start with the thesis that the earlier Mughal invasion did not destroy basic Hindu society, by which family status and occupations were determined by caste. Long before the nineteenth century the four principal divisions of Brahmins (priests), Kshatriyas (warriors), Vaishas (traders) and Sudras (farmers-peasants), with the outer-caste group of 'untouchables' had fractured into a huge number of lesser-caste divisions.[44] The defeat of the Sikhs in the Punjab in 1841 was but the last of a series of invasions which had left basic Hindu society intact.[45] The caste system served to reduce living standards of the most socially debased groupings, to restrict or divide consumer markets on the basis of sumptuary rules, to suppress social interaction, competition, learning and risk-taking behaviour, and to carry very specific work skills in a traditional and hereditary, rather than an achievement-motivated mode.[46] In addition, because of its very powers of resistance, the effect of greater foreign intrusion in the Raj period was not to create new conditions for innovative behaviour, but to increase uncertainty and instability, and thus such behavioural attributes as hoarding and risk-aversion.

In contrast, Toynbee emphasised the essential continuity of the 'cultural heritage' despite British rule, but pointed out that Hinduism was not irrevocably concomitant with rigid caste systems, as witnessed in its transfers to Indonesia and South-East Asia.[47] We would add to this the rise of the Parsees in industry and finance, the fact that the multitude of Hindu sects were not as one in their behavioural attributes, and that a significant portion of India lay outside both Hinduism and British rule.[48] Furthermore, just as has been attempted for Protestantism in England and France, Greek Orthodoxy in Russia, and Neo-Confucianism in Japan, we might emphasise those elements of Hindu belief that would foster and even encourage industrial development. Under the dual impact of the decline of the Mughals (and therefore the removal of the provision that all wealth would be vested in the State upon death) and the introduction of investment opportunities from the West, the traditional Hindu emphasis on thrift might well have been expected to result in investment rather than hoarding.[49] Similarly, the Hindu stress on charities and 'good works' could have translated into patterns of industrial or commercial investment, and there is certainly evidence of a decline in 'other-worldliness'. If caste was a disincentive to effort amongst lower social groups, did it not also promote trade specialisation, the maintenance of skills during periods of social disruption, and conventions for production in agriculture? Kaushal has pointed to the 'mutual benefit' role of caste, which could act as a basis for capital accumulation and shelter from famine or flood.[50] Interestingly enough, it was within the most *orthodox* of religious groups (e.g. the Jains in Hinduism, the Khojas and Bohras amongst Muslims) that the greatest 'response' to new economic opportunities arose.[51] It is true that caste may have induced labour-using activity as a mechanism of social welfare, but in a sub-continent where many regions were overpopulated it is difficult to imagine any continuous culture which did not do so. The work of Morris and Das Gupta combines to suggest that the Indian labour force was indeed responsive to industrial employment opportunities when they arose. The growing sub-culture of the *badlis*, the temporary cotton workers of Bombay, suggests that, if anything, there was an oversupply of labour for manufacturing industry in India, and that this was not hindered in its mobility by religious or cultural factors.[52] In his careful study of the period approximately 1870–1920, Morris illustrates how the Bombay cotton industry could draw upon a low-wage and disciplined labour force utilising labour-intensive technologies, many of whom had come from distances of up to 100 or 200 miles away from the city. Once in the mill there was little evidence of caste clustering. Finally, the development of the *dastari* work system, whereby labourers made money payments to jobbers who then negotiated employment for them, suggests that the *effective* labour supply was plentiful and competitive. Perhaps, then, Milton Singer had it right when, prior to the work of Morris and others, he argued that 'the traditional Indian philosophy of renunciation is not a major obstacle to economic development'.[53]

8.3 TRANSFER, DIFFUSION AND THE BRITISH

It remains, then, to examine features of the transfer and diffusion activities of the Raj, and to see whether this provides further reason to look towards British economic and social policy when explaining the failure of industrial modernisation in nineteenth-century India.

With its primary emphasis upon topographical, geological, and exploratory scientific effort, the East India Company set a pattern for British science policy (if such it could ever be named) which was to continue with modifications into the twentieth century. The object of the Company was to transfer the European 'natural history enterprise' to India for purely commercial purposes. The establishment of the Royal Botanic Gardens at Sidpur in 1787 was representative of a programme of research in economic botany, mineralogy and zoology. The task of the first official botanist, Robert Kyd, was to 'establish a stock for disseminating [botanical] articles which may tend to the extension of the national commerce and riches.[54] Kew Gardens were to serve as both depot and laboratory.[55] Similarly, the Geological Survey of India, established in 1851, directed its attention to the commercial potential of coal, iron, manganese and gold deposits, but employed only Europeans in its higher offices. In addition, throughout most of its years the Survey's policy was to concentrate research activity upon the discovery of resources to be refined or processed in Britain, rather than to investigate any possibility of the manufacturing usage of resources in India itself.[56]

If the scientific research of the Company was confined to its commercial goals, then its activity in the realm of education was dictated by racist attitudes as well. Westerners brought theories of race to wherever they travelled, including such successful industrialisers as Meiji Japan. The difference in India was that such *attitudes* were with relative ease applied in policy, i.e. racism became prejudicial. In its higher reaches EIC education policy emphasised medicine and administration rather than engineering or commerce, and settled around a philosophy of 'downward filtration'. The creation of a class of intellectual *compradores* required their immediate and selective induction into the wider ways of the West. Eventually this might filter down to the masses. The main purpose of British education of natives was to effect a transfer of loyalties as part and parcel of what Professor Ambirajan has recently termed a 'philosophy of occupation'.[57] The racist, elitist, as well as social-control arguments for neglect of mass and technical education were reiterated at the point when the British government was taking over the economic role of the Company:

> Do not you consider that the fact of England holding so large a portion of India proves that the natives are not capable of the same extent of intellectual culture as a European is? – In point of cleverness and mental power for acquiring sciences, they are quite equal to Europeans; but they want the energy and enterprise of Europeans.
>
> Do you consider that the energy and enterprise which are wanting in the Asiatic will always be found to enable the European to keep his place in positions of very high importance? – Yes, I do.
>
> Do you know of any way in which the Asiatic is likely to be educated, so as to give him the same energy and enterprise that the European has? – I do not.[58]

The policy line which resulted must surely be estimated against the background of the general negative cultural and institutional effects of Company rule. The land tenure reforms of the 1790s saw a fall in the social position of *pandits* and *manlaris*, the traditional scholars in Sanskrit and Arabic, and a general decline of Muslim education.[59] Macaulay's famous Minute of 1835, by which 'Anglicanists' won their battle with the 'Orientalists' over the form which British educational intervention should take, meant that the situation was never rectified. From 1837 British schools developed in order to give general education to the Indian upper classes, who were increasingly drawn into government employment as a result of depressed agricultural conditions. The curriculum was English and literary and entrance was only gained through the payment of fees. In tune with economic policy, education must pay for itself.[60] The official neglect of mass primary education was worsened by the contemporaneous economic decline of those artisan groupings who had previously maintained a reasonably comprehensive vernacular system.[61]

Contrary to the repeated claims by the British that the educated classes of India showed no interest in manual, technical or scientific instruction, there is much evidence of a considerable stirring of indigenous Indian activity in the second half of the century, despite the equally evident lack of scientific–technical employment opportunities offered by the Raj.[62] Hindu attempts at establishing their own institutions for European studies can be traced back to Delhi College (founded 1772), the Anglo-Indian College and the Hindu Sanskrit College in Calcutta.[63] In 1864 Syed Ahmad founded the Aligarh Scientific Society, and this was followed in 1868 with the establishment of the Bihar Scientific Society by Syed Indad Ali. Composed of a mixture of *zamindars* and Indian bureaucrats, the purpose of societies such as these was to translate European works on science and technology, to provide an intellectual basis for Indian industrial and agrarian improvement.[64] The later Indian Association for the Advancement of Science (founded during 1869–76), which attempted to derive its membership from lower social groups, suffered the antagonism of British officialdom. Acting as pressure groups for reform in the provincial education systems, such societies could not but come to the attention of policymakers whose main concerns were with control and commerce.[65] Despite this, the *Naturys* persevered with their work of translation, popularisation and, increasingly, the publication of Indian-based research.[66] During 1870–90 British attempts at introducing the mainly inappropriate techniques of turbine waterwheels and machine cleansing of flax and cotton confused Indian attempts at the development of agricultural education. After that date, the nationalist-inspired Bengal Association for the Advancement of Scientific, Agricultural, Industrial and Commercial Education did begin to gain some ascendency, and by the turn of the century was independently sending its students abroad for further studies.[67]

The British response to all this was minimal. Dr J. A. Voelker's appeal of 1890 for relevant and socially inclusive education in scientific fields, and his informed opinion that instruction in paring and burning, in the use of sulphate of ammonia and the fattening of animals was of little relevance compared to an improved knowledge of canal and well irrigation or the use of oil cake, still had little impact on British official opinion.[68] Between 1871 and 1882, of 3300 Indian college graduates, two-thirds entered some form of Raj employment rather than industry or agriculture. Indeed, only 53 graduated in engineering. By the early 1890s, the 16 000 students enrolled in all colleges represented a small fraction of the potential number. The higher colleges which did exist (e.g. Shibpur Engineering College, Calcutta University, Allahabad University) offered meagre scientific and technical provisions and demonstrated little sensitivity to the specific, highly-localised needs of the Indian economy. As late as the 1930s a mere 63 000 students were enrolled in adult and industrial training establishments.[69]

The British interest remained with commerce, acclimatisation and the natural history enterprise,[70] and Raj policy did not erect institutions which could provide even a significant cadre of skilled workers for service in either Indian or Western industrial enterprises. Furthermore, general socio-economic policy had reduced the ability of progressive indigenous groups to compensate for their loss of traditional systems of instruction. At most, as far as technical instruction was concerned, the British believed only in the 'educational' function of successful industrial projects. Our key to the ultimate failure of the Raj as a 'transfer mechanism' for Western technologies may be found in a government resolution of 1888:

> the extension of railways, the introduction of mills and factories, the exploration of mineral and other products, the external trade, and the enlarged intercourse with foreign markets, ought in time to lead to the same results in India as in other countries, and create a demand for skilled labour and for educated foremen, supervisors and managers. It may be conceded that the effect of these various influences on an Asiatic people is very gradual, and that it would be premature to establish technical schools on such a scale as in European countries, and thereby aggravate the present difficulties by adding to the educated unemployed a new class of professional men for whom there is no commercial demand.[71]

Here was 'downward filtration' extended to the economic process itself. But we would argue that Raj industrial policy did not, in fact, lead to either an increased demand for skilled *Indian* labour or the generation of educational servicing institutions.[72] Furthermore, several aspects of 'project' development under British rule combined to severely reduce their overall developmental *impact*. The transplantation of techniques within enclavist projects did not represent an effective transfer of technologies into the Indian economy.

In the case of Japan and other successful industrialisers of the nineteenth century we have noted that a major initial repository for the *introduction* and *adaptation* of advanced technology lay in existing agrarian and handicraft sectors. Where such sectors suffered from international economic competition, as in the case of the *kustarni* industries of Russia, the ability of such sectors to act as receptors of transplanted modern technique, and their subsequent contribution to the overall development process was greatly reduced. Indian economic history provides a vivid scenario of failure at this level. The varied activities of the IEC resulted in a manifold negative impact on the handicrafts sector: monopoly production by the Company reduced the remaining internal market directly, industrial production in Britain itself reduced the export markets, control over production retarded the entrepreneurial behaviour of small-scale Indian producers, free trade, transport and communication improvements increased the cost-effectiveness of British goods indirectly, and a rise in the demand for Indian *raw* materials encouraged the production for export of non-processed products rather than handicrafted goods. The *actual* impact of such seemingly negative forces is much debated. In an industry such as silk, which spanned the agricultural–industrial sectors (from mulberry-cultivation and cocoon-rearing to silk winding, collecting and merchandising) the total effect of IEC presence is very difficult to estimate. On the one hand, as R. K. Gupta's study of the district of Birbhum in Bengal suggests, the introduction of factory filature and the growth of production provided employment for large numbers of labourers and skilled workers, who probably gained new skills as a result. On the other, the curtailment of the British market due to protectionist policy, the competition of factory filature with the continuing country-wound silk, the fall in the quality of the silk produced, and a possible redistribution of income towards the IEC and private merchants may have served to reduce the wider, positive impact of the industry's undoubted growth.[73]

We may consider Morris D. Morris' general point that Western production or importation of some products (e.g. spun cotton) might have stimulated production in servicing handicrafts, (e.g. cotton weaving). Applied to the later nineteenth century, the point deserves some thought. However, Twomey has taken Morris at face value and has attempted a test for cotton production. Based on his *estimates* of labour productivity, he concludes that 'there was an absolute decline in handicraft textile employment over the century [with the period 1850–1880] as that of strongest displacement of domestic handicrafts by British exports'.[74] The British encouragement of cash crops such as sugar, rice, tea, indigo, silk, tobacco, hemp and opium might well have served as alternative employment and training grounds for labour in the use of modernised equipment. The evidence is shaky. Thus the British acclimatisation of European and other silk worms and the attempts at removing the disease *pebrime* was seemingly not handed on to the Indian cultivator as an information gain.[75] Technical improvements in the production of indigo were sharply curtailed by the invention of new methods of synthetic dye production from the 1850s to the 1890s.[76] Similarly, the Bengal Iron Company, founded in 1875, suffered an immediate setback from improved methods of British pig iron production, which brought down the price of imported pig to 50 per cent of its former level.[77] The Department of Agriculture, established in 1871–9, served not as a vehicle of information diffusion but of imperial revenue collection.[78] Such examples, however brief, may indicate that the introduction of Western technologies into handicraft production did not suffice to compensate for the loss of employment and redundancy of skills in areas which suffered from British production and import.

The fully modern sector projects became the last resort of those who sought a transfer mechanism. Contemporaries noted the heightened expenditure of the EIC in infrastructural development during 1834–53. One Company employee, the novelist Thomas Love Peacock, stressed the importance of irrigation, navigational and transport improvements, and went on to argue that the introduced technologies of iron and steam boats, and of steam-powered foundries and works for the manufacture of steam engines and manchinery had significant economic impacts.[79] But another influential commentator of 1859 admitted that little of this required the services of *Indian* engineers or foremen, even at the less prestigious levels of Company activity. Also, such an ambitious project as the proposed Scinde Railway was seen as a means of lowering the costs of imports of grain, oil seeds, flax and hemp to Britain and strengthening *Indian* production of raw materials rather than as serving to encourage their Indian manufacture. Indeed, the only employment problem of such large works was in the procurement of a sufficient number of British military engineers.[80] At the same time, Westernised coalmining utilised low-caste labour and women in the collieries. European owners and managers secured *zamindar* rights over the land which gave them great control over the workforce. The establishment of 'service tenancy agreements' permitted owners to grant land to faceworkers in return for working in the mines. This semi-feudal contract was unlikely to induce any skilling process.[81] Generally, the establishment of the Public Works Department seemed to have little effect on the employment of indigenous skills. Until the interwar years, the potential transfer role of a major producer-goods complex – cotton machinery, plant and stores – was rendered null by British policy which determined that the overwhelming source of such goods would be Britain herself. British machine-manufacturing firms exported over one-third of their total machinery exports to India, representing some 90 per cent of all cotton-weaving machinery utilised in the sub-continent.[82]

In contrast, the introduction of the railway system has been seen as the high mark of British technological achievement in India. The first line, from Bombay to Thana, was constructed in 1853. By 1869 mileage completed reached 4255, by 1880 this had risen to 9000 miles, and by 1900 total length exceeded 25 000 miles. In the first year, 1853, Karl Marx had written in the *New York Tribune* that 'when you have once introduced machinery into locomotion of a country, which possesses iron and coals, you are unable to withhold it from its fabrication'.

Reservations abound.[83] The pattern of railway development often took the form of radiating trunk routes, with a relative absence of lateral links between trunk lines. Spatially, this reduced the positive external economies to a small number of producers. Despite ranking fourth in the world in railway mileage in the 1950s, the effective *accessibility* of population to rail transport in India was only 50 per cent that of the United States at that time. Secondly, although generating private returns, in areas where there already existed efficient haulage by water, the railway system did not necessarily possess any considerable technological advantage in reducing costs of haulage, though it did certainly increase speed.[84] Third, from the time of Dalhousie's Railway Minute of 1853, the trunk lines were constructed primarily for political reasons and by private companies. Until 1869 the British government guaranteed rates of interest between 4.5 per cent and 5 per cent Britain's financial guidance had a twofold impact. Railways were constructed at a cost above that which would have been the ruling cost without guarantees. Payment of interest on capital invested was demanded upon deposit, not upon construction. As the gaps between deposit/instigation, construction and completion on such large projects could be large, these two features tended to lead to a debt trap. In 1868 thirteen-and-a-half million pounds sterling had been paid by the Raj from *Indian* revenues to pay interest on foreign capital invested.[85] Fourth, the period of state investment and lower-cost production (1869–82) was associated with the introduction of the famous British 'narrow gauge' line.[86] Fifth, the years 1882–1900 which saw the revival of private companies, reinstated the guarantee system (at 3.5 per cent) and fostered a great complexity of administrations. In 1900, 96 different lines were controlled by 33 railway authorities.[87] Sixth, British financial policy, which determined that the railways should be commercially profitable and export–import orientated, meant that rates on haulage were not only high – reducing the *general* cost effects of transport improvement – but also favoured foreign trade rather than internal haulage. This might not have deterred the export of materials and foodstuffs, but might have made indigenous entrepreneurs very wary of manufacture, e.g. sugar transported internally paid a rate of 40 per cent above that of the sugar imported.[88] Lastly, until well into the twentieth century, British producers supplied the great bulk of railway stock, machinery and intermediate inputs.[89]

8.4 CONCLUSIONS

As an interim summary we might admit that the overall financial impact of British rule is difficult to estimate.[90] However, our subsequent analysis has at least strongly indicated that (a) general British economic policy retarded *structural* change in the economy of India for a great variety of reasons; (b) there seems to have been no *overwhelming* socio-cultural retardative factors operating dynamically within Indian society – so-called 'barriers' to development (see Chapters 5 to 7) may transmogrify *during* initial industrialisation in such a way that they determine the *character*, rather than the level, of development in particular nations; (c) the wider scientific–technical environmental impact of the British was at best neutral, but in the context of (a) above probably acted to retard any economic development which might have stemmed from innovative learning and training institutions; and finally, (d) even the more spectacular infrastructural and industrial projects instigated or encouraged by the British rulers may be questioned as either generally developmental or specifically *conduits* for the generation of skills and learning processes. Growth in India was enclavist.

9. Science, Technology and Imperialism: (2) China and Beyond

9.1 THE FOREIGN PRESENCE

UNLIKE India, China was 'the victim of imperialism without annexation',[1] and a proving ground for a variety of industrial powers. Consequent upon the first opium wars, a major break in China's relations with the West was the Treaty of Nanking of 1842, which ceded to Britain the island of Hong Kong and opened the ports of Amoy, Foochow, Ningpo and Shanghai to foreign trade. The treaty further reduced China's economic sovereignty by depriving her of the right to fix her own tariff levels. During the 1860s further treaties were signed which expanded the treaty port system greatly. The British alone held 'concessions' in Canton, Amoy, Chinkiang, Kinkiang, Hankow, Tientsin and Newchang, ports within which foreign consulates exercised legal jurisdiction over their own people, who were not subject to any Chinese laws, and whose continuity of administration was maintained by foreign police forces and taxation systems. By the 'most favoured nation' clause the system expanded in frontier fashion; all the powers – Britain, France, Prussia, Denmark, Holland, Spain, Belgium, Italy and Austria-Hungary – in treaty with China automatically received any additional privileges wrested by any one of their number, and by

1900 some 90 treaty ports contained about 350 000 foreign residents. In many ways the foreign elements in the body politic served to unsettle Chinese society and administration, by acting as a base for the missionary assaults on the interior, by interfering both for and against Peking in several peasant-based rebellions, in providing a natural focus of antagonism for the revolutionary activities of Chinese dissidents and secret societies, and by merely existing as a proximate demonstration of what China might become and what China might fear. All of these elements were discernible in the Taiping Rebellion of 1850–64, which saw the distruction of central government control over the nation.[2] With the fall of Nanking, China became an open economy; new treaty ports were established, some of them inland, the trade in opium was legalised, foreign imports were subject to favourable inland transit duties, missionaries were free to wander the interior, and foreign gunboats abounded.

Technological progress in the form of the Suez Canal and the telegraph cut the effective distance (cost and time) from Europe and facilitated a more rapid movement of troops and gunboats. Until 1895 Western interaction with China was dominated by European and American nationalist competition. The Sino-Japanese war and the treaty of Shiminoseki brought a new power to the fore and the loss of Formosa, the Pescadores and the South Manchurian peninsula, as well as a considerable flow of capital outwards.[3] The longer-term disadvantage for China was a renewed surge of interest amongst the existing powers, particularly Russia, who with the aid of French capital began work on the construction of the Trans-Siberian Railroad in 1891. Here at least was a project for technological transfer of world magnitude instigated by imperial ambition.

9.2 ROOM FOR DEBATE ONCE MORE

> How should the traditional society react to the intrusion of a more advanced power: with cohesion, promptness and vigour, like the Japanese . . . by slowly and reluctantly altering the traditional society, like the Chinese?
>
> *W. W. Rostow, 1960*

As with India under the British, the history of China's economic 'retardation' in the nineteenth century has been much debated.

Although some historians have questioned the fact of retardation itself, most have accepted that China's 'failure to respond' requires some detailed explanation. Stephen Thomas has divided the pertinent historiography into two opposed strands.[4] On the one hand are those writers who point to internal barriers or resistances to development, singling out such features as Confucianism, the emperor system, demographic trends, the power of officialdom in the class structure, a lack of entrepreneurship or of firm government policy as sufficient explanations of China's failure to industrialise. Such 'domestic-limitation theories', whether single or multi-factored, tend to hold the contribution of the Western impact as either neutral, insignificant or positive.[5] In a distinct minority on the other side, Thomas notes a 'foreign intervention' approach. As with the Raj in India, the economic impact of a variety of Western nations in nineteenth-century China was in essence negative, not because of inherent, unchangeable barriers to development within Chinese society, but because the patterns of trade, investment and technology transfer were such as to effectively *disable* the Chinese economy up to and possibly beyond the climacteric of 1911.[6] Thomas's own favouring of the second approach and his subsequent analyses of the non-development impact of Western industrial 'projects' on China make his book of relevance to historians of technology and development. However, Thomas fails to note the room for discussion between his two extremes. For example, in contrast to China, the Japanese industrialisation programme may well have benefited from *both* certain *internal* features of her society and economy that were somehow conducive to industrial modernisation *and* from the fact that Japan escaped 'incorporation' into the dominant industrial capitalism of the West. Furthermore, the one might have led to the other. Whilst Thomas is content to label (and then dismiss) Frances Moulder as simply a 'foreign limitations theorist', there is in fact more to her analysis than this.[7] Moulder centres upon different 'degrees' of 'incorporation', and concludes that the retention of *economic sovereignty* was the necessary precondition for Japan's success within the international economy of the late nineteenth century. It was retained sovereignty during circa 1840–70 which then permitted Japan to 'respond' to the West by dictating the terms of her subsequent interaction with the West, especially during the 1890s and beyond.[8] As a prelude to the discussion of technology transfer and diffusion, we have space to note only the most salient features of the modern historiography on Chinese development.

In the year 1880 Mons de Thiersant commented upon the seeming inability of China to respond to the enlightenment of the West, and in so doing captured many of the themes of the subsequent 'domestic limitations' historiography.

> The central government, without money, and we might say, without the power of repression, is at infinite pains to retain the obedience of its four hundred millions of subjects, who lay on its shoulders the blame of the disasters they have brought upon themselves. Moreover, it has to reckon with their superstitions and their time-honoured prejudices. In the provinces the governors exhaust every contrivance in order to procure the funds which are required of them every instant from Peking for the general needs of the State: whence come the traffic of offices, the sale of justice, the arbitrary raising of the customs (of which the collectors absorb the profits), and consequently general discontent, which is fostered by the ceaseless intrigues of secret societies, as well as by the words and writings of the literary men – that frivolous, ignorant, and vain class which takes egotism to be patriotism, and only thinks of upsetting everything, instead of using its intelligence and influence for the good of the country. As to the common folk in general, given over to its instincts . . . it trembles as it thinks of the calamities which are in store for it in the future.[9]

Of the influence of Confucianism in the social structure of China there seems to be little doubt. In Toynbee's massive interpretation, the Confucian education of the administrative gentry (cf. the Hinduism of the Brahmins) forged the 'invisible citadel' from which the bureaucracy could resist the ravages of physical invasion and ideological challenge from without. Even in eras of successful takeover by outsiders this remained true: 'Sooner or later, all rulers of China before the Westernising revolution of A.D. 1911 had to call in the Confucian-educated native Chinese gentry to make the wheels of public administration go on turning, however unsympathetic the rulers of the day may have been towards Confucianism, and however reluctant they may have been to place themselves in the gentry's hands.'[10] Apart from forming the base for the high social esteem in which the gentry were held, Confucianism also bred a high-cost economy characterised by a rapacious level of agrarian extraction, conservatism, corruption and an environment within which risk-taking in material (not necessarily intellectual) matters was regarded with abhorrence.[11] Such a traditionalistic value system was carried over into business matters and was aided by China's possession of an internal frontier into which it could expand without challenge from the West or need to increase the efficiency of usage of land or labour. In addition, the maintenance of a basically hydraulic economy, dependent upon the movement and use of water, meant that any net capital formation which escaped the conspicuous consumption of the hierarchy was quickly filtered into the paddy-fields, embankments, dikes, irrigation canals and equipment for raising water.[12] Within this 'bamboo-economy' change could certainly occur, but was forever absorbed and contained within a continuing institutional fabric characterised by partial inheritance, fragmentation of smallholdings, ceremonial consumption patterns and rigid class divisions between the *literati* and the peasantry. The cultural weaponry of the elite groups was further forged by a complex ideographic script and the Confucian distinction between hand and brain workers. Within this vigorously extractive hierarchy emerged a group of middlemen, of original *compradores*, whose focuses were the increased indebtedness of their superiors, gentrification and short-term financial gains, rather than any ambition towards improved methods of agricultural or industrial production.[13] Over the many years during which control of the State rested in the hands of non-indigenous invaders, such *comprador* groups served as go-betweens, translating the demands of the military power into the needs of the scholar-gentry.[14] Resistance through soft-rule and compromise became the order of the day.[15]

Any 'progressive' developments under the early years of the Manchu or Ching Dynasty (1644–1911) were halted by increases in land or product taxes, wreckage caused by famines, a decline of shipping and a reinforcement of a brand of Confucian thought which laid great stress upon both cultural superiority and separateness. The Manchu and the bannermen required a substantial military establishment and instigated the development of a form of urbanism which served administrative rather than commercial purposes.[16] Under such a system merchants and merchant guilds could do little else but hoard their gains, or channel them into the licensed monopoly trade. It was this backdrop which dictated the failure of the *Shansi* remittance banks to become credit agents for industry and which forbade the growth of any significant level of foreign trading.[17]

But there were many-facets to the Chinese economy of the nineteenth century, and each facet may be viewed from a selection of angles.[18] The notion that the very rigidity and complacency of Chinese society somehow 'explains' its failure to 'react' to the West faces some real problems. Eckstein's China, 'in the throes of dynastic decline, political disunity and partial dismemberment' sits uncomfortably with his own argument about the failure of *the* Chinese government to respond to modernising forces, or with the general claim of a persistent continuity.[19] At the same time, and as with arguments about the similar 'failure' of the French or Dutch economies within Europe, estimates which clearly show the self-sufficiency and relative efficiency of the total economy are not at all convincing as explanations of failure to industrialise.[20]

Economic historians cannot have it all ways, and argue that both relative economic backwardness and relative economic forwardness are disadvantageous to a programme of national industrial modernisation. No country would industrialise. A more general argument might be that the first condition (backwardness) tends to promote government intervention in response to demonstration effects as was detailed in Chapters 5–7, whilst the second (forwardness) allows wealthy merchants, landlords, business and artisan groupings to broaden their inventive activity and investment portfolios, and in some cases to exploit the new processes and products available from either within their own societies or as transfers from other nations. Surely, the problem within China during the nineteenth century was that *neither* level of response occurred at a rate or level sufficient to create significant structural change in the economy (except perhaps in Manchuria), but that both levels of response did occur at some, lower level?

We are left with two simple sets of facts.

Late eighteenth-century and early nineteenth-century China demonstrated signs of general economic efficiency relative to other nations. Yields per acre were high. China was self-sufficient in most handicrafts and manufactures (hence the opium imports), and possessed a low-cost system of water transport over large areas. These together *may* have outweighed the effects of agricultural taxation. Certainly, income per head was possibly equivalent to that found in many parts of Europe in the nineteenth century and the distribution of income, despite the heavy-handedness of the upper classes, was probably more equal in China than in most other nations at this time.[21] The prosperity of the merchant class arose from a truly massive interregional trade, again equivalent in per capita terms, to that found in Europe in the nineteenth century, and in absolute quantities far greater. The economy was highly monetarised and the peasantry, though hard-pressed, was not locked into feudal structures, in contrast to the peasants of Prussia until the early nineteenth century, or of Russia until the emancipation of 1861.[22] Eckstein insists upon the absence of an agricultural revolution as a vital 'missing link' in the Chinese chain of development.[23] But we must note that the notion of a discrete, easily identifiable agricultural revolution in Europe is now much debated.[24] Furthermore, the introduction of new crops with completely different consumer qualities from the old, represents technological change of a sort, as does the incorporation of greater amounts of one factor, labour, into the production process.[25] Surprisingly, and on the basis of his own assembled data, Eckstein suggests that the productivity of both land and labour in nineteenth-century China was *higher* than that of Meiji Japan.[26] Therefore, in order to settle on a 'factor' explaining the peculiarity of China's response, Eckstein moves away from net productivity differences and resorts to fairly ill-defined claims as to a lack of infrastructural or servicing developments in China. These are difficult to define and, without completing his argument in any way, Eckstein points only to literacy differences between China and Japan as evidence of a lack of support mechanisms in Chinese agrarian society.[27] But once it is admitted that product per capita in China *might* have been higher than in Meiji Japan, then a huge weight falls upon any 'socio-cultural' form of explanation, and it is yet to be shouldered successfully. Finally, there is a great variety of evidence to suggest that from the proclamation of the *Taiping Tien kuo* (the Heavenly State of the Great Peace, 1851) to the broadening of the perspective of China's 'self-strengtheners' throughout the 1890s, there existed in China major groupings who were as aware as any Prussians or Japanese of the need to modernise their economy in order to protect themselves and their beliefs from further Western aggression.[28]

The second set of facts relates to the character and degree of Western military and economic penetration of China. Put very simply, if most of the material points are admitted, then historians who dismiss the impact of the

West as insignificant or positive must put enormous weight on major, internal factors operating *dynamically* during the nineteenth century to hold back industrial change. Major discussions have arisen around the role of government, the limitations of the 'self-strengthening' movement, and the rigidity of the class structure, which served to retard Chinese entrepreneurship.[29] The drawback of such research is that many of the so-called 'internal' characteristics of Chinese society and economy were increasingly determined by 'external' forces. For instance, the 'failure' of the Chinese government to mount a development programme after 1840 or 1870 can be seen as a reflection of underlying socio-cultural factors. But lack of government action might have been much more the result of its *loss* of effective economic sovereignty through war, concessions and treaty ports.

Until the 1840s China's foreign trade was dominated by the Canton system. Tea and silk exports flowed to the West as part of a triangular trade linking Britain, China and India. The system was fundamentally flawed. The absence of demand in China for the manufactured goods of the West meant that the substantial export trade – £20 million of tea arrived in Britain annually – was serviced by illegal imports of opium from India.[30] During the years 1817–30 the value of opium imported multiplied threefold, moving from 20 per cent to 60 per cent of total imports. In the same period the exports of raw silk and tea reached some 90 per cent of total exports. Thus the initial pattern of trade was a typical one for an underdeveloped, non-industrial nation at the periphery of an increasingly industrialised international economy, with the notable exception of an absence of machinery or consumer-manufactured imports. The serious disruptions of war and internal struggle, the first of which was wholly, and the second of which was at least partially due to aggressive actions on the part of the West, held back fundamental structural change in the character of foreign trade.[31] In addition, apart from the short period 1872–77, China showed a consistent trade deficit with the West. Trade did not act as a leading sector for growth in the internal economy, for two reasons. Because China trade was locked into the export of a few major raw materials, she suffered a decline in her *terms of trade*. Secondly, as in the case of India under the Raj, because of its commodity composition, the trading sector did not serve to stimulate (via demand or supply) structural change within the Chinese economy. On the one hand, consumers in China did not demand the manufactured products of the West to any great extent. As late as 1879 the greatest single import was opium from India and Persia (44 per cent of imports value), followed by cotton piece goods (27 per cent) and a range of basic sundries such as kerosene oil, seaweed, matches, raw cotton, sugar, coal and rice. Indeed the only product of advanced technique imported was sulphuric acid, for use in gold–silver separation processes.[32] On the other hand, even in the modernised areas of the economy, China's exports remained concentrated on raw materials. The export of silk goods illustrates the dual effect. From the 1870s to 1900 the share of manufactured silk exports to total silk exports, hovered around the 20 per cent mark, declining slightly through the period. The *volume* of manufactured silk to total silk exports during the early part of the period remained at the very low level of 7.5 per cent (1870–74) to 8.3 per cent (1875–79).[33] Thus in this second major trading period China did not appear to be gaining any of the advantages of relative industrial backwardness through the trading sector *per se*.

On the investment side the position was even gloomier. In the long period 1865–1937 of all loans to China 28.5 per cent went to military and indemnity expenditures, 28.5 per cent to administration and 37 per cent to railway construction. By 1902–31 there was a significant *outflow* of capital from China, with the overall inflow–outflow ratio at around 57 per cent. By the turn of the century the origin of such investments was much more diffuse than that of trade, with Britain accounting for 33 per cent of the total, Russia 31 per cent, Germany 21 per cent and France 12 per cent.[34]

The net *impact* of foreign economic relations on the Chinese economy is very difficult to measure. For both trade and investment the overall level of foreign economic activity was small relative to the size of the Chinese economy itself.[35] But this may have been of less relevance to Chinese development than the exact form of the economic relations which developed with the West. Because of the 'imperialist' character of the Western presence, economic relations were by no means encapsulated in trade and investment alone, for they included the activity of foreign traders and investors in the Treaty Ports, the effect of international treaties upon internal customs, monopolies, guilds and other regulations within China, and the ramifications internally of the establishment of spheres of influence and project concessions, e.g. Germany's suzerainty in

Shantung, the British at Weihaiwei and Kowloon, the Russians and Japanese in Manchuria. In addition, the size of trade and investment in *per capita* terms may have been far less in China than in the advanced nations of Europe, but in terms of the population engaged in the industrial and commercial sectors of China, and in comparison to all of Europe, such relations loom larger.

Although government expenditure was small in absolute terms, its direct investment in modern industrial projects represented some *12 per cent of total revenue* during the 1890s. In the years 1868–81 the *Meiji government of Japan* invested *only* some 5 per cent of its ordinary revenue in industrial projects, prior to the famous 'selling-up' process.[36] We would argue, in summary, that the Japanese government was more effective than the Chinese because of (1) its high absolute level of investment and because (2) such direct investment took place within a wider policy framework which included a great variety of indirect (non-capital) aid to industry and commerce. We would add that the reduction of economic sovereignty in China reduced the possibility of process (2), and that trade effects, investment patterns and foreign and civil wars reduced the absolute level of industrial investment (process (1)). In the years 1895–1905 indemnity payments alone accounted for one-quarter of total expenditure of the central government in China.[37] It becomes fairly obvious that many of the supposed 'negative' features of Chinese government activity may be attributed to the wide-ranging 'negative' aspects of foreign political and economic relationships. We seem to be left with only the possibility that transport improvements *may* have increased private returns to some Chinese exporters.

9.3 TRANSFER MECHANISMS

Contemporary opinion and subsequent historical analyses have promoted the idea that the Chinese Treaty Ports were centres of change and demonstration similar to foreign settlements in Russia, Japan and elsewhere in Asia.[38] In his history of Shanghai, Hawks Pott described the city's 'industrial revolution' of the 1890s as almost entirely caused by foreign activities. Especially after the 1895 Treaty of Shiminoseki had legalised foreign production in the ports, trading firms such as Jardine, Matheson and Co. and Ilbert and Co. moved from their trading ventures into cotton manufacture, flourmilling, the factory production of silk filature, chemical works and shipbuilding. For Pott, the effects of such projects upon employment and training were highly significant, and represented a conduit for technology transfer.[39] Supposedly, in Shanghai Confucian social relations were laid waste. In addition, as early as the 1870s inter-Treaty Port trade had risen to perhaps one-third the value of all foreign trade and the official transit trade from the Ports to the interior was similarly developing. Such trading networks could be seen as avenues of technology transfer.[40] The public works of the foreign municipalities provided further demonstration effects: Shanghai, by far the largest of the Treaty Ports, established its Gas Company in 1863–4, its improved waterworks in 1880, and an electric power plant in 1882.[41] Eckstein points to the even earlier extension of trading-company activity to the provision of shipping, insurance and commission services and the employment of Chinese *compradores*. The foreign settlements provided training grounds because they attracted Chinese investment and Chinese labour. Their failure as change agents lay in the resistance of the traditional order.[42] This view of the transforming role of the Treaty Ports is supported even further in the general analysis of Allan and Donnithorne, and in the recent work of Rhoads Murphey. In the foreign havens were to be found the 'new men' of China, the allies of the West who had moved beyond the bounds of traditional Chinese commerce.[43]

We may judge this view as overdrawn rather than incorrect. In contrast to Japan, the Chinese Treaty Ports contained only a very small proportion of foreign to Chinese residents (around the 1 per cent mark), and such foreigners were loath to decipher the complex language and norms of their trading partners.[44] Inter-port commerce seems to have had little effect on technology transfer, for it was dominated by the movement of such traditional products as sugar, paper, raw cotton, cuttlefish, bean-oil, hemp-bags, soap and tallow.[45] The resultant need to reduce the costs of transactions in the marketplace promoted the power of the *compradores*, the Chinese merchants who became the effective business managers of foreign firms. As Rawski has noted, such a system tended to channel foreign and indigenous capital into established product lines and services (the salt trade, native banking), and acted as a bulwark between the Western influence and the official class. The learning process was truncated. Rawski adds two further important points. The *comprador* system served to secure the developing relations between Chinese merchants and Chinese officialdom.[46] Secondly, *comprador* dominance meant that the introduction of Western business methods and ideas into Chinese industry and commerce was retarded. Prevalent trends towards nepotism, and the indigenous contract and currency systems were not threatened, and, if anything, Western traders were adopting *Chinese* business customs rather than stimulating attitudinal or institutional changes.[47]

We may note other, more diffuse forces operating within the foreign settlements which further restricted the role of the Treaty Ports as focuses for change. During periods of civil violence the foreigners of the Treaty Ports designed to protect their own interests by playing both sides.[48] The influx of Chinese into the Treaty Ports during years of unrest tended to lead to a rise in rents, speculation and a search for short-term profits, which diverted many funds from business activity. In 1862, a year of disturbance and influx, Shanghai rents rose 200-fold over those of the early 1850s. Exodus, as in late 1864, could quickly lead to the opposite, with immediate falls in prices and rises in bankruptcies. The Treaty Ports were also the centre of direct challenge to the Imperial government. The small municipal budgets of Shanghai's foreign settlements (and other areas) depended upon the sale of licences to opium 'divans' and brothels. The American settlement was described as a 'refuge for the criminal class' and the few educational institutions produced by foreigners privately were accused of 'republican' and anti-Imperial sentiment.[49] Nevertheless, the Chinese residents of the settlements were treated very much as second-class citizens. As residents they were taxed equally with the foreigners, but unlike the latter secured no voice in municipal affairs. At the same time, Chinese residents escaped general Chinese taxation (from 1862 a percentage of the land tax went to the government), and so effectively became 'outsiders' in the traditional system. Furthermore, although the Treaties provided for a judicial system which entailed separate courts for Chinese and Western disputation, during 1864–5 it is evident that infringements occurred, as representatives of foreign consuls became party to judgements between Chinese and Chinese within the settlements.[50] Other features of Treaty Port life were hardly demonstration of the advantage of Western styles. Cholera, smallpox and typhoid abounded. Perhaps understandably, when the modernised water system was extended to Chinese residents in various ports during the 1880s the 'improvement' was met with suspicion. Similarly, native Chinese residents believed that the danger of electric lighting outweighed its utility![51]

In this light, the sometime negative attitudes of Chinese officialdom to Treaty Port innovations becomes more understandable. Moreover, the authorities often had other plans. During the 1860s and 1870s the refusal of the Chinese to officially sanction the costly dredging of the Woosung Bar, Shanghai, was based on their opinion (partially derived from that of foreign experts), that future trade would see Chinkiang and Hankow grow at the expense of Shanghai.[52] Again, the tardiness of officialdom in aiding foreigners in their designs for a railway linking Shanghai and Soochow was based not on Confucian stubbornness, but on their considered policy that no such project would take place until it was agreed that it be underwritten by the Chinese themselves, that it should not involve large movements of foreigners in the interior, and that it should not cause friction amongst Chinese landowners who would have to sell tracts of land to developers.[53] Because such stipulations were not met, the Chinese government effected the closure of this early attempt in 1877. Thus, it may be that the attitudes of Chinese government and business were formed out of the realities of Treaty Port culture, rather than out of the quixotic nature of Chinese culture.

Despite the partial breakdown of certain aspects of the transfer process, the later nineteenth century witnessed ambitious efforts to create a modern industrial infrastructure amongst both Chinese and Western investors. These were particularly concentrated in the years following the war with Japan. Hou calculates that over the period to the 1930s indigenous Chinese investment in the modern sectors of the economy grew at a rate equal to that of the Westerners. Indeed, in such strategic areas as factory production of cotton and railway extension, Chinese investment grew at a faster rate than Western investment.[54] As in Japan, Chinese government investment was often in the form of priming or demonstration projects, such as the modernised textile mills at Shanghai or the coal and iron complex at Hanyang.[55] Such direct investment by the Chinese was associated with reform in the institutional framework of business activity. Government banks were established in competition to the financial networks of the *comprador* system and were designed to deal directly with the indigenous merchants of the interior. Such Chinese projects were seemingly aided by a high rate of *reinvestment* within the foreign financial sector itself. Thus Jardine, Matheson and Co. reinvested their profits from trade into such ventures as silk reeling, refrigeration, engineering, shipping, cotton textile, and brewing enterprises, as well as in loans to the central government. As Hou suggests, the last years of our period seem to have been dominated by a fervour of modernising activity and an economic partnership between Chinese ventures and Western capital. For Hou, such strategic projects, possessed a learning function: 'The Schumpeterian innovator (in this case, the foreign firm) was indeed followed by a 'cluster' of Chinese innovators'.[56]

Was Chinese innovation able to follow in the tracks of Western railway-builders? The development of the railway system in China absorbed the bulk of central funds for industrial investment after 1895 and accounted for 37 per cent of total foreign investment in the years 1865 to 1937. A line construction of 2708 miles in 1903 increased to 6000 miles in 1911 and thereafter the rate of growth declined somewhat.[57] It has frequently been argued that the modest performance prior to 1895–7 was a fault of Chinese cultural conservatism, a failure of vision.

We must agree that prior to 1897 the railway system was held back as a leading edge of growth because of 'negative' attitudes amongst Chinese officials. The limited development of the early years was monitored by officialdom. From the 1860s even those 'self-strengtheners' amongst the Chinese who were strong advocates of modernised trade, currency, mining and general manufacture, remained sceptical of either Western or Chinese involvement in railway development.[58] Following the terrors of the Taiping period officials saw the extension of railways as a threat to Chinese sovereignty. Thus the Governor of Kiangsi, Lie 'Kun-yi, argued that a combination of railways and telegraphs would expose confidential information and hence weaken China's attempts at defence against Western encroachment.[59] Only in 1882 was there any sign of a tentative policy of Chinese support. Approval for projects would be gained if they satisfied certain criteria. Railways should open up resources, employ Chinese labour and management and remain independent of foreign capital. Such policy stipulations combined with a difficult terrain, a very low revenue base and the existence of efficient water transport systems, meant that development was slow during the years 1870 to 1890. But there is really little evidence to show that a conservative bureaucracy hindered railway development once the stipulations were satisfied. The Tangshan railway was encouraged from 1882 as a principal means of transporting Kaiping coal. This led to the formation of the Kaiping Railway Company, which extended the line to Lutai in order to circumvent the growing inefficiency of the existing Hsikechaung–Lutai canal.[60] Similarly, the need to stimulate indigenous Chinese trade lay behind a proposal for the Peking–Hankon line. It was objective factors – a small budget, internal disruption, terrain, and alternative transport systems – which prohibited a speedier development of the system.

The railways' immediate absorption of large fractions of fixed capital also precluded more extensive government activity. The central government's lack of *economic* sovereignty and its consequent inability to raise customs duties or tax foreign trade and production reduced severely its ability to accumulate capital. Simultaneously, the refusal of the Powers to lend money for Chinese railway projects in the absence of general concessions came up against China's reasonable fear of foreign control. So China's railways were built literally piece-out-of-piece. The income from partially completed sections was used to pay for the construction of later sections. Labour was used more intensively than in the case of railway systems in Europe. The Chinese were attempting to forge an appropriate technology. Between 1889 and 1897 the growth of the Imperial Railways of North China did depend upon small amounts of foreign borrowing.[61] But from that date the real 'scramble' began. After the defeat of China by Japan, indemnities, treaty concessions and fear of further aggression converted the Chinese railway system into a haven for foreign capital. The considered stipulations of the 1860s to 1880s period broke down under the pressure of necessity. During 1895–1911 £34 million of foreign investment entered the railway system. By 1911 some 41 per cent of railway mileage was constructed through 'concessions' and the bulk of the government-controlled lines were financed by foreign investment.[62] But during these years popular opinion began to demand a return to Chinese control over the railways. Resentment against any increase in salt or rice levies led to government attempts to raise scarce capital by an issue of stock to the indigenous population. Capital was not the only problem, for the government faced the recurring difficulty of negotiating between competing Western interests for control over the lines.[63] Stringer, who roundly condemned the 'greedy backstairs scheming' of the Powers noted that 'Railway development has been localised mainly by spheres of influence, which empire or necessity has dictated.'[64] Railways served enclaves, not hinterlands.

The negative effects were even greater in the case of concessions. The leased areas of the concessions were entirely controlled by foreign investors and with the spheres of influence acted to reduce China's control over projects on her native soil.[65] Within such areas China's technical capability was completely bypassed.[66] Because of transport and commission charges, together with a natural tendency amongst the foreign controllers to inflate costs in order to justify further lending, the total cost of foreign railway construction was 50 per cent above that of comparable Chinese railway projects. Concessions filtered out the potential positive externalities of railway projects, leaving only nominal 'growth' as their result. Whereas the emergence of some Chinese technical capacity may have served to reduce the cost of Chinese railways to a level below that of foreign railways after 1897, such infrastructural gains were lost in the years prior to 1911.[67]

We may summarise that in the years before 1895–7 the 'developmental' impact of the railway system was insignificant because of its small size. After 1897 the greater growth of the system had little positive impact on Chinese economic development because of the control exerted on the system by foreign capital. But we might remember that the small size of the indigenous system prior to 1895 may have been dictated not by Chinese 'cultural' attitudes, but by a lack of revenue, a proper fear of Western control and doubts about the rectitude of a policy which shifted scarce resources into projects the potential social returns of which were to be doubted. As with the Indian case, the emergence of the 'high technology' sector of railways seems to have had very little positive effect on economic growth, economic development, or sustained technology transfers.

9.4 'ENLARGED INTERCOURSE' – INDIA, CHINA AND JAPAN

The machines used in the cotton manufactures were, up to the year 1760, nearly as simple to those of India.

Edward Baines, *History of the Cotton Manufacture*, 1845

In the long sweep of history, India and China had much in common. Both were large (with populations of around 100–130 million in the sixteenth century, when Europe's population equaled perhaps 70 million), both suffered threat and invasion, and both developed military despotisms protective of coherent 'civilisations'. Their large size meant that both were disadvantaged by heavy transport costs and the random or dissenting activities of provincial power-brokers. But geography meant more than this – numerous languages and dialects, internal customs, periodic pillage, superstition and ignorance, all hindered the emergence of either nationalism or entrepreneurship.[68] However, when all this is written down, it may be more or less repeated for early modern *Europe*. To treat of either India or China as entities whose 'characteristics' may be labelled and then generalised wholesale, is to do for Asia what most Western historians would never do for Europe.

By the nineteenth century neither space nor technological backwardness *forbade* industrialisation. They did condition it. The case of Russia during the 1880s and 1890s suggests that industrialisation in a large nation of extensive land frontiers could take the form of a disjointed, expensive and military-orientated but nevertheless dramatic industrial drive. At the same time it might be noted that the sceptical attitude of Nicholas I's Finance Minister, E. F. Kankrin, towards railroads was founded upon just the same fear of rampant change and social dislocation as could be found amongst China's scholar gentry: was resistance to railways as much a Slavic as a Confucian virtue?

The history of railway projects illustrates the importance of the spatial relations considered in Chapter 1 above. Contemporaries estimated that the efficient *maintenance* of existing Chinese railways required the expenditure of over £3 million annually by the early years of the twentieth century. But the Chinese railways serviced 100 000 people and 400 square miles territory per mile, the Indian system served 8600 people and 40 sq. miles, the US system 3800 and 12 sq. miles per mile of constructed line.[69] Despite tremendous efforts in the interim years, by the mid-twentieth century China possessed a mere 2 miles of rail per 100 000 of its population, at a time when northwest Europe boasted 50, and the USA 250 per 100 000. Smaller nations inhabited a transport and communications world of their own – kilometres of line per 100 square kilometres reached 13.5 for Britain, 7.5 for Japan, 6.8 for Italy.[70]

Under the right conditions, *regions* industrialised in a manner analogous to advancing areas of Europe or Russia in the late nineteenth century. Murphey has made the point that China was so large as to 'absorb' into itself without trace the modernisation which stemmed from the Treaty Ports.[71] We might comment that, if China north of the Great Wall is *excluded*, the hinterlands of modernising regions were areas of population density and high levels of land irrigation, productivity per acre and urbanism, i.e. contexts for response as much as absorption. Again, the history of Manchuria in the early twentieth century suggests that there was little in Chinese culture which prohibited economic modernisation. From 1902 the region accounted for an increasing proportion of foreign investment in China, and witnessed a tenfold increase in its foreign trade between 1907 and 1927.[72]

What Japan, Russia or the USA did not share with India or China in the nineteenth century was the *loss of economic sovereignty*. Relative economic backwardness in 1800 or even 1850 did not *prohibit* industrial advancement thereafter. In themselves, spatial and demographic factors seemingly did not *preclude* industrialisation. Loss of inner control did.

Although a 'sharp stimulus' of the type identified by Rostow may engender a response, traumatic shocks to existing economic and political systems merely reduced the efficiency of social control mechanisms and, in the case of China and India, led to the loss of sovereignty over decision-making in the economic sphere.[73] Perhaps one of the most immediate impacts of loss of control was the disestablishment of existing rural by-employments in both China and India. We have seen, and Thorner has emphasised, that existing rural industries may act as centres of technological *transfer* and *diffusion* (see especially Sections 5:2 and 7:3–4).[74] Whereas in nations such as Russia, Germany or Japan such by-employment could be protected or nurtured by sovereign governments, in India and China they were opened up to all of the negative forces of 'imperial' competition. At the same time, scarce resources were channelled into 'projects' which absorbed modern technologies but employed little in the way of indigenous techniques, skills or labour, and satisfied little in the way of existing patterns of consumer and producer demands.

In per capita terms, by 1913 the level of foreign investment in Japan was significantly higher than in China, and 50 per cent of

Japan's government debt was held by foreigners.[75] Yet there is no sense in which it might be stated that Japanese industry was a mere adjunct of Western imperialism. In the crucial years of the industrial drive (circa 1868–81) technology had flowed into Japan under Japanese auspices and firm measures had been instituted for its control and utilisation.[76] From this phase emerged the entrepreneurship and responsive vigour of the Japanese industrial revolution. Who would not buy into the government's Shingawa Glass Factory when it was offered for sale to private enterprise in 1883 at an optional downpayment of 4 per cent and a low capital valuation, to be paid for in interest-free deferred payment over 55 years? Intellectual debates over Westernisation and economic change wilted in the climate of successful, state-monitored industrialisation. Would this have been true if Japan had been caught in the net of the Opium Wars of the early nineteenth century? Japan's sovereign choice included the construction of social and economic overheads, a military programme, the application of measures of direct social control, the government operation of enterprises and demonstration plants and the beginnings of regional planning.

The Japanese government concentrated its efforts upon the wholesale use of Western techniques and personnel as a prelude to the indigenisation of knowledge and decision-making. In the 1880s, after a phase of Western influence and government ownership, the local mines of Takashima were taken over by Mitsubishi, who introduced Japanese water wheels for drainage and established techniques for excavation.[77] The introduction of Western technologies into the 43 government metal mines during 1868–75 redefined the effective resource base, e.g. discarded ores from the tin-mines around Kagoshima became sources of gold and silver when F. Coiquet reported on the appropriateness of Western extraction methods in 1874. Although mining technique often remained Japanese (labour-intensive), the processes of crushing, pulverising and concentrating utilised Western methods.[78] In this environ new technologies transferred quickly; the water-jacket blast furnace developed at Detroit in 1882 was being used at Ashio in 1890; the first successful application of the Bessemer steel process to copper-refining was in America in 1884, was searched by Japanese in America in 1890 and was operating on a large scale in Japan in 1893; electrolytic refining processes developed in America in the late 1880s were installed in Japan by the mid-1890s, even though their use required a support structure of hydroelectric power plants, electrical generators and water pumps; prior to the First World War, the government-operated Yawata Steel Works reached a level of technical sophistication comparable to that of its German tutors.[79] Such transfers were induced by the war economy, especially after 1894.

Kozo Yamamura argues that it was strategic *demand* resulting from the military plants which was responsible for 'assuring the survival and for aiding the growth of often financially and technologically struggling *private* firms in shipbuilding, machinery and machine tools industries'.[80] It should be emphasised that the activity of the Japanese arsenals was discontinuous with that which had gone before. Thus the change from wooden to iron ships was no isolated affair – it involved steam hammers, docks, riveting machines, blast furnaces, skilled workers, cranes, ores and cokes all imported, adapted and brought together on a significant scale of production. By the early 1880s the arsenals were employing 10 000 workers in comparison to the 3000 in small private establishments. Examples of *diffusion to non-military enterprises* are numerous. The Osaka arsenal produced machinery used by several private firms, including a large-scale cotton textile company from 1882. Six of the ten private cotton textile firms (capital-intensive and linked to a labour-absorbing sector) established during the early 1880s used spindles imported from the West and powered by steam engines made at Yokosuka arsenal.[81] Military establishments were involved in mining, machine tools, bridge construction, the repair of private ships, lighthouses, government buildings, private factories, the construction of roads and harbours, and boilers and related machine parts for mining, textiles and railroads.

From a comparison of India, China and Japan, it appears that the retention of economic sovereignty during the industrial drive is vital, not only in general financial terms but as a necessary condition for the successful transfer of modern technologies into large industrial projects. Furthermore, we argued in Chapter 7 that the retention of sovereignty is not simply a function of the nature of economic and political penetration from outside, but is also conditioned by internal spatial and demographic factors.

Kurt Mendelssohn has claimed that 'Western man' made his true intellectual departure from the rest of the world when he formulated general notions of 'energy as a physical quantity of paramount importance'. He goes on:

It was a creation of Western thought for which the others utterly lacked the basis. Energy, its conservation, and the usefulness of the notion of conservation, were ideas that had grown out of more than two centuries of natural philosophy. And this was a field of development in which none but the white man had participated. . . .[82]

Whether there is anything at all in this approach is open to question. If such stale distinctions do explain features of *intellectual* divergence between Europe, India and China, do they help in understanding the 'failure' of Western *technology* in China and India during the nineteenth century and its 'success' in Japan? Of the three nations, China seems to stand as an intellectual contender to the West.[83] On the other hand, Japanese science was basically Chinese science for most of the nation's history, yet Japan used the artefacts of Western science in order to escape from remaining a 'second-class citizen of the world',[84] and does not appear to have been hindered in doing so by its cultural inheritance.[85]

On the one hand Japanese culture (and its 'science') was non-Western. On the other hand, within that cultural foundation, specific Western scientific knowledge was imbibed by key intellectual groups. During the Meiji years Western scientific knowledge was absorbed *not* by Japanese culture, but within the micro-cultures of small, strategically-placed and powerful elite groups.[86] Such groups wielded the nation's sovereignty. In a world in which machines could be transferred at an economic cost, and within which understanding of technologies could be taught, the nature of the 'culture' into which they came (itself a dubious category) may have been of considerably less importance than the character of the social and political structures which were maintained during the phase of 'Westernisation'. As in the cases of China and India it was only when the power of established elites was effectively *removed* that the question of the *cultural appropriation* of techniques arose. A close 'fit' between host culture and introduced artefact would then represent a massive coincidence. At the same time, even the existence of such a 'fit' might have been irrelevant to the successful utilisation of advanced technique when other, particularly economic, conditions remained unsatisfied.

10. Centre and Periphery: Science and Technology in America and Australia

10.1 SCIENCE, TECHNOLOGY AND RECENT SETTLEMENT

Like the transplant of any tissue, the organic structures of science may well be rejected by inappropriate hosts.

Charles Rosenberg, 1974

IN different periods, America and Australia shared the distinction of being both formal colonies and areas of recent settlement.[1] From the imposition of the Navigation Acts in the 1660s – whereby American trade was directed and American manufactures were discouraged – to at least the revolutionary wars of 1775–8, the economic development of the new world was conditioned by isolation, migration, resource endowments and external trading relations, as was that of Australia throughout most of the nineteenth century.

The scientific and technological cultures which emerged in such nations were unlikely to merely *duplicate* those of either established or newly-industrialised nations in Europe or elsewhere, nor that of their *heavily populated*, traditionally-settled colonial acquisitions. Here there existed no powerful indigenous political systems resisting the encroachment of Western ideas, artefacts and technologies, for the indigenous people were sufficiently disabled by force, and eventually disestablished by time and circumstance. So the combination of cultural resistance and political suasion which did result from socio-economic clashes in densely populated nation-states was of far less importance in the history of America or Australia than in the histories of Japan, India or China. In colonies of recent white settlement the transplantation and modification of Western knowledge and technique in the hands of European cultural activists and entrepreneurs and in the context of restricted manufacturing development became 'the history of science and technology' in those nations. Fairly rapidly, all other models for the scientific enterprise became illegitimate and powerless.

In both continents, the process of early settlement was associated with the activities of groups of European intellectuals whose energies were directed to the establishment of a European scientific culture which addressed itself to the economic demands of staple exploitation and the political requirements of colonial rule and security. However able, such scientific intellectuals were 'peripheral' in a threefold sense: in their marginal positions at the edge of European culture; in their partial commitment to scientific enquiry (forced upon them by the immediacy of survival and development); and in their role as agents for the exploitation of natural resources of economic significance to Britain, itself the focus for sanctions and rewards.

From its beginning, science at the periphery stressed the development of natural resources, the collection and reportage of natural artefacts and specimens, and the application of European scientific knowledge and methods to such immediately practical pursuits as agriculture, mineral exploration and metallurgy and to the public services required for efficient drainage, water supply, transport and public health. Colonial pressures joined with the natural imperatives of recent settlement to forge a utilitarian and initially subservient scientific programme. Finally, the work of Fleming, Basalla and Shils combines to suggest that the intellectual haven provided by natural history investigations offered up psychological rewards to the scientific intellectuals once settled at the periphery.[2]

10.2 FROM COLONIES TO INDUSTRIAL ECONOMY – THE AMERICAN CASE

I always feel America as well as Canada is on the periphery of the circle.

Ernest Rutherford, 1907

In America, the colonial, revolutionary and Napoleonic periods were characterised by a substantial rate of population growth, with absolutes moving from around 330 000 in 1710, to 2 million in 1770, 2.8 million in 1780 (25 per cent of the English population), to something less than 4 million in 1790, by which time slaves numbered 700 000 individuals and urban settlement involved only 5 per cent of the population. The expanding frontier of the new West permitted a further population growth of 8.5 million in 1815. For the colonial years alone the population had grown at the startling rate of 34 per cent per decade. Here was the first example of rapid European settlement into a relatively open space. Amidst this the immediacy of survival promoted an approach to 'knowledge' which was fairly utilitarian and specific and reminiscent of the 'philosophy of improvement' which developed contemporaneously in Ireland and Scotland, discussed in Chapter 2. Writing in the mid-eighteenth century, Jared Eliot pleaded for a local journal or information source which 'might serve to increase *useful knowledge*, giving a faithful account of the success of all the experiments and trials that may be made on various Sorts of lands, and of divers Sorts of Grains, Roots, Grass and Fruit, not only such as we have in use, as also what we have not yet introduced among us.' And writing of the settlers themselves Eliot recognised that 'in a Sort, they began the World a New'.[3] But the new-world scientific intellectuals had been pursuing an acclimatisation programme for some time. John Winthrop, who had been a member of the original Gresham College group, researched into the growing of maize, the extraction of sugar from cornstalks, and began a collecting programme in mineralogy and botany. Just as Eliot was writing, Benjamin Franklin was helping to found the American Philosophical Society, whose original scientific agenda included botanical acclimatisation, archaeology, palaeontology, mapping and surveying, and animal breeding. The society first came to European prominence through its local observations of the transit of Venus in June 1769. With one eye to fame in the sky, the other focused on mundane matters of the earth – silk cultivation, the reclaiming of marsh lands, medicine and public health and the preparation of an American pharmacopoeia.[4] From 1780 the work of the American Academy of Arts and Sciences was of even more utilitarian a nature, including original investigations into such a variety of pursuits as the culture of wheat, distemper and iron-working. The natural history enterprise resulted in international rewards as well as local utilities; prior to the Revolution Benjamin Franklin and Cotton Mather were but two of fifteen American Fellows of the Royal Society.

Nevertheless, before the 1840s it is difficult to write of an American 'scientific movement'. Following but adapting Pred, we might argue that a truly interdependent, national urban cultural system would see significant changes in one city or group reflected in similar changes in others.[5] But for most of this period Philadelphia remained a cultural enclave, with Boston a tardy second-runner. The population of Philadelphia grew from over 28 000 in 1790, to a population in an expanded area of 68 000 in 1800, 160 000 in 1830, and 400 000 in 1850. Until this demographic boom Philadelphia remained the 'city of entrepreneurs' and professionals.[6] The social dominance of merchants, lawyers and gentlemen gave rise to a 'polite' science allied to the work of the American Philosophical Society, founded in the city in 1769. At a more popular level the public lecture thrived. 'The love of Science' was enshrined in the lecture courses of John Ewing, Benjamin Rush, Ebenezer Kennedy and David Rittenhouse. The city's biographers stress the general popularity of science in Philadelphia in these years:

> Interest in natural philosophy had preceded that in other fields of culture at Philadelphia, and throughout the eighteenth century kept pace with other phases of intellectual endeavour. Beginning with the amateur dilettantism of the contemporaries of James Logan, it came by the end of the period to command the professional respect of its notaries and the avid curiosity of hundreds of eager and interested laymen. While actual accomplishments in the field were considerable, perhaps the most significant achievement of these years was the very generality of popular interest in the mysteries and potentialities of nature.[7]

During the later years of the century a prolonged attempt was made at science popularisation by Charles W. Peale (1741–1827), who opened his Philadelphia Museum in 1784, from which he and others delivered regular lectures on chemistry, natural philosophy and electricity. Peale saw his role as a lecturer in 'impressing on the mind of the citizens the importance of the study of Natural History and the presence of a museum'. What is more, 'Natural History is not only interesting to the individual, it ought to become a National Concern, since it is a National Good – of this, agriculture, as it is the most important occupation, affords the most striking example.'[8] His activities included correspondence with individuals both within and outside America, and expenditure upon works of reference, chemicals and apparatus was considerable.[9] But generally, until the 1820s, Philadelphian science was the cultural property of the educated elite. Only during the 1820s did a more purposeful, popular and applied science emerge.[10] Between then and the 1840s the science culture of the city, as in other cultural centres, reflected more the patterns of development in Britain than those of eighteenth-century America.[11]

From the 1840s a 'scientific movement' on a more national scale did develop and was strongly linked to government aid, large private donations, the technological demands of the economy, and the democratic ideal.[12] Bell has argued that the American Philosophical Society began to exert itself as a national academy during this period, a 'Congress of Philosophers as well as of Statesmen'.[13] The stress of the APS was on both technology and natural history, on expeditions (Michaux, Lewis and Clark, Wilkes), and upon the maintenance of intellectual contacts with Britain, Europe and Latin America. Although some applied science had begun at Harvard from 1814, and at Pennsylvania and Columbia, the decade of the 1840s witnessed a far greater specialisation and expertise in the applied fields and the beginnings of the movement of higher education away from teaching only and towards research and the establishment of higher degrees. In addition to the work of Rensselaer Polytechnic Institution, engineering was introduced into the teaching programmes at Harvard, Yale, Dartmouth and Pennsylvania. In 1859 Yale established a professorship in industrial mechanics and physics.

The decade which boasted the foundation of the American Association for the Advancement of Science (1847), an association at least loosely analogous to the British Association for the Advancement of

Science, also witnessed the beginnings of a promotion of national science and technology in a manner quite different from that of Britain. Something of this difference has been pointed out in Chapter 4. The utilitarian, specialist and research-orientated programme was far more dominated by endowment and direction from outside of the scientific community itself. Against the backdrop of a considerable machine-engineering culture, the expenditure of the individual states from 1845 became less devoted to transport and other infrastructural projects and more focused upon educational provisions, which by the 1870s took up 30 per cent of all expenditure at this level. In addition, the land grants of the Morrill Act of 1862 represented a tremendous resource transfer to university and engineering, agricultural and mechanical institutions.[14] Further boosts to research originated from vast philanthropic grants from Johns Hopkins (which developed a graduate school on an initial endowment of $3.25 million), Ezra Cornell, Andrew Carnegie and John Rockefeller. The number of doctoral degrees grew from 44 in 1876, to over 560 in 1918, to over 1000 in 1924, and the per capita expansion in the number of higher and technical students was at the rate of 100 per cent for the years 1872–1900, whilst the per capita production of trained scientists and engineers increased fourfold during 1880–1900. The land-grant schemes encouraged not only extension training, experimental stations, and research on such specific 'applications' as rust-resistant wheat, corn-hybridisation and geological surveying, but stimulated the establishment of engineering facilities in existing universities and colleges (e.g. at Wisconsin in 1870). In addition to the development of domestic research, such officially encouraged centres of learning as Purdue and Ohio State became important as *entrepôts* for the transfer of recent research findings from Europe. This bolstered the transferal-effects of the tendencies towards foreign study and travel which had been established at mid-century.[15] The Board of Agriculture (1862) and the Bureau of Animal Industry (1884) were perhaps the foremost official institutions for the rejuvenation of the acclimatisation tradition, e.g. in the development of entomology from the early 1880s. By 1888, the year in which a much smaller Australian scientific community was making its first tentative steps towards a federation of effort, the American government was investing $0.75 million in its agricultural experimental stations alone. Yet, as Fleming points out, this was also the decade during which the intellectual dependence of young American scientists on Europe for *general* scientific training and leadership reached a peak.[16]

10.3 THE 'AMERICAN SYSTEM': SCIENCE, TECHNOLOGY AND ECONOMIC GROWTH

> A free people will in due time produce anything useful to mankind.
>
> *Joseph Priestley, 1794*

Despite the continual expansion of cotton textile production in New England, American economic growth up to the 1840s was still based on agriculture, the exploitation of natural resources, and the development of specialised external markets. As late as 1879 the contribution of agricultural production to total output was equivalent to the total outputs of mining, manufacture and construction. Indeed, until well into the later nineteenth century America remained a rural nation whose economic growth was a direct function of immigration and population growth, foreign capital importation and associated transfers of European technologies.[17] As such, a very large portion of efficiency increases in the economy resulted from trade expansion (with related developments in railway and shipping services) and positive economies of scale. Although the factory production of textiles expanded remarkably after the invention of the cotton gin in 1793, growth in the Napoleonic period had, in fact, been dominated by an expansion of staple production for export – to cotton were added sugar, coffee, cocoa, pepper and spices, staples which gained their income effect from positive movements in the terms of trade, and not from improvements in the methods and organisation of production.

In summary, despite its large absolute size, the growth of the manufacturing sector of the American economy was but one aspect of the expansion and diversification of a major trading economy, whose regional characteristics were determined by variations in the pattern of land and resources exploitation. Scarce resources were mainly shifting into the servicing of the trading sector – shipping, canals, railroads, telegraphs and public works – rather than into the industrial manufacturing sector.[18] The significant falls in overseas freight rates between 1815 and 1850 were associated with the *sailing* ship, not the steam engine. Again, the steam engine played its role in reducing internal transit charges initially through the steam ship and the canal, rather than the railroad.

Nevertheless, it was within this pre-Civil War setting that a distinctive American system of manufacture began to emerge, and this was to be of decided significance in instigating the important institutional and technological changes of the last decades of the century. The special character of the 'American system' was the early emergence within the industrial sector of an *organisation* of sequential operations on specialised machines which produced interchangeable parts for an array of often new products.[19] From the International Exhibition of 1851, through the Crimean War and the American Civil War this novel character of American industrial production was noted in Britain and copied in the manufacture of small arms, a process of emulation encouraged by influential reportage from Joseph Whitworth, a Manchester machine-tool manufacturer who toured America, and through the active support of John Anderson, technical expert to the British Board of Ordinance.[20] The process of technical transfer seemed to have come full circle.

Chapter 5 has noted the importance of war as a stimulant to technical change. Throughout the early nineteenth century the US War Department was committed to French theories of uniformity in arms production. American mechanics, aided by government support and contracts, attempted to develop standardised machine-based interchangeability in arms production. These efforts led to such important innovations in machining as the development of bearing points and fixtures by John Hall at Harper's Ferry from the early 1820s, and a series of interrelated advances at the federal ordinance factory at Springfield. The early cost-reducing effects of such production engineering improvements were minimal, the chief advantage of the new system being the more efficient assemblage and repair of arms on the battlefield. By the 1850s the new ordinance production system was being utilised by private armoury producers such as Colt, Robbins and Sharpe, all of whom employed machines in aiming at 'thorough identity in all parts'.[21]

Prior to 1850 there is some evidence of a spread of what was to become known as 'the American system' into the fields of clockmaking, flourmilling and textiles. During the next decade or so, although still centring on the production of small arms (encouraged by the Mexican War from 1848), the system diffused through a range of products. However, the history of both the Singer (sewing machines) and the McCormick (agricultural machinery, in particular, the reaper) corporations between the 1840s and 1880s illustrates the considerable difficulties inherent in full application of the principles of machine-based interchangeability.[22] Even famous locations of the new approach, such as Harper's Ferry, did not in fact maintain the model system and at times resorted to the use of craftsmen and hand-tooking at very important points of production.[23] The failure of the system to spread into such areas as furniture production was clearly not a result of any shortfall in technique, but is more likely evidence of the importance of large, homogeneous markets to the successful adoption of mass production technique. The early 'American system' was costly, encountered labour resistance and technical inefficiences (e.g. loose 'fits' in assembled products), and so prior to 1880 there were relatively few examples of the diffusion of the new complex of technique and its organisation into areas other than armaments. Several key technical elements were still missing. For instance, precision grinding, which made it possible to machine parts after the metal was hardened, was an innovation of the 1880s.[24] Again, new management systems for the direction and control of labour and production were a feature of the 1880s and 1890s, and prerequisite to the highly-developed, interchangeable, assembly-line production of the early twentieth century.

Integral to even this rudimentary growth of the system was the American capital goods sector, essential to the crucial process of initial technology transfer from Europe, a process in turn associated with the large migration of skilled Europeans into America during these years. Habakkuk has argued that labour scarcity, natural resource abundance and the ability of the economy to attract foreign investments, led to capital-intensive production and a frequent scrapping and replacement of machines which incorporated the recent advances in knowledge. Entrepreneurs attempted to maximise the productivity of labour and reduce its usage in maintenance and repair.[25] This combination of forces led to the growth of a machine-tool industry based initially on advanced 'core' technologies from abroad. As Rosenberg summarises, the early American experience with industrialisation 'did not necessarily involve an inventive process but, rather, one of transferring technologies that had already been developed elsewhere . . . [this] involved frequently complex questions of selection, adaptation and modification'.[26] In his view, interchangeability and machine production of specialised parts may be seen as a substitute for both general and highly-skilled craft labour, a substitution induced by the demand resulting from an ever-expanding, increasingly wealthy and urbanised national market. Within this metal-based, adaptive industrial sector was harboured a learning process concentrated in an increasingly specialised machine-tool sector. This sector in turn acted as the major focus of further internal transfers of technology and skill from one industry to another, aided throughout by declining metal, power and transport costs. As Rosenberg puts it, 'the nature of the technological innovations was a system of interlocking changes that fed upon themselves'.[27]

It would seem that the major dynamic of the American manufacturing system prior to the 1860s or even 1880s cannot be properly grasped without some understanding of the essential nature of 'machinofacture'. The division of labour and of machinery is limited by the extent of the market. A growing population, rising incomes, and improved transportation of goods by river, canal and railroad, set the scene for the emergence of a metal-based industrial sector benefiting from rapid technical change (much of the 'core' of which was imported from Europe) and positive returns to scale, and characterised by the machine production of interchangeable parts, a process which after 1860 was to spread through a range of industries. Such successful machine emporiums as the Lowell Machine Shop of Massachusetts and the Amoskeag Manufacturing Co. of New Hampshire accepted orders for not only textile machinery as such but also turbines, mill machinery and a range of tools and equipment and, from as early as the 1840s, locomotives. Out of the initial requirements of the textile and railway sectors emerged a capital goods industry revolving around the major ingredients of metal-based manufacture – lathes, planers, boring machines. Added to these were the new techniques emerging from the arms enterprises – drilling and filing jigs, fixtures, taps and gauges, die-forging, the stocking lathe (to reproduce irregularly-shaped objects), milling machines (substituting for hand-filing) and the turret lathe (allowing a sequence of operations without a resetting of the part), all of enormous significance to the growth of the American system and applicable to a wide range of products, from clocks and agricultural machinery to typewriters and other office equipment. Machinofacture, that group of industries centring on the working of metal by machinery, dominated the industrial world of the late nineteenth century, and thus the huge diversity of final products which emerged was dependent upon a relatively small number of universal operations – turning, dieing, boring, drilling, milling, planing and polishing. A relatively small number of machines rather than men could perform these functions once a common group of technical problems associated with power-transmission, control, feed-mechanisms and friction-reduction had been resolved or reduced. The American system may be viewed as the technical and institutional cost-reducing solution to the central problem of mass production of a wide range of commodities utilising a relatively narrow range of processes. Rosenberg depicts this as a common technology with shared technical solutions.[28] It was the machine-tool manufacturers who, in essence, *did* 'normal technology', i.e. solved the day-to-day puzzles involved in improving the machinofacture operations of cutting, planing, boring and shaping of metal parts. This was at the very heart of the indigenous technological system.

From the early years of the century, when Francis Cabot Lowell introduced power-weaving into the Boston Manufacturing Company as a direct result of information gained in Manchester and elsewhere, through to the loosening of British restrictions on machinery exports in the mid-1820s and the subsequent increase in the flow of machinery, the American textile sector benefited from a long-term ease of supply. Between 1815 and 1860 perhaps 80 per cent of the decline in cotton-cloth prices resulted from technological progress, the bulk of which derived from Europe. In textiles, the use of the power loom reduced weaving costs by around 75 per cent. From the early years of the century historians may point to several specific innovations and their diffusion – original borrowings led to American improvements in flour-milling, the application of small water power sources to the timber industry and in the production of edged tools and equipment. Metal manufacturing by machinery was really a feature of the 1840s. From then, the enormous increase in population (from around 17 million in 1840 to 63 million in 1890) makes it difficult to separate out the effects of new machinery upon production-costs from the impact of positive returns to scale in newer industries.[29]

Sliding skilfully into and out of the Habakkuk thesis, Douglass North claims that 'progress was made whenever a machine could be devised to economise on high-priced labour by utilising richly abundant natural resources such as waterpower or wood. It was this ability to innovate and to modify the British improvement, coupled with the growing size of the American market, that reduced our manufacturing costs as compared to England's. The mills and factories could now begin to produce on a large enough and efficient scale to compete with foreign imports.'[30] Furthermore, North goes on to point out that indigenous knowhow plus formal education provided the human capital which was needed for the production of 'significant innovations' within industry: 'Human capital accumulation provided the essential skilled labor force that was so important for high levels of productivity in both manufacturing and agriculture.'[31] From this immediate backdrop, rather than from formal provisions for science and technology, came the new organisation of technique (continuous flow, interchangeable parts) and new innovations of technique (e.g. water turbines in the textile mills of New England, the standardised steam engines of the 1850s and onwards).[32] The diversification of production from cotton into steamboats and railroads provided the backward linkages for the development of technology-intensive sectors – general engineering services, steam-engine production, iron and steel, machine-shops. American historians emphasise the importance of the 'artisanal' culture of the machine-shops, within which was located the 'human capital' of the industrial economy. Until the 1860s this machine culture was unspecialised, and the transferable skills of lathing, drilling, boring and planing serviced the repair, design and 'parts' demands of the textile and other industries. Even the rise of a more specialised, factory style of heavy machine production during the 1870s did not prohibit the constant sharing of information within a coherent culture, the mechanisms of which were informal contacts, journals and periodicals and the apprenticeship system. Henceforth the culture of the practical engineer was increasingly linked to the growing communities of scientists and academic technologists. The machinist was transformed into the engineer, whose value in production lay not only in his possession of specific skills but of general organised knowledge.[33]

Hounshell has shown that the interrelated technologies of the American system spread from armaments to the production of sewing machines, woodworking, agricultural machines, the bicycle and, eventually, the automobile. Important to this diffusion was the network of skilled mechanics whose activities gradually moved outwards from the federal armouries to the large private enterprises which began to monopolise such fields.[34] The American system of manufacturing was forged out of the concern of American production engineers with the production of specialised machines, interchangeable parts, carefully coordinated sequences and materials flows and new methods of metal working. True mass production awaited the assembly line of the Ford Motor Company (1913) and the development of systematic marketing.

Chandler and others have argued that, especially from the 1880s, technical change was of overwhelming importance in shaping the highly centralised, functionally departmentalised and vertically integrated American firms of the late nineteenth century. The operational requirements of the new technologies demanded 'managerially operated business enterprises'.[35] In particular, it was the multifold impact of the railroads which created the qualitative shifts in the relations between science, technology and institutions which were in evidence from around the 1880s. From 2818 miles in operation in 1840, the railway network grew to 30 626 miles in 1860, 93 262 miles in 1880, to 166 703 miles in 1890.[36]

Until railroad expansion, most large industrial enterprises serviced the predominant agricultural sector. But henceforth industry itself became a prime consumer of machinery and equipment and the fastest-growing parts of manufacturing became dominated by a few integrated, bureaucratic enterprises.[37] Of most significance, the reliable movement of goods and passengers on the railroads, the maintenance and repair of stock, 'meant the employment of a set of managers to supervise these functional activities over an extensive geographical area . . an administrative command of middle and top executives to monitor, evaluate, and coordinate the work of managers . . . the formulation of brand new types of internal administrative procedures . . . the creation of the first administrative hierarchies in American business'.[38]

Low-cost railroad transport, combined with other communication technologies such as the telegraph and telephone, allowed enterprises to extend the geographical stretch of their markets and realise greater scale economies, and this was particularly so of those businesses whose products were amenable to the 'American system' of manufacture. A resulting need to integrate manufacturing and market functions was one of the forces which led to a general process of forward integration through merger. In particular, production of the high technology new-product lines of consumer durables (especially sewing machines and automobiles), and electrical and office machinery, was characterised by integration from production into the retail function.[39] It was precisely in such areas that the problems of property rights – over scientific, technological and enterprise-specific commercial knowledge – were greatest, and where enterprises were, therefore, that more likely to establish their own facilities for industrial research and development and their own methods of protecting the knowledge or information thus generated. Section 4.4 has shown how enterprise research and the patenting of technologies were of central importance to the generation of a substantial scientific enterprise in the USA during the late nineteenth century.

From the 1880s strategies of vertical integration amongst large scale enterprises associated with the 'American system' (i.e. those who employed continuous-process machinery) were common enough, and increasingly embraced the pooling or capturing of formal research and development capabilities. Perhaps merger and departmental organisation were a response to technical change and opportunity, but in turn new organisation furthered technical change and mass production. The firms of Singer, Westinghouse and McCormick appear to fit this approach, with merger necessitating the adoption of new management systems. In such strategies, the nurture and protection of high-technology intellectual property became an important aspect of enterprise policy.[40]

North concludes that in the years approximately 1860–1914 the growth of the economy was more significantly related to the efficiency increases generated by organised knowledge. The relatively high incomes of a large population 'made possible all potential economies of scale inherent in the technologies of the individual industries', even if they were mainly adaptations and augmentations of techniques originally transferred from Europe. As earlier chapters of this book show, adaptation is by no means a natural or necessary process in a 'receiver' nation and the high level of human capital formation may well have been the requisite of efficiency growth. As

North is quick to note, 'it is difficult to separate entrepreneurs from their environment and to credit the organisers with much responsibility for this expansion'.[41]

Patent data provide some indication of the emergence of an American inventive capability. The number of patent applications rose from 268 in the decade 1790–99, to 911 in 1800–1809, 1998 in 1810–19, 2697 in 1820–29, 6808 in 1830–39, to the 11 869 applications of the 'industrial' decade 1840–49.[42] Patents *granted* per annum grew from an average of 500 in the 1840s, to twice that in the 1850s, to some 4000 in the 1860s. Between 1860 and 1890 440 000 patents were granted. From 1865 George Westinghouse (1846–1914) registered no less than 400 patents, a proportion of which represented the stock of knowledge upon which the Westinghouse Electric Company was founded in 1886.[43] At this level of total patenting, which fails to capture the comparative *quality* of the American or other innovations, of 155 000 patents granted by 15 nations in the years 1842–61, some 22 per cent originated in America. Of more significance are figures suggesting the quality of innovative activity. For the years 1842–64 only 56 per cent or so of all patents filed in America were actually issued by the Patent Office, compared to a figure of around 65 per cent for Britain in the same period.[44] This suggests that even when compared to Britain, the American system was able to select from a host of new ideas emerging from the machine culture; during the 1840s America overtook Britain in the absolute number of patents granted. Furthermore, Table 10.1 suggests that there was an American patent invasion of the leading industrial nations.[45] Of 38 major foreign cities whose residents lodged 2057 patents in Britain during the three years 1867–69, only Parisian inventions outnumbered those of New York. Within Britain, only London and Manchester outranked New York, and only five British cities outranked Boston. Of all 38 foreign cities, the 11 American cities produced 40 per cent of patents in Britain.

We might reasonably assume that such foreign lodgements represented at least potentially competitive techniques and argue that such data adds to the work of Pred and others which emphasises the importance of urban-based innovation in America from mid-century.[46] Furthermore, by 1870 foreign patent lodgements in *America* had reached an all-time low – of 12 000 patents issued in America in that year only 5 per cent originated with residents of

TABLE 10.1 American urban patenting in Britain, 1867–9 (11 cities; 824 patents)

City	*Number of patents*	*American rank*	*International rank (38 cities)*
New York	393	1	2
Boston	140	2	3
Philadelphia	83	3	4
Brooklyn (NY)	54	4	5
San Francisco	33	5	8
Newhaven	30	6	9
Providence	25	7	10
Chicago	18	8	14
New Orleans	17	9	15
Washington	17	9	15
Cincinnatti	14	11	19

foreign nations. From this time the American patent invasion extended beyond the 'Atlantic economy'. Of the 172 000 patent applications lodged in *Germany* between 1877 and 1894, 30 per cent originated with foreign residents, amongst whom Americans ranked second only to the British.[47]

From this limited account we may draw some very tentative conclusions. Until at least the 1820s the relationship between the pursuits of science and national economic growth was forged *primarily* at the level of the natural-history enterprise. Between then and the 1840s the enterprise of science became the property of urban-based professional and intellectual groupings at a time when the economy continued to rely for its growth upon a heady mixture of expansionism (migration and new areas of settlement) and technological borrowings from Britain and Europe. For a time the technological capability of the American economic system, although dependent upon European innovation, ran ahead of provisions for scientific culture and research. From the 1840s there seems little doubt that the American machine culture, itself some function of earlier transfers of human capital, began to act as the centre for an inventive and adaptive technological capacity, and that this was largely independent of the contemporaneous movement towards the large-scale endowment of organised research. Exceptions to this unweildy generalisation were the innovatory institutions, such as the Franklin Institute, which undertook specifically 'applied' investigations.[48] But even here, attention was directed to 'civic', agricultural and infrastructural applications, rather than to manufacturing industry *per se*.

Significant contributions to *national* productivity from technological progress in manufacturing became a leading feature only after mid-century, but this was yet set within the context of an expanding, trading economy. One might postulate that in these years the investments in formalised scientific and technical investigation ran ahead of their contributions to economic growth. Until the 1860s America remained a technologically dependent economy, and new technologies arose within a machine culture which shared many of the features of the British model as outlined in Chapter 3. From then, when America emerged as a challenger to European technologies, the endowment of science was probably of greater economic import. Until then, American scientific culture was conditioned by its 'peripheral' status. Cultural activists were greatly motivated by their desire to capture some of the prestige gained by science in Europe. For such reasons, the history of the American economy and of its technologies is not readily explainable in terms of the history of its own scientific enterprise.

10.4 LOCAL IMPERATIVES AND THE NATURAL HISTORY ENTERPRISE: AUSTRALIA

> One main object we had in view in asking the assistance of Colonial Societies, was that they might inform us from time to time of the discovery of new sources of raw materials likely to be useful to our manufactures at home, and so extend the trade and commerce both of the Colony and the Mother Country.
>
> *P. Le Neve Foster, Secretary of the Royal Society of Arts, London, 1855*

Prior to 1890 the growth of the Australian economy was derived from the exploitation of natural resources and new lands, population growth, staple expansion and the inflow of British capital.[49] The early impress of administrative and intellectual authority from London had permeated many aspects of Australian life, and was by no means at odds with the arguments produced by the dissident Edward George Wakefield in his *Letter from Sydney* (1829) and *England and America* (1833): policy from the centre should create 'communities' rather than mere 'scatterings' of British people abroad, settlement of whom should benefit both themselves and the mother country.[50] In the words of D. N. Jeans, such components as the Colonial Land and Emigration Commissioners (founded in 1840) 'acted as a ministerial conscience in preserving the fixed principles of systematic colonisation'.[51] The gold-based economic boom of the 1850s, centred in Victoria, created a population explosion on a very small base (from 438 thousand in 1851 to 1168 thousand in 1861) and a measurable increase in both voluntary and government-aided scientific and technical institutions. This was the decade of the Geological Survey, the Museum of Economic Geology, the Victorian Institute for the Advancement of Science and the Victorian Science Board, all of whose membership was dominated by professionals and experts employed in the bureaucracy. European scientists such as G. B. von Neumayer and Ferdinand Mueller were attracted in by the opportunities advanced by official employment. Jan Todd argues that the coming of self-government in the individual colonies from the 1850s 'implied a new self-perception on the part of the colonies – one of permanence and self-sufficiency', and that this stimulated the development of universities (Sydney 1852, Melbourne 1855), official surveys and observatories, museums and the bureaucratic employment of a variety of scientific experts, all of which meant a larger and more ambitious Australian scientific community.[52]

Nevertheless, in the main the character of scientific discussion and investigation remained that of the natural-history programme of earlier years. Colonial grants for establishment and maintenance meant that the more applied institutions for acclimatisation, geology and botany overtook the earlier attempts at the provision of general intellectual forums; by 1873 the nearly 500 members of the Victorian Acclimatisation Society alone outnumbered the total membership of Australian 'philosophical' societies.[53] The scientific and commercial acclimatisers devoted themselves to research on a variety of potential staples – flax, silk, cotton, game and fisheries were all grist to the mill of applied Australian science, and were subjects of trial and experimentation from the earliest years of the acclimatisation societies to the initial discussions of the Australasian Association for the Advancement of Science, founded in 1888.[54]

The 1880s saw the emergence of specialist societies in the fields of engineering, geology, geography, meteorology and anthropology, together with the development of more specialist research endeavours in university departments. Melbourne established its doctoral degree in 1887. But the research programme had been led by men such as Richard Trelfall, previously of the Cavendish Laboratory, William Haswell, a fòrmer pupil of Huxley, and David Masson, whose 1880s research was an extension of the work of European intellectuals such as Ostwald, Arrhenius and Vant Heff.[55] The bulk of Australian academic scientists were fully engaged in state service more generally, from the direction of museums and observatories to the provision of advice and expertise for a variety of Select Committees.

It was the immediacy of depression and economic isolation as much as the ideal federation which forged new relationships between science, technology and economic growth. As in America, officially supported scientific endeavour had led to a variety of good research. The Departments of Agriculture acted as centres of research and employment for chemists, plant pathologists, entomologists, bacteriologists and biologists, some at least of whose findings represented work of international repute. Such structures as the technological museums spawned fruitful investigations, including that of Richard Baker and H. G. Smith in phytochemistry.[56] But the 1890s depression, resulting from a disastrous combination of the collapse of Melbourne's land and construction boom, the stoppage of capital inflow from Britain and a fall in wool export prices, stimulated a restructuring and diversification of economic and scientific enterprise. Recovery was conditional upon the application of new knowledge or techniques in food processing, wheat production, refrigeration, urban and marine services and gold production. Reginald A. F. Murray, FGS, formerly government geologist for Victoria, who had delivered a paper on 'Mining in Victoria' before the 1891 AAAS, now advertised his services as a consultant chemist/engineer for new gold processes anywhere in Australia.[57] New alluvial mines at East Gippsland received advice from H. H. Howitt, FGS. E. M. Krause, FGS, FLS, Professor of Geology at the School of Mines, Ballarat, was appointed manager of the General Gordon Mines in Western Australia, which was evolving new methods for gold processing.[58] Such engineers as T. H. Houghton, MICE, used the facilities of the scientific community to advertise the importance of new chemical processes of gold extraction.[59] Such examples, which may be multiplied with ease, suggest that de-

TABLE 10.2 Patenting in New South Wales and the Australian Commonwealth, 1888–1904[63] (Selected years; 6112 patents; percentages)

Patent category	*1888–89 New South Wales*	*1894 New South Wales*	*1904 Commonwealth*
Manufacturing	15	11	25
Mining	2	2	6
Food processing	5	8	4
Agriculture	8	8	10
Transport	13	10	12
Other	57	60	43
All patents	1538	720	3854

pression induced a redeployment of scientific expertise away from direct state employment into the agricultural and primary sectors of the economy.[60] Redirection of informed interests resulted in information dispersals, e.g. as early as 1896 J. C. F. Johnston of the Australasian Institute of Mining Engineers had sold some 10 000 copies of his *Practical Mining*, mostly within Australia.[61] A result of a complex of such activities was not only an expansion of gold production, but a general economic recovery led by increased *productivity* in the rural sector.[62] Table 10.2 suggests that the process of recovery involved a shift of inventive activity toward manufacturing and mining in particular. This was concurrent with a move of both patenting and scientific efforts from Victoria towards New South Wales, a process which continued in the years up to the Great War.[64]

Thus, by the early years of the twentieth century, nearly 60 per cent of Australian patenting was located in areas of production associated with recovery. Two further points might be made. Firstly, by the early years of the twentieth century the registration of inventions per capita in Australia compared very favourably with that of large, industrial nations at that time.[65] Second, as suggested in Table 10.3, inventive activity in Australia became less and less dependent on foreign registrations during the years 1904–18. Seemingly, Britain and America retained their relative shares of Australian patent lodgements only at the expense of other nations in the international patenting system.

There seems to be little doubt that a combination of Australian

TABLE 10.3 Patent registrations by foreign residents in Australia five-year periods, 1904–18, (percentages)[66]

Period	*All patents*	*All foreign patentees*	*British patentees*	*American patentees*
1904–08	13673	34	10	11
1909–13	18645	26	9	10
1914–18	16246	22	8	10

inventiveness and foreign transfers was sufficient to supply the economy with the specific information required for increases in productivity over a range of sectors. More precise breakdown of the patenting data suggests that innovations in manufacturing tended to cluster around areas which served the trading and agricultural economy – agricultural machinery, refrigeration equipment, urban services.[67] Efficiency increases arising from such applications peaked around 1910, and from thence economic growth was associated with large capital inflows, railway development, irrigation works and further expansion of settlement.

10.5 SUMMARY AND CONCLUSIONS

At different times in both America and Australia, a concentration upon staple exploitation led to an emphasis upon natural history and acclimatisation. This was seen as eminently suited to both the economic needs of the regions and the psychological requirements of the scientific workers concerned, whether temporary visitors or permanent settlers. Only when market demand was sufficient to generate a significant manufacturing sector did the character of scientific investigations alter at all dramatically. Whatever the phase of economic expansion or settlement, the development of the scientific culture was set against the backdrop of a technological capability highly dependent upon initial borrowings from Europe.

Differences between the two regions of settlement arose as a result of differences in the temporal, spatial and demographic characteristics of their overall growth. Timing was of importance insofar as it determined the nature of the technologies available from external sources. A scientific culture capable of absorbing and transforming the textile techniques of the 1820s would not necessarily provide the expertise or information needed to settle the metallurgical or chemical technologies of the 1880s and 1890s. Spatial factors are not only important in explaining the emergence of national markets (which require some reduction in effective distance) and therefore the timing of the demand for industrial technologies, but are also relevant in encouraging or hindering the communication and dispersal of discrete information. The supply of the latter is an ingredient in raising the technological capability of any economy at any time.

Much of the observed difference between the cultural and economic experiences of America and Australia may be explained in demographic terms. Around 1860 the population of the Australian colonies stood at just over one million, approximately one-thirtieth of that of the United States. If Melbourne was Australia's Philadelphia, then it was only in the 1850s that its population size approximated that reached by the latter city in 1800. Furthermore, the Australian colonies were far more urbanised than was America. As a result, in Australia urban communities spawned scientific endeavour in the context of a small demand for manufacturing technologies and slim colonial budgets; in America urban culture and invention arose amidst the artisanal homesteader and a sturdy machine culture, all of which combined to generate the 'American system' once demographic change and transport improvements had lifted the level of effective demand for manufactured industrial products.

In both America and Australia, official support for either science or technology was of greater significance than in Britain. Once freed from British interference and injunctions, the American states, and eventually the American government itself, boosted the scale and quality of American scientific and technological research. Smallness of size combined with colonial direction to reduce the extent of official support in Australia. The sparsity of the scientific audience and the weakness of the market for manufactures meant that Australian scientific development was intermitent, heroic, dependent and pragmatic.

It seems that in areas of recent settlement, the links between scientific enterprise and technological systems in the nineteenth century were not as strong as those found in Europe. In the United States prior to the 1880s the major driving forces behind technological change appear to have been migration and the ever-expanding market, which together yielded a high rate of skill-formation and a homogeneity of taste in the marketplace which encouraged the growth of the 'American system'.[68] From then the complex demands of city growth, the new organisation of business generally, and the internalisation of research and development capabilities within the firm led to a rapid rate of technical change, yet associated with the efficient adaptation of European core technologies.[69]

It was the absence of such market features rather than the colonial connection or a backwardness in science which dictated the slightness of Australia's technological efforts within manufacturing at this time. Formal colonialism was of little importance here. By the 1890s the proximity of Canada to the United States and large investment flows had led to a far greater degree of technological dependency there than was discernible between Australia and the United Kingdom.[70] At the same time there was a sufficient strengthening of the Australian technological system to permit an increase in adaptive and innovative capacity over a range of industries and some weakening of the technological ties with Britain. But the very small scale of events did not lend itself to wholesale improvements in machinofacture along North American lines. Rather, Australian enterprise quite naturally focused on staple export or services, on machine-importing for a range of sectors and on staple-processing, and very few enterprises in the latter category were either of a nature or a size to justify formal scientific or engineering research establishments.[71] Although both Australia and the United States were at different times regions of recent settlement and frontier economies, by the late nineteenth century their paths had diverged entirely. Massive differences in scale, location and distance spelt similarly great contrasts in the relationship between science, technology and economic development.

11. Twentieth-century Aftermaths: Science, Technology and Economic Development

11.1 DEVELOPMENT AND UNDERDEVELOPMENT

CONCENTRATING on the eighteenth and nineteenth centuries, this book has illustrated that the analysis of economic development must embrace perspectives still unconventional to economics as a modern academic discipline, but common enough to students of economic history, social history and technological change. In particular, analyses of the bounded rationality of economic agents, of knowledge or information, of institutions and of technology, might be at the centre of the study of the development process.[1] All of these are themes introduced in Chapter 1 and illustrated at various points throughout this book. The present chapter expands on some of these themes through brief consideration of research and development, institutional structures, technology transfer mechanisms and alternative patterns of development in the twentieth century.

Gunnar Myrdal and many other writers argue that continued growth and change in advanced economies creates underdevelopment in poor nations as a dynamic, objective, interacting process. For Myrdal, the international trading economy works in reverse of classical free trade doctrine: trade means that industrial nations with long technological leads and developed marketing facilities and superior information (external economies) merely out-compete existing small-scale industries in poor nations, who are then faced with high unemployment and chronic balance of trade problems – underdevelopment. A *co-existence* of rich and poor nations is converted into an *interaction* through the pervasive workings of the international economy. The expansion of trade

> strengthens the rich and progressive countries whose manufacturing industries have the lead and are already fortified by the surrounding external economies while the underdeveloped countries are in continuous danger of seeing even what they have of industry and, in particular, small scale industry and handicrafts priced out by cheap imports from industrial countries.[2]

It is possible to identify two general perspectives on the problem of underdevelopment in the twentieth century. The first accepts that underdevelopment exists and that it is associated with the past and present relationships between poor and rich nations. Sutcliffe summarises this: 'Many of the conditions which are generally associated with the state of underdevelopment derive directly or indirectly from the relations of underdeveloped with developed industrialised nations.'[3] Indigenous production systems are challenged by manufactured imports; protectionism and politics curtail the export promotion attempts of many underdeveloped countries; available techniques are produced within high-knowledge frameworks and are transferred by institutions which are inappropriate to the existing resources of poor countries. Moreover, given that population growth is a fundamental obstacle to development, even where growth exists this demographic feature may also be attributed to contact with already industrialised countries – modern medical practices and products have been successfully transplanted where technologies 'more relevant to economic development have not'.[4]

The second perspective is that, for some set of reasons which varies between writers, underdeveloped nations have failed to 'respond' to the development potential offered by interaction with the advanced industrial nations. Without in any way agreeing with it, Brett has summarised this position as it manifests itself in optimism about the recent success of the Newly Industrialising Countries (NICs):

> By adopting rational market-oriented strategies, and by fully exploiting both domestic and external opportunities, they have broken out of the vicious circle of underdevelopment and given the lie to those who have asserted that this can only be achieved by those willing to 'de-link' from the capitalist world economy . . . the many failures in the Third World are the result not of the necessary operation of the capitalist market mechanism, but of the 'irrational government interventions' introduced in response to incorrect theories designed to correct its supposed imperfections.[5]

Of course, inefficient or corrupt government can be interpreted in turn as an inherent characteristic of the value systems and cultures of underdeveloped nations. Elkan has suggested that 'there do seem to exist greatly varying propensities to develop among different peoples at any one time, and that these may be more easily explainable in non-economic terms'.[6]

As highlighted in Chapters 8 and 9, several of the characteristics of the present international economic system stem from relationships of power and status laid down during the long swing of high colonialism (circa 1870–1939). We have already noted that colonialism conditioned a great variety of social, political and cultural processes in the colonial nations. Unfortunately, many of such processes have now been interpreted as inherent 'characteristics' of present-day underdeveloped economies, and this has led to arguments which explain poverty in terms of 'culture' or 'value systems'. There is little doubt that in most of such colonised countries, *dualism* – that is, a state not only of distinction between but separatedness of, the old and the new, the city and the country, the modern and the traditional, the enclave and the hinterland – evolved as a social as well as an economic syndrome.[7] Colonial forces served on the one hand to regenerate the existing social hierarchy (in some cases, as with the British in India, solidifying an originally more fluid, *ad hoc* situation), and on the other to diffuse Western machinery, knowledge and values. We are not here especially concerned with the niceties or otherwise of European racial, cultural or social attitudes towards subject peoples. The point is that the inheritance of colonialism may be seen in both the present state of the international economy and the present economic and political conditions of a great many of today's poor nations. Once this is allowed, then analysis might shift to (a) evidence of a true and sustained escape from such an inheritance; (b) the strategies and policies required in order to escape continuing underdevelopment.

11.2 SCIENCE, TECHNOLOGY AND INSTITUTIONS

All too often, the noticeable shift in policy within the underdeveloped countries towards human resource development, institutional change and technological reformation, and away from a reliance upon large increases in investment only, has been interpreted as a rejection of the goals of industrialisation and economic development. This is not the case at all. Policy has simply shifted towards areas more likely to yield economic development in the manner defined in Chapter 1 of this book. At one point in the draft plan of the Sixth Indian Five Year Plan (1978–83) it is stated that 'in the next phase of development, it will no longer be appropriate, in the light of past experience, to formulate the principal objectives of a particular plan period merely in relation to a specified target of growth for the economy'.[8] The Plan targeted a lower rate of GDP growth, but did not neglect the goal of economic development in any way, for it clearly aimed at a reduction of unemployment (rather than net transfer payments to the unemployed), a rise in living standards (rather than net transfer payments to the poor) as well as the fulfilment of 'basic needs' through service sector development (e.g. the clean drinking-water campaign, now evident in Delhi and elsewhere), adult literacy, health care, rural roads and housing, all inputs into any reasonable definition of overall economic development. It is increasingly believed that the developmental impact of a sustained linkage between improved employment, education and techniques is far greater than the seemingly weak linkages forged between savings, investment, entrepreneurship and industrialisation in past development strategies.

Any strategy must take account of the international economy. The NICs of today have clearly passed through import substitution industrialisation to export-promotion industrialisation. This alone suggests that trade is of importance. The expansion of trade allows reallocation of resources to areas of greatest efficiency, a fuller employment of total resources, economies of scale, skilling and technological improvement and a variety of efficiency increases arising from increased competition. As we have seen, trade expansion does not *in itself* deliver such results. The real world is composed of differential levels of protectionism, different mechanisms of protectionism, political power and transnational corporations.[9] As such, trade theory has little to say about the spread of growth from the trading and exporting sectors into other areas of the economy, and nothing to say about how imported, better technology becomes utilised within the importing nation. Whether a trading sector is also enclavist almost certainly depends on institutional arrangements, both within and without the nation. Thus, where increased export income arose from plantation-type production, as in Java in the later nineteenth century or Northern Rhodesia in the early twentieth, the 'spread' effects of growth were strictly limited.[10] Again, as illustrated in Chapter 6 above, the nature of the export product is important. Cotton or coal had more potential than beans or bananas in the mid-twentieth century.

It would seem that any consideration of overall strategies and the place of research and development (R&D) leads to an advocation of development projects, institutional reform and technology transfer, to be overviewed by government intervention in both these fields directly and in the area of international trade and investment. In advanced nations there is evidence of a positive relationship between technology transfer and indigenous technological advancement. But even within the shared production functions of the advanced nations, transfers do require adaptive R&D if they are to be incorporated into the marketplace. Japan and Australia, in different ways, are still beneficiaries of an earlier strategy of development through imported technology. But this does not mean that complementary relationships will necessary be sustained between indigenous technology and transferred technology in today's underdeveloped nations. Much depends on the wider institutional framework.

If it is agreed that continuing economic underdevelopment is a result of 'the dominant influence of archaic social institutions and the consequent failure to utilise available technical knowledge and skill',[11] then institutional reformation becomes a leading element in development planning. Marx and Veblen both saw institutional inertia as a retarding factor in technological progress, but from our historical evidence we would wish to add three modifiers to any argument which links institutions and development, which are that technology is itself institution-bound, some technologies are institutions (what is a factory?) and that physical technology may be adapted to conform to the functional requirements of existing institutions.[12]

Of course, for two other great thinkers, Adam Smith and John Stuart Mill, the market was an institution. It was the only institution which could sufficiently provide the information, the checks and balances, required for economic expansion. The exercise of power in non-market institutions (e.g. government) was destructive of economic growth insofar as it impinged upon the market mechanism (e.g. tariffs, poor relief, colonialism). For Smith, it was the scholar-gentry institutional system and the resulting restriction of markets which held back development in China, 'a much richer country than any part of Europe'.[13]

In a recent study, Gerald Scully has tested the hypothesis that efficient resource allocation requires that 'all resources be owned exclusively by private individuals and that these resources be transferable'.[14] For 115 economies institutional variables measuring political liberty, economic freedom and so on are correlated with rates of economic growth.[15] Scully concludes that economic development requires an open, liberal institutional framework which minimises 'intervention and regulation'. Smith and Mill would feel quite happy about this. Five Year Plans do not seem to come into the picture. But we are highly suspicious of this sort of approach. Increasing GDP growth is not economic development. A reasonable quantitative fit does not satisfy the need for a proper causal argument. Nations of lowest income per head may be nations of lesser income per head growth and lesser freedoms. If so, there is a problem of identification. Does a high level of economic growth, occurring for whatever reason, induce greater freedoms? As we have seen in this book, Scully's present 'capitalist and democratic' market economies at one time passed through definite phases of industrial revolution, during which markets were undeveloped and interventionism rife. The rise of the classical economic doctrine, upon which writers such as Scully depend, was not a perfect measure of freedom's increase, but a plea for its increase as part and parcel of a belief in the increasing efficacy of market mechanisms. I would disagree with Scully in his claim that Adam Smith would reject out of hand Myrdal's argument for a 'complete transformation of the values and attitudes that people hold and in the institutions that foster those values'.[16] Radical institutional reform – in this case, towards market mechanisms in a highly conditional manner – was precisely what the *Wealth of Nations* was all about. The difference between Myrdal and Smith is surely that Myrdal is writing of nations relatively and dynamically underdeveloped and faced with tremendous political and economic exigencies. Smith was writing of the most powerful trading economy in the world at that time. Myrdal was never analysing poor nations with highly developed market institutions and wide degrees of political freedom who had to choose to maximise liberty or maximise statism. In most cases, the latter was the fact.

Institutional change may be stimulated by either technological change or by sustained government activity. Present advanced technologies pose far greater challenges to existing institutions than did the transfers between nations prior to 1939.[17] The mechanisms of technology transfer are embedded within institutional arrangements of other nations and derive from entirely different ideological frameworks. The open economy is subjected to foreign ideology as well as foreign technique. Whilst institutions may surely adapt to such imperatives, the Gerschenkron or Hirschman approaches emphasise the role of government in promoting institutional reformation.[18] Planned institutional innovation may be hampered if the costs of change are high, if rewards are low or risky, or if the authority of key elites or power-brokers were to be threatened. Traditions may work to reduce costs. It has been argued for Meiji Japan that 'the traditional moral obligation in the Japanese village community to cooperate in joint commercial infrastructural maintenance has made it less costly to implement rural development programmes than in societies where such traditions do not prevail'.[19] New information could spread through existing cooperative networks very effectively, as in the case of knowledge of silkworm breeding technology to the 16-village association or *Matsukawa-gu-mi* near Matsumoto in 1870.[20] But we have also seen that strong government-'created' legitimising 'traditions' in Meiji Japan. More recently, China has undergone programmed institutional change legitimised in cultural terms, whilst the industrial success of South Korea was clearly some function of the regulated intervention of the government of Chung Hee Park, which drew strength from a general, national acceptance of the need for industrialisation.[21]

11.3 TECHNOLOGY TRANSFER: JAPAN AND INDIA

Technology transfer dominates the twentieth century. It is a major subject of discussion for managers, academics and planners, as well as for labour movements or governments as a whole. On the one hand there are those who see the transfer of advanced technique as a panacea or a promise, as a mechanism of economic development. Alfred Marshall's identification of the international economy as the 'engine of growth', might in this perspective be translated as a realisation of the promethean power of technological transformation. On the other hand, there are those who emphasise the difficulties involved in technology transfer, the low probabilities of success, the frequent occurrence of spectacular failure. No one has uncovered a convincing model of technology transfer and its relationship to economic development, but recent history *appears* to offer examples of the profound impact of imported technology on indigenous economic and technological systems.

In Japan after the Second World War, the intent of the Occupation Forces, led by the USA, was to debilitate the great *zaibatsu* firms, which were seen as hosts or harbingers of war and imperialism. The response to this within Japan during the 1950s was to reinvent the wheel, that is to create the complex entrepreneurial combination of the *keiretsu*, which provided a novel institutional framework for the continuous, sustained interactions of producing units, banking facilities and trading companies or *sogo shosha*. It has been convincingly argued that it was such *institutional innovation* which allowed the Japanese to partake of the technological and economic boom of the 1950s and 1960s.[22] Soon the ability of the Japanese to first copy and then adapt unto themselves the best techniques of the West became evident enough, and increasingly Japanese goods sold in the West not only because they were cheap, but because they were better. Japan may well have continued to lag behind the West in the *creation* of better hardware, chemical processes, etc., but it led the world in adaptive adoption of technique, in marketing, in redesign. More recently, the NICs of Asia and South America have demonstrated a broadly similar pattern of development. Particularly in Asia, very high levels of industrial growth and productivity, based initially on fairly labour-intensive, mature forms of production (e.g. textiles) emanating from the West and from Japan, have been associated with a growing technological maturity, innovative institutions, private initiative and increasingly sophisticated government guidance.

With the outbreak of the Korean War in 1950, procurement payments in Japan joined with US aid as a major recovery device in the Japanese economy.[23] From then, the fast growth of the economy was to be closely associated with technology importing. Between 1955 and 1973 there was a sixfold increase in Japanese GNP in real terms. Measured in terms of royalty and license payments, technology-importing grew at the rate of 30 per cent per annum between 1951 and 1961, and continued to grow at over 10 per cent per annum during the 1960s and 1970s.[24] Two features should be noted at the outset. Technology transfer took place alongside a rapid growth of domestic R&D expenditures and patent registrations.[25] Secondly, the pattern of industrial growth followed that of gross fixed capital formation (measuring the expansion of investment), the large extent of which was almost certainly induced by technical opportunity.[26] By the early 1970s the overall efficiency of the Japanese economy was probably equivalent to that of the USA, although this is a very general judgement and does not entail any equivalence of *technology* between the two nations. The efficiency of the economy from that time was illustrated in the Japanese response to the 1973 oil crisis. Despite a very high dependence on oil and oil-related inputs, continued increases in productivity led to a higher industrial growth rate than in any other capitalist economy and the continuation of a buoyant export sector in the face of a general depression in world trade.

Most of the technologies transferred into Japan during the 1950s and 1960s had already passed through Vernon's monopoly stage of the product cycle, i.e. relevant technologies were standardised and generally available through trade or licensing.[27] Between 1956 and 1972 over 176 000 license agreements were signed, involving well over $3 billion in royalty payments. These were concentrated in such areas as electrical machinery, chemicals, pharmaceuticals, synthetic fibres, transportation equipment and iron and steel. In 1969 alone Japan paid $350 million in foreign licenses of technology.[28] During the 1960s perhaps one-third of Japanese industrial production was directly dependent upon foreign technology, a dependency at its greatest in the export industries, public works and in the industries nominated as of strategic importance by the Ministry of International Trade and Industry (MITI).[29]

The official targeting of industries was justified on the grounds of the high costs and risks associated with the new technologies and their long lead time, emphasised against the backdrop of market failure. That is, micro-industrial policy involving repeated interventions in the area of R&D and technology transfer was deemed necessary in order to push resources into areas where the potential for fast productivity growth was high but where the normal (unaided) operation of market forces would not sustain the necessary levels of investment and infrastructural provisions. Once a strategic industry (not a firm) was identified, MITI had available to it a package of support mechanisms, from the stimulation or funding of related R&D, to tax benefits and loans, or market research.[30] Seemingly, it was the emerging synergy between government and enterprise which permitted the quick transformation from technology transfer to import substitution to export promotion.

The underlying success of the Japanese economy and the strength of the linkage between imported technology and indigenous technical capability almost certainly lies in the original large scale of operations, the high investment rate, the continually expanding export sector and the peculiar organisation features which define the relationship between the public and private sectors. The reformulation of the *keiretsu* firms dominated the 'economic miracle' years of the 1960s. Such coordinated enterprise permitted cooperation in decision-making (e.g. in the selection of appropriate foreign technologies), the pooling of information, research and other resources between medium-size production units and the sharing of risks in areas of new, high technology. With the growth of the bank-centred *keiretsu* (e.g. Fuyo, Sanwa) the links between production, trade and financing were strengthened through cross-holding of stocks, interlocking directorships (especially between banks and their client firms), sub-contracting between 'upstream' and 'downstream' elements (e.g. between large assembly plants and small component-parts firms), by numerous lender–borrower financial dependencies and by a variety of contract buyer–seller relationships.[31]

Many commentators argued that by the early 1970s Japan had 'caught up' with the West in terms of major manufacturing technologies. The model of high growth associated with high investment levels induced by technology transfer was no longer applicable.[32] However, this particular turning-point in Japanese economic history is difficult to locate. In 1973 the ratio of technology exports to

TABLE 11.1 Technology imports to Japan as a percentage of technology exports from Japan (1971–81), in ¥ billion measures

Year	*Tech. imports as % of tech. exports*
1971	500
1975	252
1976	213
1977	204
1978	157
1979	181
1980	150
1981	148

technology imports for Japan stood at 0.13, for Britain at 1.04 and for the USA at 9.30, and a very large number of foreign patents were still being lodged in Japan.[33] There is little doubt that by the early 1980s Japan had mounted a patent invasion of other major industrial nations. Thus, by 1981, Japanese patent lodgements as a proportion of all *foreign* patent lodgements in the USA stood at 32 per cent, in Germany at 30 per cent, in the UK at 21 per cent and in France at 16 per cent.[34] Nevertheless, as Table 11.1 suggests, even as late as 1981 Japan remained a net importer of technology, although the trend towards a less dependent structure is also evident.[35]

Efficiency-increases arise from a variety of sources, of which machine technology is only one. The organisation of production in Japan may have yielded efficiency increases which brought Japanese productivity above that of the USA or European nations, but there is evidence of a remaining technology-lag in many areas. Although there was certainly a decline in the demand for foreign technology in the machine sector, where Japan probably had caught up with best-technique, there was if anything an increase in the demand for overseas technology in the high-growth areas, especially electronics and chemicals.

As we have seen, Japanese trading companies are closely tied to production processes and possess tremendous intelligence systems. In several cases Japanese firms have bought into university research in other nations and obtained control over intellectual property. Other enterprises will continue to establish their own research facilities in advanced industrial economies, where they can employ advanced skills and techniques generated originally in the technological systems of the host nations. Again, whatever the restrictive policies of the US government are on technology-transfer to Japan, individual firms will continue to allow a drain of their technologies through licensing to Japanese enterprise in exchange for access to Japanese finance or marketing expertise. Given the intense competition between large firms in Japan, then such transfer mechanisms may permit continuous technological progress in a world of restrictions and catch-up. The link between enterprise culture and technology transfer will undoubtedly be strengthened through cooperative agreements for private–government research activity.

In contrast to the already-developed industrial base of Japan at the time of the Korean War, at Independence the Indian government inherited an economy where less than 7 per cent of GNP came from factory production, which employed 2.4 million people, or less than 2 per cent of the population. Within such sectors as textiles, machine tools and iron and steel there existed a very wide range of technique.[36] Technology policy was seen as an integral part of overall planning and was directed to upgrading of existing techniques, promotion of new techniques and adaptation of selected techniques to the needs of the Indian economy. From the early days it was recognised that technological transfer and progress necessitated changes in the wider institutional structure. The First Five Year Plan stated clearly that: 'The problem is not only merely of adopting and applying the processes and techniques developed elsewhere, but of developing new techniques specially suited to local conditions. Certain forms of economic and social organisation are unsuited to or incapable of absorbing new techniques and utilising them to the best advantage.'[37] At this stage, transfers of technology would flow directly into the modern sector, as 'technological filtration' would mean that superior technique trickled down to small-scale suppliers or users of equipment produced in the modern sector.[38]

It is almost certainly the case that technology transfer processes have failed to result in effective linkages between Indian industry and Indian government-financed R&D institutions. Whilst much expertise exists and many facilities are made available, a common complaint is that R&D is highly-centralised and bureaucratic and has little relationship to the basic needs of the Indian economy and its people. Up to 90 per cent of all R&D is financed by government;

a very high proportion of this is in the area of 'pure' scientific research, and perhaps 70 per cent or 80 per cent of all government R&D finds its way into a small elite group of industries – atomic energy, space and defence.[39]

As a result of such tendencies within the system, many Indian commentators have advanced strategies which are focused on the basic needs of the economy, the development of the indigenous technology system and the control of foreign influence. Foreigners have invested hugely in India since 1948, taking advantage of arrangements which permitted foreign firms an equity share of up to 74 per cent in Indian enterprise. By the early 1970s the remittances abroad of such foreign firms had reached very large proportions, composed of profits, dividends and interest payments, as well as payments on royalties and technical knowhow.[40] It was also claimed that many technical collaborations actively competed with existing available Indian capacity. The Public Undertakings Committee found several instances of obvious incompatibility. Indian Oxygen Ltd obtained foreign technology for oxygen plants when Bharat Heavy Plants and Vessels had the relevant knowhow. Texmaco were involved in a similar collaborative arrangement for boiler manufacture when BHEL had the relevant knowhow.[41]

The basic critique of technology transfer through TNC activity in India is twofold. First, it is argued that indigenous capabilities are significantly underutilised, as user industries turn towards foreign sources for machinery and designs. Secondly, it is maintained that capabilities are diverted from basic needs, 'to link with the global network of multinational corporations'.[42] Thus 30 per cent of the research expenditure of Hindustan Lever is 'corporate' activity, undertaken in the commercial interests of the TNC Unilever as a whole. Similar trends are found in the operations of Union Carbide, Philips and Chloride India. Yet such enterprises are in receipt of government subsidies and incentives.[43] In summary, it is argued that the TNC mechanism does not generate independent capacity but does co-opt existing capabilities, drains the nation of resources and brainpower and directs attention away from basic needs.[44]

P. M. Pillai has suggested that 'one may be outwardly impressed by the remarkable degree of adaptation of imported technology to local conditions and needs but the situation leaves much to be desired when one looks into the process more closely'.[45] There is evidence of adaptive capacity in the engineering sector, where techniques have been scaled down and adapted to the use of local material and skills. Menon and others have found evidence of technological innovation associated with the decreasing dependency on foreign technology in the fertiliser plants in the public sector.[46] Again, this sort of effect seems strongest where foreign technique did not involve foreign ownership or control. In the capital goods sector more generally, foreign-controlled firms remained more technologically dependent than the domestic firms, who on the whole seem to have been ready to enter a new technological world on the expiry of original agreements with foreigners.[47]

The Indian case provides very mixed evidence about the utility of technology transfer generally, and its relations to indigenous technological capacity in particular. It should be remembered that the Indian economy is a very large one. The Indian case differs from that of Japan in major respects. The status of industry in the economy and the level of industrial investment were and are quite different. The bargaining power of Japanese firms (bounded by a sturdy Japanese government presence) was far greater than that of Indian firms, and TNC activity was of far less importance. Technology transfer into Japan was predominantly through trade and licensing. Income per head in Japan was far closer to that of the technology 'donors' than was income per head in India.

11.4 ALTERNATIVES: APPROPRIATE TECHNOLOGY AND THE CHINESE MODEL

There are other recent models, perhaps at a polar extreme to that of Japan and very different from that of India. After the 1950s and the great Russian interlude, China chose to close her doors. The combination of a Communist ideology with the vast resources of a large nation, the relatively slight importance of foreign trade in total economic transactions, and a lack of foreign exchange, together meant that this somewhat drastic strategy was not entirely unrealistic.[48] It is, indeed, interesting and relevant to note that from around 1984 World Bank and other distant but informed reportage has reassessed the growth performance of the Chinese economy during the 1950s to 1970s to yield a very respectable industrial and overall result.[49] Indeed, basic sectoral data suggest that China underwent an industrial revolution. In China, the emphasis was on making the most of what was available, regionalism, employment, reskilling and resource-saving. Institutions were devised in order to induce technical change and information-diffusion – the most infamous of these being the rural and urban communes – and a form of planning emerged which, for all its faults, confusion and disasters, was less top-heavy and centralised than the counterpart Russian model. By the later 1970s the Chinese technological system could begin to cope with an influx of advanced technology, and the four modernisations ideology emerged.[50] Since then, from the mid-1980s, the economy has increasingly opened up to the world at large, and massive high-tech transfers have been negotiated with the US, Europe and Japan. Although this more recent trend has been often seen as a denial of the past, it may also be interpreted as an outcome of the reasonable success of the earlier 'alternative path' phases of development. Despite the personal tyrannies associated with Maoism and much else, the transformation of the enormous Chinese economy does seem to have proceeded relatively quickly (in historical terms), and the stage is set for a considerable period of technology transfer.

Until at least 1978 the institutional structure for Chinese science policy followed certain ingrained assumptions. Amongst these were the notions that agricultural growth should not draw resources from other sectors, that intellectual work must not expand at the expense of productive labour, and that decentralised industrialism would boost employment, encourage rational resource-usage, small-scale industry and labour mobilisation.[51] At the level of the county or below, intermediate and traditional technologies would be supplied by means of the diffusion of known information more efficiently, rather than by the formal application of R&D resources. Under this schema, formalised ST resources were directed at the modern, large-scale industries of the centrally-planned sector of the economy. From the late 1970s the ground rules have shifted. For instance, the success of the production responsibility system in the rural sector prompted the central committee of the Chinese Communist Party to mount a programme of urban economic reform in October 1984. These reforms opened up industrial enterprise to a far greater degree of choice, competition and incentive. The impact of the latest reforms on the institutional character of Chinese R&D has yet to be clarified. Generally it is true that the movement towards indirect administration, financial incentives and competition in the economy at large has spilled over into the management and pursuit of science.

Until recently, central government agencies and ministries commanding large resources in major research institutes have been considered the appropriate mode of organisation for formal R&D. Typically, the State Scientific and Technological Commission selected 38 key research projects for sustained financing under the Sixth FYP. Certain central agencies have acted as information, feedback and diffusion centres for large enterprises or bureaus in the planned sector. But, given the lack of market signals in the economy, this system may not have been so successful in providing well-placed support to enterprises growing out of the sub-county level, an important phenomena of the late 1970s and 1980s. In addition, lack of cooperation between institutes in the provision of components and ancillary services necessary for the integration of new technologies into production in the modern sector of the economy may well have reduced the overall appropriateness of the system.

In March 1978 the National Science Congress convened by the Central Committee of the Communist Party successfully proposed a 'Draft National Outline programme for Scientific and Technological Development 1978–1985', intended to usher in a 'spring-time of science' by increasing rewards and professionalism. Even more importantly, Hu Yaobang's speech at the Second Congress of the China Association for Science and Technology, held in March 1980, isolated the need for increased specialised, intellectual skills, large-scale training and research programmes and greater freedom of scientific enquiry.[52] From that point, scientists were increasingly appointed as advisers to provincial and central governments. In water conservancy, economic botany and demography, provincial governments responded to the findings of research teams.[53] In 1980 scientific and technological workers in Shanghai put forward 42 major proposals to government, over half of which were adopted. The Scientific and Technological Commission of Liaoning Province set up 22 groups to serve as advisory bodies for the provincial leaderships. A 'brains trust' in Zhuzhou city, Hunan Province, submitted to the city authorities a whole series of proposals for the application of technique to local industry. In Yichun County, Jiangxi Province, a scientific and technical service company was set up.[54]

The most significant adjustments to new conditions relate to those designed to further the links between formal R&D and *all* sectors of industry. Just as industrial enterprise has been opened up to competition and incentive, so too research institutes have been encouraged to form closer links with production through contract research and the setting-up of joint research–production teams in a manner reminiscent of the Japanese system. A large number of factories and mines have set up their own research institutes. Liaoning Province claims an expansion in the number of research institutes and laboratories run by enterprises to more than 800, with 13 000 staff members.[55] Autonomy in the research institutes was first experimented with in Sichuan and influenced by changes in the organisation of productive enterprises. Increasingly, institutes could plan their own research projects and undertake tasks assigned to them under contract by production units to which they were not otherwise affiliated. Institute-enterprise contracts included those for scientific research projects (designing and testing), for transfers of research results, technical services, training of personnel and product development.[56]

It is not clear in exactly what manner the reformulated goals of 'readjusting, restructuring, consolidating and improving' the economy have impacted on the established institutions for scientific research. The National Science Council remains the paramount scientific and technological executive agency, recommending to the Premier national policy and consolidating research and development activities. As a funding agency the NSC promotes a range of research, supports personnel training and monitors the institutes of the Academia Sinica, the Industrial Technology Research Institute and the Science Industrial Park.[57] The NSC has been particularly concerned with the promotion of basic research in the fields of energy resources, agricultural production, medical and public health, defence and high technology industry. In 1984 the NSC claimed to have implemented some 1301 research projects in universities and research institutes.

China's new 'open policy' or *kaifang zhengce* of the 1980s provides a good example of the great variety of modern technology transfer mechanisms. The first half of the decade witnessed over 2000 Sino-foreign joint-venture agreements between state enterprise and foreign companies, but these were in the main simply the organising component of a package of mechanisms which involved licensing, technical assistance grants and trade.[58] At the same time, many of the foreign 'donors' of technology argued that there were a host of obstructions on the transfer process. Dealing with China meant difficulty in repatriating profits, unclear legal restrictions, over-bureaucratic procedures and so on. Of course, such 'difficulties' were evidence of the more general conflict of interests between parties which is inherent with transfer processes in a complex world. The reason that such conflicts were more acute in China was because of the speed with which new policies were implemented and because China's transfer arrangements were based on the firm belief that the strategy must benefit the state overall.[59]

11.5 THE DISTRIBUTION OF SCIENCE AND TECHNOLOGY RESOURCES

There can be absolutely no doubt about the skew in the world distribution of scientific and technological resources, away from underdeveloped nations and towards the advanced economies. The maldistribution goes beyond that accountable for in terms of differing incomes, incomes per capita or industrial outputs per se. The ratio of R&D facilities to total sales for Indian manufacturing enterprises lies around 0.7 or 0.8 per cent at best, far below the 3 per cent and more of nations such as America.[60] Japan, the USA or Germany commit six times the proportion of their GNP to their R&D resources than do Venezuela or Argentina, two of the more successful underdeveloped nations of the 1960s and 1970s. In addition, it can be argued that the institutional context of such expenditure is far more refined, complex, sustained and efficient in advanced nations, and more responsive to the needs of the manufacturing industrial sector. In underdeveloped nations highly formalised R&D resources are also highly centralised, and – if only because they are frequently the major form of employment and career advancement for large numbers of talented scientists and engineers – are responsible, if at all, to the political requirements of the centre rather than the intermittent production needs of heterogeneous industrial sectors. If relevant R&D is indeed essential to the sustained economic transformation of large underdeveloped nations, then the distribution and institutional structure of the world's R&D enterprise is not well suited to the needs of all but the most advanced nations.

By 1980 the developed nations possessed 23 times the R&D personnel per capita of the underdeveloped world and five times the physical R&D resources even when national incomes are used as the deflator. In absolute terms underdeveloped nations possessed only one-fifteenth of the physical resources devoted to ST in the advanced industrial nations. Japan and Germany, the two fastest-growing of the major industrial nations during the 1960s and 1970s, were in the mid-1960s yet to establish R&D facilities on a scale comparable (in per capita terms) to that of the two major powers. Both employed far less of their ST personnel in productive enterprise.

But of all nations, Japan showed the fastest growth of gross expenditure on R&D (GERD) as a percentage of GNP during the 1960s (with no real growth in the USA or UK), and in absolute terms Japan had overtaken France and Britain by 1970–72.[61] At the same time, a high proportion of Japan's GERD was financed from the private enterprise sector. The proportion of government financed GERD to total GERD in Japan by 1971–72 was around 16 per cent, in UK and the USA the figure was around 55 per cent, in France 62 per cent, in Germany 48 per cent.[62] Thus by the early 1970s Japan was already identifiable as a fast-growing R&D nation with a low level of direct government-financed input.

In contrast, even over a period of a decade or more's effort, an underdeveloped region's relative ST resources position might remain little changed. Thus, despite the tremendous efforts of individual Asian governments, the rise of Japan and the NICs, the relative production of R&D scientists and engineers in Asia remained almost constant throughout the 1970s, moving from 17.4 per cent of the world total in 1970, to 18.6 per cent in 1975 to 18.5 per cent in 1980.[63]

Clearly, the least-developed nations depend more on government financing of R&D than do the NICs or the major developed nations. All nations, however low their income per head, attempt to distribute a significant proportion of R&D expenditure to 'fundamental research', although the usage of this term varies greatly. Again, the wealthy or growing economies spend a significantly high proportion of R&D resources in the productive sector of the economy.[64] Vast amounts of R&D in the Philippines or India are directed to 'general services' as defined by UNESCO, i.e. 'various private or government establishments serving the community as a whole'.[65] High amounts of expenditure on the services sector is almost certainly equivalent to a large fraction of 'productive sector' spending in Pakistan. All of the developing Asian nations have utilised higher education as a tactical sector for the production of engineering, technical and agricultural skills, far more so than either the USA or Japan.

All Asian nations devote a very high production of R&D facilities to agricultural development, industrial development and energy exploitation – in the case of India these three directly economic categories take 54 per cent of total funding, in the Philippines 64 per cent and in Singapore 88.3 per cent. In the USA these three categories in combination take up only 9 per cent of funding. At a world level, Japan is the exceptional industrial market economy, devoting some 64 per cent of its R&D expenditure to these key economic aims.

Generally, and with the exception of the complications of military expenditure, it might be suggested that underdeveloped nations attempt to throw their severely limited R&D resources into the goal of economic development against all odds. On the basis of stark aggregates and highly generalised categories it may not be argued that R&D resources in underdeveloped nations are focused on non-economic goals. General indicators suggest that the problem for the underdeveloped world is lack of funds rather than wholesale misuse of them.

11.6 THE INSTITUTIONAL CONTEXT (1): JAPAN AND THE USA

In Japan, over three-quarters of all expenditure on R&D is raised in 'productive enterprise', but only 65 per cent of the total is expended in that sector. If there is a redistribution effect it is from the private to the public sector. In contrast, in the US only half of all R&D expenditure is formed by productive enterprise, but that sector spends 73 per cent of all financing.[66] The income transfer is from the public sector to the private. We would hypothesise that the prime reasons for this contrast are: (a) that the Japanese spend less on R&D for defence and space and, (b) that government–business institutional links and 'enterprise culture' ensure that relatively small amounts of strategically-placed public finance yield large amounts of private enterprise activity.

Any explanation of differences in the economic import of American and Japanese R&D requires a somewhat closer look at the contending institutional framework. This is of particular concern at the present time. In the 1960s American institutions, character and decision-making were seen as models for Europe.[67] In 1990 Japanese institutions, character and decision-making are seen by many Americans as vital components of Japanese industrial success, to be emulated if at all possible.[68]

In the USA, anti-federal philosophies led to a lack of overall control over the massive government R&D effort which was combined with relatively close relationships between universities and industry and between research and development.[69] These inherited characteristics, together with the military-inspired origins of science policy, means that a great amount of US government-financed research is actually located in the private sector itself, rather than in public research institutions.[70] Apportioning a very large proportion of its budget to atomic, space and defence objectives, the US government provides large resources through specialised agencies (the most famous being the National Science Foundation, established five years after the end of the Second World War) and into selected industries by means of formal contracts with individual enterprises. This is a less common mechanism in, say, Japan or West Germany, where defence is a smaller item. The nature of contracts vary with industry, task and administration, from simple fixed price contracts to costs – plus fixed-fee to costs-plus-incentive-fee contracts. The best known of government agency funding of private industry research on a massive scale was the Apollo Project, which established NASA and ensured that Americans were the first men on the moon.[71]

From 1969 the guidance of federal funding into industry has been dominated by the NSF through the industrial programme of its Division of Intergovernmental Science and Public Technology. Its task has been to encourage work of the large-scale, long-term, interindustry and high-risk variety so avoided by individual corporations! However, a complex combination of subsequent spending cuts, an increased concern with social acceptability, a fall away in economic growth set against the surge of the Japanese economy (influencing the US trade imbalance) and a belief that existing institutions were failing to provide for even fundamental research, led to renewed attempts at institutional rejuvenation during 1985.[72] Increased centralism might, it was thought, reduce the resource replication involved with research regulation between agencies, gain economies of scale, set standards for excellence and direct R&D towards both fundamental research and selected applications. A 1985 report by the Presidential Commission on Industrial Competitiveness recommended that agencies should be rationalised in order to create a Department of Science and Technology to administer all non-defence/space government-funded R&D.[73] However, neither history nor contemporary comparison with Japan or the European nations provide robust evidence of the economic efficacy of centralism. The barriers to a closer relationship between US resources for R&D and the US industrial economy and those which reduce feedback *from* the latter *to* the former, are almost certainly founded in the enormous emphasis on space/defence (three-quarters of all federal R&D financing in 1986) together with the workings of the agency-contract system in the civil sector.[74] It is possible that dispersements of super-contracts from a super-agency would merely complicate and therefore further stultify the institutional context of R&D in the US.[75]

Between 1980 and 1985 expenditure for R&D in Japan rose at an annual average rate of over 11 per cent, amounting to over 3 per cent of GDP in 1985.[76] 79 per cent of this was raised by the private sector and 67 per cent was spent in that sector. The surge of spending during 1985 came not from government but from the private sector (a 15.6 per cent increase over 1984).

Major agencies contract with individual corporations in a similar manner to that of Europe or USA. Thus Mitsubishi heavy industries were the main contractors for the STA space-station project.[77] The difference lies with the strategic position of MITI. Some loss of status by MITI during the 1970s and a need for industrial restructuring led to the MITI policy document *Visions for the 1980s*, which emphasised high technology as a key to restructuring of the entire economic system.[78] Through combinations of protective legislation and research organisation, MITI has targeted 14 high technology industries for accelerated growth by means of ten-year R&D projects (e.g. in superphosphates, robotics, ceramics, biotechnology, semi-conductors).[79]

Sheridan Tatsuno had depicted MITI's role as the regulator of a product cycle. That is, MITI enters the infant industry stage of a new high technology area using legislation, research cartels and so on (e.g. satellites), withdraws from industries at maturity in order to allow market forces to operate (e.g. fibreoptics, precision instruments) and re-enters to monitor a regulated decline in the industry (e.g. steel, mining, petrochemicals).[80] But, financing and contracting directly is only a portion of MITI's agency influence over industrial R&D. Other institutional strategies which may be isolated include incentive allowances, the technopolis concept, telecommunications provisions, the encouragement of venture capital and selective import promotion. Much of such activity, although directly related to the stimulation of private industry R&D, is not in itself captured by formal R&D aggregates of the sort summarised in Section 11.5.

For both MITI and Japan, a key industrial R&D strategy for the future lies in the idea of the *technopolis*, variable and decentralised locations for the generation of economies of scale and specialisation across the whole range of R&D activities, from pure science to technology adaptation and diffusion.[81] The projected new cities symbolise the major contrasts between Japan and other nations' arrangements for the generation of knowledge in the ST system and its application to industry. Schumpeterian-like leaps in technology arising from formalised R&D may or may nor generate significant increases in industrial efficiency and output. An efficient 'soft' infrastructure for the transfer, adaptation and diffusion of technique, is seemingly vital to sustained increases in industrial productivity.[82] Again, even MITI's ascendency over the new approach does not spell overwhelming government financial involvement. Each technopolis will be financed by a combination of municipal funding (gathered from taxation and corporate donations), central government depreciation allowances (30 per cent on machinery) and subsidies to joint research projects between corporations and prefectural industrial research laboratories.[83]

Finally, it might be considered that Japan's success in financing a large proportion of total R&D from private industrial funding lies as much in distribution as in the minute workings of institutions. That is, Japan spends a large amount of money on a relatively small clutch of sectors or projects. Despite Western assertions to the contrary, the Japanese government is not financially omnipresent throughout industry; most of industrial R&D comes from *very* large corporations and is highly concentrated in electronics, chemicals and the automobile industries. Restructuring of the economy entails more the selection of new targets than the broadening of the government's R&D enterprise.[84]

From technicians to managers to scientists, human capital formation in Japan takes place primarily within the enterprise system itself. It is here that the swiftest response to dramatically changing conditions is located. An American firm which introduces Japanese-type quality control circles into its structure soon finds that it must do more. Similarly, any attempt to graft attributes of the Japanese educational system on to that of the US will involve further changes. These should not to be made at the expense of the taxpayer if it may be reasonably argued that the success of Japan in human capital formation owes more to private enterprise culture than to formal educational systems. It may be more productive to use US public funding in order to stimulate private enterprise activity, not in the field of external education but in that of internal training. Nevertheless, it would still have to be acknowledged that in the US this strategy would have to be formulated in the absence of important institutions which, in Japan, forge the almost indefinable synergies between government and business.

11.7 THE INSTITUTIONAL CONTEXT (2): THE NEWLY INDUSTRIALISING COUNTRIES

It does not require much imagination to see some identity between the Japanese model of development via sustained technology transfer and that demonstrated in the very recent past by the NICs of Asia and Latin America. Indeed, the recent coincidence of a high industrial growth rate in Asian NICs and strong trading links between that group and the developed nations (the so-called export-promotion strategy for industrialisation) has called into question earlier core–periphery or 'dependency' theories which postulated a dynamic process of underdevelopment in the Third World nations as a necessary outcome of their incorporation into the world economy.[85] Industrialisation-through-dependency seems to suggest that transfers of technology do indeed act to the betterment of the technology receivers. Success in the NICs can be interpreted as resulting from close trade–technology links with advanced economies, in turn a cause of the revamping of NIC technological systems, a vigorous export-orientation and a purposeful and sustained intervention of national governments in the overall planning process. Although the effects of formal colonialism in the past may still be debated, it is now increasingly argued that institutional changes in today's international economy (particularly the rise of the TNCs) may have induced the sustained industrialisation of nations receiving both standardised and advanced technologies. An original (say 1950s and 1960s) predominance of standardised technique gives way to an export-orientation and to a more sophisticated indigenous technological system capable of not only absorbing but also adapting and further developing sophisticated technologies. Sustained economic development (industrial revolution) through technology transfer is again on the agenda.

Particularly successful NICs, such as South Korea, seem to illustrate the replaying of the Japanese model.[86] The direct impact of the Japanese economy on the Asian NICs is also of especial importance – by 1982 Japan accounted for 27 per cent of the total exports of the Association of South East Asian Nations (ASEAN) and 22 per cent of the group's imports. More noteworthy, Japan supplied 56 per cent of all the technology imported into Korea during 1962–83, the USA only 23 per cent.[87] Foreign technology was then centred on standardised product lines (in the Japanese manner) and patents, therefore, tended to be neglected in favour of importing and direct investment.

But by 1983 Korea was paying out $US150 million in royalties alone, representative of approximately 1.5 per cent of world technology trade.[88] Government policy undoubtedly influenced the extent and direction of technology transfers. Prior to 1978 Korea controlled technology importing through the authorisation of import licenses, given on condition that the import fulfilled fixed development criteria.[89] After 1978 trade and technology were liberalised as a result of the efforts of the Ministry of the Economic Planning Board to stimulate the heavy and chemical industries.[90] As high technology lines emerged in Korea the transfer system was further opened up, and in 1981 the Foreign Capital Inducement Act lifted most restrictions and offered positive incentives to the high-technology areas of the economy. Between 1977 and 1983 the import of technology grew at an annual average rate of increase of 11.5 per cent.

All of the major NICs (Hong Kong, Taiwan, South Korea, Singapore, Indonesia, Malaysia, the Philippines, Thailand, Brazil and Argentina) have exhibited higher-than-average growth rates despite recession. This achievement is marked by export-led growth and the workings of a variety of technology-transfer mechanisms. Yet in almost every other respect NICs vary enormously. It is all but impossible to identify a NIC in terms of personnel skills, resources, income per head, geographical location or investment levels. During the 1970s the features which identified them, apart from their growth rate, were the high proportion of direct foreign investment in manufacturing sectors, export-orientated manufacturing and an increase in State economic interventionism. As in Japan, planning serves as a guideline for action in the *private* sector, and policies result from collaboration with large companies – the so-called 'concensus' effect. The 'state sector' has also taken on a distinctly Japanese aspect. As Andràs Hernàdi has put it recently,

> The so-called state enterprises act in the manner of private enterprises. In contrast with many highly industrial countries, it is not considered as inevitable that they should operate at a loss and, in fact, where such is the case, they are made profitable by transferring them to the private sector. Nor are they created in the way that is customary in the highly developed countries, that is, in a move to 'save' a big company in a contracting industry. The motivation is much rather the desire to launch an industry that does not yet exist in the country. Similarly, while the basis for the inflow of foreign capital has been the market conditions, the opportunities for the return on capital, i.e. the attraction of the economy, the State also played a considerable role through offering various preferences, the establishment of export enclaves and free trade zones.[91]

Technology transfer in NICs has involved definite adaptation capabilities. Imported machinery is used on longer or over more shifts, at a greater speed, with shorter down-times and with greater labour inputs for maintenance and repair. In some cases the setting of the machine has not been a 'factory' at all, and quite sophisticated machinery has been erected within proto-industrial environments; 'individual machines were set up in the households and by using workers and families as home workers it was possible to achieve large savings in fixed capital, while at the same time using a large mass of cheap manpower'.[92] The labour-using bias is measurable. In cotton spinning and weaving the number of workers per spindle or loom is decidedly higher in NICs than in, say, Japan. The Korean and Taiwanese electronics industries use far more labour than in the

West or in Japan despite the predominance of transferred technology.[93]

By the 1970s the first group of NICs had clearly transcended their original place in the product cycle (above) and economic activity had become more centred on telecommunications, iron and steel, shipbuilding, office equipment, watches and instruments and so on. In South Korea the leading edge of industrial growth had passed to electronics, cement, oil refining, artificial fibres and petrochemistry. By the 1980s computers and other technology-intensive industries were introduced quickly. A similar 'maturity' of economic structure is seen in Singapore. Yet such nations were heavily dependent upon imports of oil, which may indeed have encouraged technological change in search for substitutes. From 1978 South Korea's Ministry of Energy and the Energy Research Institute were engaged in research directed at conversion from oil to coal and nuclear energy.[94]

Benefits involve costs, and even a success story such as Korea did not escape this truism. Industry might well have grown at a faster rate than did the indigenous technological system. The transfer of technologies by transnational corporations in many cases involved exclusive rights to supply materials and intermediate goods or prohibited the rights of the receiver to develop the same technologies themselves. That is, they forestalled certain feasible 'downstream' effects.[95] Of a large number of Korean firms surveyed in the early 1980s, 32 per cent reported evasion of technology transfer commitments by TNCs and a large number complained of insufficient information.[96] The indirect costs of technology transfer may include overpricing of inputs, loss of profit on capitalised knowhow in the case of equity participation, commissions paid to licensers for use of their distribution and sales services and so on.[97]

11.8 EXITS: ON PURSUING THE PAST

Eighteenth-century and early nineteenth-century European history is a story of national conflicts, social conflict, revolution, compromise and the rise of social class.

In Britain, the emergence of a strong trading class and a phalanx of industrial capitalists was the harbinger of a spirit of individualism, competition and entrepreneurship. It might be recalled that

Adam Smith's *Wealth of Nations* was primarily a plea for the removal of mercantilist restrictions and a move towards market institutions.[98] Transfers of capital, information and institutional innovations towards the industrial sector created the 'industrial revolution from below'. The resulting ideology of 'liberalism' – which in its pursuance of market forces, often went well beyond the greater caution of Smith – legitimised and clarified the world of the businessmen. From the middle of the nineteenth century the need to find new markets and raw materials, the emergence of high expectations amongst commercial groups, and the development of new technologies of travel, communication, and conquest led to the commercial expansion which was to so disturb the pattern of economic development in Asia and elsewhere.

Industrialisation under such conditions of relative 'economic forwardness' (in Europe) depended upon a prior assemblage of a host of 'givens' – demographic change, commercial expansion, institutional change, social adjustment. Industrialisation resulted from the coalescence of many activities performed by many individuals in a 'liberal' market environment of social and economic choice. Choices were made by both consumers and producers, and both exerted their sovereignty within the context of retained national sovereignty. Although not to be entirely neglected, technology transfer did not loom very large until the industrialisation of Germany, the USA and of Russia in the last decades of the nineteenth century. In contrast, industrialisation under conditions of relative economic backwardness was replete with tensions and transfers, conflict and the tortuous decision-making and control devices of the State itself. Technology transfer together with the threat of political encroachment (through colonialism) created traumas in both economies and societies. It is in this international context that the Japanese industrial experience belongs. Without a retention of economic sovereignty, the exogenous shocks of international trade and investment, colonialism and knowledge-flows, could not lead to national economic development.

Sovereign governments intervene in economies under pressure or threat from the outside world. The behaviour of the State is not necessarily *primarily* dictated by the direct economic costs and returns of its interventions. Given a threat to the entire system, governments will act if they calculate that industrialisation will yield less social and economic costs than would inaction. They will view the cost of State inaction primarily in political terms. Similarly, we may stylise the conditions for entrepreneurial behaviour in the private sector, assuming more than rudimentary markets for inputs and outputs and a degree of commercial competition. Here the cost of inaction (failing to follow a leader) is loss of profitability. Now, we may postulate that the behaviour in the political system will directly impinge upon behaviour in the business system. The very retention of control and sovereignty by government during early industrialisation will serve to reduce risks for the entrepreneur. During revolutions businesspeople do not look to the long term, but rather emigrate, place their capital overseas, invest in land, or hoard. By their activities governments may also reduce the costs of production and the costs involved in scrapping the equipment of the pre-innovation production system. Without interventionist decisions in the political system under conditions of sovereignty (e.g. under colonialism, civil war, foreign war or revolution) decisions in the business system are likely to be adverse to industrialisation.

Hence the claims about lack of 'responsiveness' or 'entrepreneurship' in colonial or post-colonial poor nations. As a historical example, the emergence of entrepreneurial activity in Meiji Japan occurred precisely when the government had completed its first phase of interventionism during the industrial drive (i.e. around 1881–84). The costs of production in the private sector were lowered by government provision of information, physical infrastructure (especially transport) and capital. Risk was reduced by government guarantees, social controls and political continuity. Thus, the environ for innovative investment decisions was a reasonably good one by the late 1880s and 1890s. The explanation of the rise of Japanese entrepreneurship at that point lies not in a *long tradition* of business-merchandising or profit-seeking, but in the short phase of government interventionism.[99]

It might be too easily concluded that Asian historical experience suggests three things only: that a loss of *economic* sovereignty (not equivalent to, although strongly associated with actual political sovereignty) is not compatible with sustained economic development via industrialisation; that technology transfer, widely defined, was central to the development of Asian economies; and that *therefore* a certain economic sovereignty is required if nations are to secure to themselves the great *potentials* (and they are only ever that) of the international economy. Industrialisation requires contracting, negotiation, arbitrage between buyer and seller, government and government, firm and transnational corporation, politics and people. In particular there must be a continuity of *institutions* and policies in order to settle new technique into an indigenous economic framework and to allow the development of an indigenous technological system.

But there is more to our story than this. Such messages are the outcome of an *interpretation* of economic history. What *does* seem universal is that 'intellectuals' must be satisfied, for they are behind the policies and ideologies of many political movements. Politicians or military elements may at times wield power, but that power is most truly secured or legitimised by argument and by ideas. A policy programme based on a coherent vision is more likely to survive an industrial drive than a series of abrupt, pragmatic policy lines. Cultural suasion and legitimacy lasts longer than the charisma of individuals.

Explicit political ideology may be of less real importance in intellectual debate during the industrialisation of an underdeveloped nation than the very wide variety of conflicts which can arise from different positions on basic economic issues. Therefore a fundamental *developmental* (that is, an economic as much as a political) task of central authorities is to provide a mode of industrial change which will allow intellectuals to accommodate themselves to new circumstances. Intellectuals must be gainfully employed.

If the relation between decision-making at the centre and implementation in the regions is not close, a large amount of effort and expenditure will be involved in controlling the divisive forces unleashed by development through industrialisation, especially if such forces are linked with non-metropolitical political opposition. Pre-industrial social groupings must be acknowledged and integrated into the new political economy unless the armed force of the new state is sufficient to expel them altogether, and this further adds to the non-productive expense of the industrial drive. Accepting the importance of technology, as argued here, an alternative might be to consider 'appropriate technology' as more than an economic affair. Technology as imported or created may be appropriate not only to factor endowments, the extent and nature of markets, existing technique, and structural and spatial balance, but also to the political requirements of the nation concerned. In Meiji Japan, non-political economic factors forced a technology which created relatively few political strains. In today's underdeveloped nations economic criteria combine with international political factors to encourage a technology which possesses a very high political profile. If the capital cost of a technology is great, and if the capital cost of controlling the social unrest created by its introduction is also great, then the real and opportunity cost of transfer is huge and the process of industrialisation will not bear a family resemblance to that of Meiji Japan.[100]

Traditionally, too much emphasis has been placed on physical capital as a determinant of technological and industrial success. The attack upon such an emphasis is now well under way, and we have attempted to bring to the fore of analysis the enormous importance of mental capital, human capital and institutions which provide the creation, diffusion and adaptation of information.

Eighteenth-century history may not be *directly* transposed to the present day. This is so for two major reasons. First, technology is fundamentally different: twentieth-century technique embodies a far greater scale, capital, knowledge and skill content than that of the eighteenth century. Second, the mechanisms whereby technology is transferred from one locus to another have also changed radically. The role of individuals, of basic, generalised skills, of trade and of markets seem far less than the role of transnational corporations, of highly specific, scientific knowledge and of non-market contractual obligations. Nevertheless, as suggested at the outset of the book, technology does transfer and must diffuse into existing systems if it is to have any significant impact on the overall process of economic development. *Enclave* technology is as wasteful (and wasting) an asset in the twentieth century as it ever was in the eighteenth century. Of greater importance, any emphasis on technology diffusion and institutions helps us to perceive the realities behind the rhetoric of the twentieth century.

The place of Japan within the framework of modern world technology is a case in point. An important debate is taking place in Japan on the problem of restructuring, of generating a creativity which will lift the technical capability beyond that of other nations, given that presently Japan appears to have 'caught up' with Western best-technique and therefore can gain little from continued transfers in.[101] But we would suggest that, in fact, Japan may lead the industrial world without a cultural revolution. Greater world-level creativity in Japan might well lead to the existence of a series of Japanese best-techniques.

But let us instead assume that Europe and the USA retain their seeming leadership over best-technology. As long as Japan retains its ability to penetrate that technology and to transfer it, then the ultimate industrial outcome depends on the institutional context. Here, Japan has enormous advantages, for the nation retains a definite lead in the area of organisations designed to diffuse information, adapt technologies and relate technology to capital and to markets. This may well ensure that Japan continues to benefit from the sustained development of new technologies in the USA or in Europe.

Perhaps the most important lesson of the historical record is that technology transfer from regions of high income to regions of low income has never been a panacaea, an all-healing and final remedy for poverty and distress. What might be suggested is that technology transfer fails very frequently, that institutions and ideas are as important to technology transfer as is foreign exchange, and that any form of technology transfer which merely swamps and does not integrate into existing indigenous technological systems is likely to fail. Today, economic policy in poor nations is to an extent influenced by this recent historical experience of rich nations, and policymakers must view the promise seemingly inherent in technology transfer against an historical background of very mixed results. As is well-known, technology transfer is increasingly dominated by large transnational corporations, whose ultimate goals may or may not coincide with (or at least agree with) the needs of large, complex technological systems such as that of India. But this dominance of corporations is a relatively recent phenomenon. In our more detailed and distant examples of eighteenth-century Europe or nineteenth-century America and Japan, the mechanisms whereby knowledge and technology flowed from nation to nation were highly individual, variable and, at times, quite effective in transporting industrial modernisation from one location to another. Yet most instances of technology transfer failed. Whilst the transfer mechanisms and the nature of the technologies being transferred prior to 1914 were quite other than those dominating the world economy in the late twentieth century, it is nevertheless possible to construct a series of historical tendencies and relationships which are not without interest in the present day. After all, then as now, transferred technique does have to be understood and utilised, does require *diffusion* within receiver nations, does have to relate in some functional way to existing skills and, indeed, to help redefine the basic economic and industrial resources of a receiver nation. In both the eighteenth century and our very recent past, men, manuals and machines moved.

Derek de Solla Price once wrote that 'there is much more past to live in if you discuss politics and wars than if you discuss science'.[102] The present book has tried to place the history of science and technology within a fairly broad canvas, one which certainly allows for the importance of both politics and wars. Because the stress is on industrialisation, then certain second-level emphases appear quite naturally. There is more here on technology transfer and diffusion than on initial invention. There is some acknowledgement of the importance of institutional and technological changes which are not embodied in large amounts of fixed capital. The physical imperatives of commanding technologies are themselves seen as conditioning the pattern of economic development. The art of organisation is visualised as of at least equal importance to the act of 'creativity', however that term may be defined. Industrialisation is interpreted as a phenomenon which appeared in sporadic form, but which, where successful, is then sustained within those wider socio-economic environs which permit experiment and provide continuity. Sovereignty is vital, for it must be at least feasible to interpret the present as better than the past, the future as better than the present.

Appendix 1: Inventions and Innovations – Britain c. 1700–1880

1709 Abraham Darby, coke smelting, slow diffusion, widely used in 1760s and 1770s

1720s Newcomen's pumping engine (first working 1712)

1730 Increased momentum to parliamentary enclosure

1733 Kay's 'Flying Shuttle'

1736–7 Ward's sulphuric acid process by combustions (patented 1749)

1746–9 John Roebuck's sulphuric acid plants in Birmingham and Prestonpans (bleaching), (lead chamber process)

1748 Lewis Paul's carding machine (cotton) patented

1749 Huntsman steel smelting

1750–70 Fivefold increase in turnpike road mileage

1759 John Harrison's chronometer (£20 000 prize awarded)

1760 Bakewell's stockbreeding

1761 Bridgewater Canal (completion cost £250 000)

1764 James Hargreaves' jenny (patented 1770)

1769 James Watt steam-engine (lower fuel costs, separate condenser)

1769 Richard Arkwright's waterframe

1770 Ramsden's screw-cutting lathe (Maudslay improvement 1797)

1771 First patent to obtain mineral alkali from common salt

1773 Manchester–Liverpool Canal

1774 John Wilkinson's device for boring cannon (applied to cylinders)

1774 James Watt's improved steam engine (1776 applied to blast furnace) in Birmingham

1775 Boring mill (for Boulton and Watt cylinders)

1779	Samuel Crompton's mule
1779	Cast-iron bridge of Abraham Darby at Coalbrookdale
1780	Hornblower's compound engine (high pressure cylinder to Watt engine) first working
1782	Watt's 'parallel motion' (beam and piston-rod of steam engine)
1782	Jethro Tull's geared seed-drill (slow diffusion)
1784	Cort's puddling process (puddling achieved from 1779, not taken up until c.1810), for wrought-iron manufacture
1784	First mail coaches; 1780s stage-coaches generally.
1784	James Watt – patent rotary motion (invented 1781), and put in operation in cotton-spinning factory
1785	Edmund Cartwright's first power loom, first patents 1786–8, adoption deferred to investment booms of 1823–5, 1832–4. 1833 = 100 000 power looms in UK, and 250 000 handloom weavers
1786(c.)	Mechanical power introduced to calico printing in printworks of Liversey, Hargreaves & Co.
1786	William Murdoch's steam carriage (roads)
1787	Wilkinson's first iron boat
1788	Chlorine bleaching in Manchester
1788–95	Application of new techniques in road making (Macadam, Telford)
1780–90	Development of high-pressure engines
1790	First steam-rolling mill erected in England
1790	Cartwright's wool-combing machine
1793	Cotton-gin invented in USA
1795	Joseph Bramah's hydraulic press
1797	Hen Maudslay invents carriage lathe
1797–99	Tenant's process for bleaching (chlorine over lime)
1800	Richard Trevithick's high-pressure steam-engine
1800	Hen Maudslay screw-cutting lathe
1801	General Enclosure Act
1802	Bramah's rotary wood-planing machinery
1803	Woolf's compound/high-pressure engine
1803	Begin building Caledonian Canal
1804	Trevithick's first successful railway locomotive
1810	Development of food canning (tinplate, sealed by soldering)
1812	Central London Streets first lit by gas
1812	Chapman's invention of bogie-truck
1813	Hedley's steam locomotive on smooth rails at Wylam
1814	Power printing
1814	Stephenson's steam locomotive
1815	Humphry Davy's safety lamp
1815	First efficient colliery locomotive
1818	Blanchard's copying lathe
1819	Steamship crossing of Atlantic
1820	Patent for rolling wrought-iron bars into edge-rails

1824	Portland Cement, invented at Wakefield
1825	Stockton and Darlington Railway (horse-drawn)
1828	Neilson's hot air blast furnace (iron-smelting)
1828	Brunel's Thames tunnel
1830	Liverpool and Manchester Railway
1830	Richard Roberts automatic mule patented; £12 000 spent on second patent; fully self-acting (one man could work 1600 spindles) Mid-1830s widely adopted
1831	Phillips' contact process (sulphuric acid) patented, Bristol
1836	Gossage's absorption towers (for hydrochloric acid in Leblanc process)
1837	First railway telegraph
1834–43	Commercialisation of superphosphate manufacture by J. B. Lawes, Rothamsted
1839	Application of electric telegraph to railway system
1839	Nasmyth's steam hammer
1840	Penny Post
1842	Ransomes' application of steam power to threshing
1842	Establishment of Rothamsted Experimental Station
1843	*The Great Britain* first screw-steamer to cross Atlantic
1845	First Agricultural College, Cirencester
1845	First successful compounding of Watt engine, McNaught of Bury
1840s	Innovations for drainage of heavy land (cylindrical clay-pipe, pipe-making machines)
1848–78	Joseph Swan's development of carbon-filament lamp
1851	Great Exhibition
1851	Channel Cable
1855	Paris Universal Exhibition
1856	Perkin-aniline dye (first synthetic)
1856	Bessemer Converter
1861	Machine riveting patented
1864	Siemens-Martin open-hearth furnace
1865	Mushet's use of tungsten and vanadium in steel manufacture
1860s	Steam-roller (roads)
1869	Suez Canal opened
1870	Deacon's catalytic oxidation of hydrochloric acid gas (to chlorine), bleaching
1872	Mond's introduction of Solvay (ammonia–soda) process, by patent right
1875	Gilchrist–Thomas basic steel process
1876	Commercial working of contact process, W. S. Squire
1879	Automatic screw-making machine
1880	Frozen meat imports from Australia (canned from 1847)

Appendix 2: Key to Patents Subject Typology 3.6

1. Includes fuel, and covers extraction and substitution processes.
2. Generation, conversion, new uses.
5. Includes substitution, dyeing, cleansing, all non-metals.
6. For example, reducing friction, boiler-improvement, heat conduction, levers, rollers, etc.
7. Including for canals, navigations, incorporating dredging, weighing, ropes and cordage, cranes, separating.
8. And some 'parts'.
10. and 11. Including both chemical and non-chemical (e.g. metal) *products*.
13. For example, brewing.
14. Non-motive and non-moving, including pumping, cutting etc.
17. Including watches, clocks, nautical.
20. Incorporates public building and includes furnishing, heating, etc.
21. Mostly design, decorative but some substitution of raw materials, cost reductions, etc.
22. Medicines, drugs, apparatus, including dental.
23. For instance, locks, musical instruments and doubtful categories.
24. Includes some civil engineering (minor), machinery, chemicals.

Notes

1. INTRODUCTION: SCIENCE, TECHNOLOGY AND ECONOMIC DEVELOPMENT

1. D. Lardner, *A Treatise on Silk* (London, 1831); G. R. Porter, *Treatise on the Origins, Progressive Improvement and Present State of the Silk Manufacture* (London, 1831); F. Warner, *The Silk Industry of the United Kingdom* (London, 1921).

2. Francesca Bray, *The Rice Economies, Technology and Development in Asian Societies* (Oxford, 1986).

3. M. Abramovitz, 'Research and Output Trends in the United States Since 1870', *American Economic Review, Papers and Proceedings*, May 1956; R. Solow, 'Technical Change and the Aggregate Production Function'. *Review of Economics and Statistics*, August 1957; S. C. Gilfillan, *The Sociology of Invention* (Cambridge, Mass., 1963); A. Fishlow, *American Railroads and the Transformation of the Ante-Bellum Economy* (Cambridge, Mass., 1965).

4. See essays by Enos, Mueller and others in National Bureau of Economic Research, *The Rate and Direction of Inventive Activity: Economic and Social Factors* (Princeton, 1962); S. Hollander, *The Sources of Increased Efficiency: The Study of Du Pont Rayon Plants* (Cambridge, Mass., 1965).

5. Robert M. Solow and Peter Temin, 'The Inputs of Growth', in Peter Mathias and M. M. Postan (eds), *The Cambridge Economic History of Europe*, Vol. VII, Pt I (Cambridge, 1978).

6. E. F. Denison, *Why Growth Rates Differ* (Washington, 1967); *Accounting for Slower Economic Growth: The US in the 1970s* (Washington, 1979).

7. Alexander J. Field, 'On the Unimportance of Machinery', *Explorations in Economic History*, *22* (1985).

8. J. D. Gould, *Economic Growth in History* (London, 1972), Table p. 299.

9. For essays which consistently illustrate the importance of incremental changes see Nathan Rosenberg, *Perspectives on Technology* (Cambridge, 1966).

10. Denison, op. cit. (n. 6), pp. 299–301.

11. Joel Mokyr, review of Paul David, in *Economic Development and Cultural Change*, 1977, p. 236.

12. G. Mensch, *Stalemate in Technology: Innovations Overcome the Depression*, English transl. (Cambridge, Mass., 1979).

13. M. J. Peck, 'Inventions in the Post-War American Aluminium Industry' in NBER, op. cit., (n. 4); D. Hamberg, *R and D: Essays on the Economics of Research and Development* (New York 1966); J. Jewkes, D. Sawyers and R. Stillerman, *The Sources of Invention* (New York, 1969).

14. C. Freeman, *The Economics of Industrial Innovation* (Harmondsworth, 1974).

15. Organisation for Economic Cooperation and Development, *Impact of Multinational Enterprises on National Scientific and Technical Capacities: Computer and Data Processing Industry*, Directorate for Science, Technology and Industry, Paris (Restricted) DSTI/SPR/7739-MNE, Paris 27 December 1977.

16. Jacob Schmookler, *Invention and Economic Growth* (Cambridge, Mass., 1966), p. 173; C. F. Carter and B. R. Williams, *Investment in Innovation* (London, 1958), p. 10; J. A. Kregel, *The Theory of Economic Growth* (London, 1972), p. 50.

17. See especially op. cit., Gilfillan (n. 3) and Rosenberg (n. 9).

18. Kendall Birr, *Pioneering in Industrial Research: The Story of the General Electric Research Laboratory* (Washington, 1957).

19. S. Agurin and J. Edgren, *New Factories: Job Design Through Factory Planning in Sweden* (Stockholm, 1980).

20. C. Babbage, *On the Economy of Machinery and Manufactures* (London, 1832).

21. E. Zaleski et al., *Science Policy in the U.S.S.R.* (OECD, Paris, 1969), Table p. 475.

22. NBER, op. cit., (n. 4), p. 318.

23. C. M. Cipolla, *Before the Industrial Revolution* (London, 1976), quote p. 172.

24. For those wishing for an introduction to the economics of technological change, the following should be read in order of listing; Solow and Temin, op. cit. (n. 5); Gould, op. cit., n. 8, Chapter 5; F. R. Bradbury, 'Technological Economics: Innovation, Project Management and Technology Transfer', *Interdisciplinary Science Reviews*, *6* (1981); Rosenberg op. cit., n. 9; Arnold Heertje, *Economics and Technical Change* (London, 1977); Paul Stoneman, *The Economic Analysis of Technological Change* (Oxford, 1983).

25. E. Mansfield, 'Technical Change and the Rate of Imitation', *Econometrica* (1961), p. 752.

26. Ibid., p. 762.

27. A. C. Gatrell, *Distance and Space: A Geographical Perspective* (Oxford, 1983).

28. L. A. Brown and E. G. Moore, 'Diffusion Research in Geography: A Perspective', *Progress in Geography*, *1* (1969), Diagram p. 147.

29. T. Hagerstrand, *The Propagation of Innovation Waves* (Lund, 1952); 'Quantitative Techniques for Analysis of the Spread of Information and Technology' in C. A. Anderson and M. Y. Bowman (eds), *Education and Economic Development* (Chicago, 1965), pp. 244–80.

30. E. Katz, 'The Two-Step Flow of Communication: An Up-to-Date Report on an Hypothesis', *Public Opinion Quarterly*, 21 (1957); E. Katz, M. Levin and H. Hamilton, 'Traditions of Research on the Diffusion of Innovation', *American Sociological Review*, *28* (1963).

31. Nathan Rosenberg, 'Technology' in Glenn A. Porter (ed.), *Encyclopedia of American Economic History: Studies of the Principal Movements and Ideas*, 3 vols (New York, 1980), Vol. I, pp. 294–308.

32. For the first see G. von Tunzleman, *Steam Power and British Industrialisation to 1860* (Oxford, 1978); Stoneman, op. cit. (n. 24), pp. 141–8.

33. Peter Temin, *Iron and Steel in Nineteenth Century America* (Cambridge, Mass., 1964); Nathan Rosenberg, *Technology and American Economic Growth* (New York, 1972).

34. R. Minami, 'The Introduction of Electric Power and Its Impact on the Manufacturing Industries' in H. Patrick (ed.), *Japanese Industrialisation and its Social Consequences* (Berkeley, 1976); S. Ishikawa, 'Appropriate Technologies: some Aspects of Japanese Experience', in A. Robinson (ed.), *Appropriate Technologies for Third World Development* (London, 1979).

35. T. Ozawa, *Japanese Technological Challenge to the West, 1950–1974* (Cambridge, Mass., 1974).

36. D. C. North, 'Transaction Costs in History', *The Journal of European Economic History*, *14* (1985).

37. Joseph E. Stiglitz, 'Information and Economic Analysis: A Perspective', *Economic Journal*, Supplement, *95*, Conference Papers, University of Bath 1984 (London, 1985).

38. A series of examples of this may be found in the very thorough, W. T. Jackman, *The Development of Transportation in Modern England* (London, 1962).

39. Nicholas Leblanc (1742–1806) patented his process in 1791, which treated common salt with sulphuric acid, producing sodium sulphate which was roasted with coal and limestone, from which soda was extracted. The process had initially transformed from France to Britain fairly quickly, but commercial exploitation came in the 1820s, and general adoption in the 1840s. Ernest Solvay (1838–1922) filed his patent in Belgium in 1861, first commercial success occurred in 1867, transfers occurred to France in 1876 (Dombasle) and Britain in 1872 (Mond and Brunner at Winnington, Cheshire), Germany in 1880 (Wyhlen) and the United States in 1884 (Syracuse).

40. L. F. Haber, *The Chemical Industry in the Nineteenth Century* (London, 1958), pp. 109–121. Other improvements included new absorption towers for the removal of waste hydrochloric acid, and chemical methods to treat galligu to recover sulphur. By Henry Deacon's patent 1870 bleaching powder could be manufactured from chlorine, produced by the catalytic oxidation of hydrochloric acid gas.

41. Although it is a major element in the contemporary debate over the utility of military technologies, we do not here consider the problem of the transfer of technologies from one industry to another within the same economy.

42. Bradbury, op. cit. (n. 24), quote p. 151.

43. A. C. Sutton, *Western Technology and Soviet Economic Development*, 3 Vols (Stanford, 1968–1973); G. D. Holliday, *Technology Transfer to the USSR 1928–1937 and 1966–75: The Role of Western Technology in Soviet Economic Development* (Boulder, 1979).

44. R. Nurkse, *Patterns of Trade and Development* (Oxford, 1962).

45. Michael Chisholm, *Modern World Development: A Geographical Perspective* (London, 1982), p. 83.

46. R. Solo, 'The Capacity to Assimilate an Advanced Technology', *American Economic Review, Papers and Proceedings*, May 1966, pp. 91–7.

47. John E. Sawyer, 'The Social Basis of the American System of Manufacturing', *Journal of Economic History, 14* (1954), quote p. 376.

48. H. B. Binswanger, W. V. Ruttan et. al., *Induced Innovation: Technology, Institutions and Development* (London, 1978). On the surmounting of obstacles see: A. O. Hirschman, 'Obstacles to Development: A Classification and a Quasi-Vanishing Act', *Economic Development and Cultural Change, 13* (1965) and Paul M. Sweezy, 'Obstacles to Economic Development' in C. H. Feinstein (ed.), *Socialism, Capitalism and Economic Growth* (Cambridge, 1967).

49. Benjamin and Jean Downing Higgins, *Economic Development of a Small Planet* (New York, 1979), quote p. 87.

50. J. K. Galbraith, *The New Industrial State* (New York, 1967), Chapter 2, quote p. 35.

51. For an extension of this list of 'imperatives' see Ian Inkster, *Japan as a Development Model?* (Bochum, 1980), pp. 72–84.

52. See T. C. Smith, *Political Change and Industrial Development in Japan: Government Enterprise 1868–80* (Stanford, 1955), p. 59f; T. Izumi, 'The Cotton Industry' in special issue, 'Technology Change and Adaptation: The Japanese Experience', *The Developing Economies, 17* (1979).

53. F. R. Bradbury (ed.), *Technology Transfer Practice in International Firms* (Rijn, Netherlands, 1978).

54. R. M. Solow, 'Economic History and Economics', *American Economic Association Papers and Procs, 75* (1985), quote p. 328.

55. Ibid., p. 328 and passim.

56. Paul A. David, *Technical Choice, Innovation and Economic Growth* (Cambridge, 1975), quote p. 10.

57. Solow, op. cit. (n. 54), pp. 328–30. Of course, natural science is itself not entirely free from such observations. For an introduction to critiques of method, see P. B. Medawar, *Induction and Intuition in Scientific Thought* (London, 1969) and Michael Mulkay, *Science and the Sociology of Knowledge* (London, 1979).

58. Solow, op. cit. (n. 54), p. 329; see also the paper by Kenneth Arrow in the same issue of AEA 1985, pp. 320–23.

59. R. M. Cyert and K. D. George, 'Competition, Growth and Efficiency', *Economic Journal* (1969), quote p. 26.

60. H. Leibenstein, 'Allocative Efficiency vs X-Efficiency', *American Economic Review* (1966); 'Organisational or Frictional Equilibria, X-Efficiency, and the Rate of Innovation', *Quarterly Journal of Economics* (1969), quote p. 600.

61. David Hamilton, 'Technology and Institutions are Neither', *Journal of Economic Issues*, *20* (1986); Thomas Sowell, *Knowledge and Decisions* (New York, 1980).

62. R. Kumar Sah and J. E. Stiglitz, 'The Architecture of Economic Systems: Hierarchies and Polyarchies', *American Economic Review*, *76* (1986).

63. R. C. O. Mathews, 'The Economics of Institutions and the Sources of Growth', *The Economic Journal*, *96* (1986), quote p. 305.

64. A. D. Chandler, *Strategy and Structure* (MIT Press, 1962); O. E. Williamson, *The Economic Institutions of Capitalism* (New York, 1985).

65. See references Chapter 9 below; David Landes, *The Unbound Prometheus* (Cambridge, 1969).

66. K. Arrow, 'The Economic Implications of Learning by Doing', *Review of Economic Studies*, June 1962.

67. M. Bowman, 'Schultz, Denison and the Contribution of Education to National Income Growth', *Journal of Political Economy*, October 1964; see also G. Becker, *Human Capital* (New York, 1964).

68. Benjamin Higgins, 'Economic Development and Cultural Change: Seamless Web or Patchwork Quilt', in Manning Nash (ed.), *Essays on Economic Development and Social Change in Honour of Bert F. Hoselitz* (Chicago, 1977).

69. Irma Adelman and Cynthia T. Morris, *Economic Growth and Social Equity in Developing Countries* (Stanford, 1973), pp. 187–9.

70. Raymond Vernon has done more than anyone else to argue that the commanding imperatives of a technology will generally vary considerably over the lifecycle of a project or product: see his 'International Investment and International Trade in the Product Cycle', *Quarterly Journal of Economics*, May 1966.

71. This has been seen as a major lesson of modern Japanese economic history; see H. Rosovsky, 'What are the "Lessons" of Japanese Economic History?', in A. J. Youngson (ed.), *Economic Development in the Long Run* (New York, 1972); Asim Sen, *Science, Technology and Development: Lessons from Japan* (Ankara, 1982). Sen argues that underlying *social* factors provide the relevant explanation of Japanese success. See also Chapter 7.

72. The trend has been towards the establishment of single-purpose machine shops, manufacturing components by assembly-line techniques. To be at all competitive a newcomer would need to overcome the high comparative cost disadvantages of a smaller operation. For a statement of the issues, Alan Parkinson, 'Transfer of Nuclear Power Technology to Australia: Problems and Perspectives' [M.Sci.Soc. dissertation, University of New South Wales, 1980].

2. MENTAL CAPITAL – TRANSFERS OF KNOWLEDGE IN EIGHTEENTH-CENTURY EUROPE

1. R. W. Meyer, *Leibnitz and the Seventeenth-Century Revolution* (Cambridge, 1952). As in Russia, mining and agriculture figured largely in the German academies.

2. E. G. Barber, *The Bourgeoisie in Eighteenth Century France* (Princeton, 1955), quote p. 46.

3. D. J. Struik, *Yankee Science in the Making* (New York, 1948, 1962), quote p.58.

4. M. C. Jacob, *The Newtonians and the English Revolution 1689–1720* (Hassocks, 1976); G. Holmes, 'Science, Reason and Religion in the Age of Newton', *British Journal for the History of Science*, *12* (1979).

5. For Whiston's lecturing and the quotation see M. Farrell, 'The Life and Work of William Whiston' [PhD thesis, University of Manchester, UMIST, 1973]. My judgement that the surge of popularisation came in the 1730s is derived from a fairly complete survey of the following newspapers: *The Daily Post Boy*, *Fog's Weekly Journal*, *The Whitehall Evening Post*, *The St. James' Evening Post*, *The London Evening Post*, *The Universal Spectator and Weekly Journal*, *The Post Boy*, *The Post Man*, *The Hyp-Doctor* (satirical) and *The London Journal*. See also, M. Rowbottom, 'The Teaching of Experimental Philosophy in England 1700–1730', in *Procs of International Congress of the History of Science*, 1965, Vol. II, (1968), No. 4.

6. *Fog's Weekly Journal*, 4 January 1735.

7. Ibid., 6 November 1731.

8. *The Daily Post Boy*, 24 November 1731.

9. *Fog's Weekly Journal*, 19 February 1732; Conyers Middleton (1683–1750, MA 1707, DD 1717, from 1731 Woodwardian professor at Cambridge), in a series of publications 1731–3 defended the deistical opinions of Matthew Tindal. In 1731 Zachary Pearce (1690–1774, DD by patronage, later Bishop of Rochester) attacked Middleton as a covert infidel.

10. Ibid., 22 April 1732.

11. Ibid., 10 June 1732.

12. For example, *The Daily Post Boy*, 21 March 1732; *The Whitehall Evening Post*, 7 December 1731; *The St. James Post*, 16 January 1721; *The Daily Post*, 30 January 1721.

13. *Fog's Weekly Journal*, 1 February 1735.

14. E. S. Shils, 'Towards a Modern Intellectual Community' in J. S. Coleman (ed.), *Education and Political Development* (Princeton, 1965), quote p. 498. The most recent thorough survey of eighteenth-century societies is J. E. McClellan, *Science Reorganised, Scientific Societies in the 18th Century* (New York, 1985), of which Chapters 5 and 6 concern communications networks and common research interests.

15. W. H. E. Armytage, *The Rise of the Technocrats* (London, 1965), p. 35; J. R. Partington, *A History of Chemistry II* (London, 1961), pp. 723–9.

16. The tables are derived from figures in N. Hans, *New Trends in Education in the Eighteenth Century* (London, 1951), pp. 32–3 and K. M. Birse, *Engineering at Edinburgh University 1673–1983* (Edinburgh, 1983), p. 16, and both sets are based on only men entered in DNB, and their characterisation as 'scientists' or 'engineers' is approximate. Of course, a less-biased sample would almost certainly yield a significantly larger percentage of both groups receiving no university education. The considerable 'high' education available in dissenting and private academies is not considered here. Mostly,

such academies were attached to a particular urban culture, rather than to any national movement.

17. For a comparison with Scotland's own intellectual elite see B. and V. Bullough, 'Intellectual Achievers: A Study of Eighteenth Century Scotland', *American Journal of Sociology, LXXVI* (1971).

18. Two volumes and 50 copper plates gave details of designing, gilding, japanning, perspective and engraving and etching; working in metals and mineralogy, dyeing, and the preparation of colours, ink-making, etc.

19. Again, the surge in such publications appears in the 1730s. Most sold at around 1/-. Stephen Switzer's *A General System of Hydrostatics and Hydraulics, Philosophical and Practical* introduced steam engine construction, using theoretical material from Desaguliers, and Hauksbee, at a time when both included something of this in their London lecturing courses. The engine discussed by Switzer was being exhibited, demonstrated and lectured upon by Desaguliers at Channell Row, and its proprieters would be there 'to treat with such persons as want this engine, for supplying of great Quantities of Water, draining Mines and other uses in which this engine may be of service' (*The Daily Post*, 30 January 1721, p. 2, advertisement). Simultaneously, William Harding was giving public demonstrations of an engine which would raise water 70 feet, 'and is applicable to any Power that other engines are'. Scientists or engineers of the calibre of John James, Nicholas Hawksmoor, James Gibbs, John Harris and Desaguliers (all DNB) recommended publicly a variety of trade-based publications.

20. A model of detective-work for the eighteenth-century historian, R. Darnton, *The Business of Enlightenment, A Publishing History of the Encyclopedie, 1775–1800* (Cambridge, Mass., 1979).

21. Ibid., p. 529.

22. See Chapter 1 of R. Darnton, *The Literary Underground of the Old Regime* (Cambridge, Mass., 1982).

23. R. K. Merton, 'Priorities in Scientific Discovery: A Chapter in the Sociology of Science', *American Sociological Review*, *22*, (1957). The case of J. A. Rabaut-Pomier illustrates how the very informality of contact created priority disputes. Around 1780 in Mountpellier, Rabaut had observed that smallpox, '*le claveau des montrons*', and cows' pustules were regarded as identical illnesses under the term '*picote*'. His subsequent considerations were informally communicated to Jenner: see *Biographie Universelle, Ancienne et Moderne*, *36*, Paris (1823), pp. 475–6.

24. D. A. Kronick, *A History of Scientific and Technical Periodicals* (New York, 1962). For an example in another field see Fritz Redlich, 'An Eighteenth Century Business Encyclopedia as a Carrier of Ideas', *Harvard Library Bulletin*, *XIX* (1971).

25. D. S. L. Cardwell, *The Organisation of Science in England* (London, 1957), p. 10.

26. R. Hahn, in his paper in *Studies on Voltaire and the Eighteenth Century*, *25*, (1963); *Biographie Universelle*, *9*, Paris (1813), p. 86.

27. L. J. M. Coleby, 'Richard Watson, Professor of Chemistry in the University of Cambridge, 1764–71', *Annals of Science*, *9*, (1953).

28. A. and N. L. Clow, *The Chemical Revolution* (London, 1952).

29. J. H. Park and E. Glouberman, 'The Importance of Chemical Development in the Textile Industries during the Industrial Revolution', *Journal of Chemical Education, 9* (1932).

30. Kronick, op. cit. (n. 24), pp. 73–5, 239.

31. For sources and information about the patenting system see Chapter 3 below. Increasingly, patent specifications become more explicit and exact, and were often designed to improve quite slightly upon an existing 'major' technique.

32. The society has been analysed in two hitherto unpublished papers; R. J. Morris, 'The London Philosophical Society 1780–1788' [delivered at Meeting of American History of Science Society, Philadelphia, 28 December 1976] and Gwen Averley, 'The "Social Chemists": English Chemical Societies in the 18th and Early 19th Century', [April 1984, Department of Humanities, Teesside Polytechnic, England]. See also, H. Guerlac, *Lavoisier The Crucial Years* (New York, 1961), Chapter 2; *Gentleman's Magazine*, *99*, (1829); intellectual conflicts within such a group must have been very strong, with Kirwan and Pearson diametrically opposed over the phlogiston question, and with Kirwan complaining of Magellan's use of his work on mineralogy (1784).

33. See Hans, op. cit. (n. 16) extensively; J. Kendrick, *Profiles of Warrington Worthies* (Warrington, 1854); H. McLachlan, *Warrington Academy, Its History and Influence* (London, 1943: Chetham Society); idem., *English Education Under the Test Acts* (London, 1931); W. Turner, *Warrington Academy* (Warrington, 1957), reprinted from *Monthly Repository*, 1813–15.

34. J. Kendrick, *Profiles of Warrington Worthies* (Warrington, 1854); W. C. Henry, 'Memoir of William Henry of Manchester', *Manchester LPS Memoirs*, Series 2, *6*, (1842); A. Thackray, 'Natural Knowledge in Cultural Context: The Manchester Model', *American Hist. Review*, *79*, (1974).

35. *Observer* (London), 12 August 1804; *Liverpool Chronicle*, 18 April, 10 October, 26 December 1804.

36. J. Money, 'The Schoolmasters of Birmingham and the West Midlands, 1750–90', *Historie Sociale–Social History*, *IX* (1976); 'Birmingham and the West Midlands, 1760–1793: Politics and Regional Identity in the English Provinces in the Later 18thc' *Midland History*, *I* (1971); 'Taverns, Coffee Houses and Clubs: Local Politics and Popular Articulacy in the Birmingham Area', *The Historical Journal*, *XIV* (1971). In their paper on Whitehaven's mathematical teachers and lecturers, F. J. G. Robinson and P. J. Wallis indicate a similar acceleration; 'Some Early Mathematical Schools in Whitehaven'. *Trans of the Cumberland and West Morland Antiquarian and Archaeological Society*, *LXXV* (1975).

37. *The Dublin Evening Post*, 16 January 1798.

38. T. Rutt, *Memoirs of Priestley* (London, 1831); W. H. Chaloner, 'Priestley, Wilkinson and the French Revolution', *Trans. Royal Historical Society* (1958); N. Garfinkle 'Science and Religion in England 1790–1800', *Journal of the History of Ideas*, *XVI* (1955); W. P. Hall, *British Radicalism 1791–97* (London, 1912); W. T. Laprade, *England and the French Revolution*

(London, 1909); S. Maccoby, *The Radical Tradition, 1763–1914* (London, 1952); E. Robinson, 'The English Philosophes and the French Revolution', *History Today, 6,* (1956); R. B. Rose, 'The Priestley Riots', *Past & Present, XIV,* (1960); J. R. Western, 'The Volunteer Movement as an Anti-Revolutionary Force, 1793–1801', *English Historical Review, 71* (1956).

39. E. P. Thompson, *The Making of the English Working Class* (London, 1963), pp. 26–7, 74.

40. G. A. Williams, *Artizans and Sans-Culottes* (London, 1968), p. 64.

41. See for 1790s scientific dissent and the reaction to Priestley, Ian Inkster, 'Studies in the Social History of Science in England During the Industrial Revolution 1790–1850', [PhD thesis, Department of Economic History, University of Sheffield, 1977], especially pp. 234–55, 261–79, 391–429, 494–500, 619–32.

42. C. C. Gillispie, *Science and Polity in France at the End of the Old Regime* (Princeton, 1980); Terry Shin, *Savoir Scientifique – pouvoir social, L'Ecole Polytechnique, 1794–1914* (Paris, 1980); R. Taton, 'Sur quelques ouvrages recents concernant l'histoire de la science française', *Revue d'histoire des Sciences, XXVI,* (1973); C. C. Gillispie, *Lazare Carnot savant* (Princeton, 1971), and his note on Carnot in *Dictionary of Scientific Biography, III* (New York, 1971).

43. See Shin, ibid.

44. Colin Russell, *Science and Social Change, 1700–1900* (London, 1983), Chapter 7.

45. See ibid., and references in note 42, and L. P. Williams, 'Science, Education and the French Revolution', *Isis, 44* (1953); M. P. Crosland, 'The Development of a Professional Career in Science in France', *Minerva, 13* (1975); R. Fox, 'Scientific Enterprise and the Patronage of Research in France 1800–1870', *Minerva, II* (1973).

46. This is especially debatable, and see J. Langins, 'The Decline of Chemistry at the Ecole Polytechnique', *Ambix, 28,* (1981). Between 1783 and 1788 the Académie offered a prize for an improved method of preparing soda from salt, at which time Leblanc was receiving patronage from the Orléans family, which lasted until the duke was guillotined in 1793.

47. During 1790–94 there is some limited evidence of an emergence of 'popular' science in France – voluntarism, radicalism, the introduction of a metric system, the 1793 schemes for 'Applied Science' of the Committee of Public Safety, which included Lavoisier, Laplace and Coulomb.

48. J. Robinson, *Proofs of a Conspiracy Against all the Religions and Governments of Europe* (Edinburgh, 1797); R. Olson, 'The Reception of Boscovich's Ideas in Scotland', *Isis, 60,* (1969); J. B. Morrell, 'Professors Robinson and Playfair and the *Theophobia Gallica*: Natural Philosophy, Religion and Politics in Edinburgh 1789–1815', *Notes and Records of the R.S. of London, 26,* (1971).

49. J. Norman to Richard Bright, 15 September 1774, in 'Letters of the Brights of Bristol', Melbourne University Archives [MS, Unsorted].

50. John Cuthbertson, *Practical Electricity and Galvanism* (London, 1807), quote p. 198. See also *Nicholson's Journal* (June 1792), *Philosophical Magazine*

(March 1805); R. J. Forbes, (ed.), *Martinus van Marum. Life and Work, I* (Haarlem, 1960); T. H. Levere, 'Friendship and Influence, Martinus van Marum, FRS', *Notes and Records of the Royal Society, 25*, (1970).

51. A full account and the tracing of a complex connection is in Hans op. cit., (n. 16), pp. 213–20. For the flow of enlightenment and technology into eighteenth-century Spain see the following: J. Harrison, *An Economic History of Modern Spain* (Manchester, 1978); R. Herr, *The 19th Century Revolution in Spain* (Princeton, 1958); R. Carr, *Spain 1808–1939* (Oxford, 1966); R. J. Shafer, *The Economic Societies in the Spanish World, 1763–1821* (Syracuse, 1958); J. V. Vives, *An Economic History of Spain* (Princeton, 1969), esp. Chapter 35; and the masterful, H. Kamen, *The War of Succession in Spain 1700–15* (London, 1969), esp. Chapter 6.

52. W. H. G. Armytage, *The French Influence on English Education* (London, 1968), quote p. 15.

53. *The Universal Spectator*, 22 January 1732.

54. E. P. Oberhottzer, *Philadelphia, A History of the City and Its People, II* (Philadelphia, 1920); S. and J. Bridenbaugh, *Rebels and Gentlemen, Philadelphia in the Age of Franklin* (New York, 1962), (first pub. 1942), esp. pp. 321–40; H. Simpson, *The Lives of Eminent Philadelphians, Now Deceased* (Philadelphia, 1859).

55. A. Parry, *The Russian Scientist* (London, 1973), quote p. 14.

56. Alexander Vucinich, *Empire of Knowledge* (Berkeley and London, 1984), pp. 1–56; B. N. Menshutkin, *Russia's Lomonosov* (Princeton, 1952); V. Boss, *Newton and Russia 1698–1796* (Cambridge, Mass., 1972); M. von Lane, *History of Physics* (New York, 1950); H. M. Leicester in his introduction to *M. V. Lomonosov's Corpuscular Theory* (Cambridge, Mass., 1970), and his essay on Mendeleev in *Journal of Chemical Education, 25* (1948); P. I. Lyaschchenko, *The History of the National Economy of the U.S.S.R., I* (Moscow, 1947), Section 6; N. K. Rozhkova, (ed.), *Essays on the Economic History of Russia in the First Half of the 19th Century* (Moscow, 1959), esp. pp. 121, 153–68; J. Scott Carver, 'A Reconsideration of Eighteenth Century Russia's Contribution to European Science', *Canadian–American Slavic Studies, XIV* (1980); the essay by H. Neuschäffer in U. Liszkowski (ed.), *Russland und Deutschland* (Stuttgart, 1974). Between 1755 and 1765 Lomonosov was instrumental in the publication of Moscow University's *Monthly Essays* which served to introduce to Russian intellectuals a variety of European works in natural history and economics.

57. The Edict of Nantes, 1597, had operated so as to differentiate the two religious communities of France into separate institutions of government, education and worship. Protestants probably numbered about one million (maximum: c.1.5 million) in a population of some 20 million, and tended to concentrate into industry, commerce and finance. The flow into England was undoubtedly encouraged by an initial government relief grant of £64 000 which was followed by private subscriptions, mainly from London, of up to £200 000; see F. Warner, *The Silk Industry of the United Kingdom* (London, 1921).

58. F. C. Green, *Eighteenth Century France, Six Essays* (London, 1929), esp.

pp. 26–69; W. C. Scoville, 'The Huguenots and the Diffusion of Technology', *Journal of Political Economy, 60,* (1952), in parts I (294–311) and II (392–411); W. Cunningham, *Alien Immigrants to England* (London, 1897, 1969), esp. Chapter XIV. The first was published in Amsterdam, the second in the Hague – this should not be confused with the later Geneva publication of M. A. Pictet, C. Pictet and F. G. Maurice (1796–1815). The major edition of the last was published from Paris during 1733–1740. A. F. Prévost d'Exiles, the main editor, was assisted by P. F. G. Desfontaines and C. H. Lefebvre, and supervised publication up to the twentieth volume: *Biographie Universelle, 36,* Paris, (1823), pp. 64–71. The best account of the Huguenots as carriers of human capital is in Chapter 10 of W. C. Scoville, *The Persecution of the Huguenots and French Economic Development, 1680–1720* (Berkeley, 1960).

59. N. Hans, *History of Russian Educational Policy 1701–1917* (New York, 1964, 1931). For wider aspects of bureaucratic involvement in transfers of European knowledge to Russia, see P. H. Clendenning 'Eighteenth Century Russian Translations of Western Economic Works', *Journal of Eur. Econ. History, 1* (1972) and his 'The Economic Awakening of Russia in the Eighteenth Century', ibid. *14* (1985).

60. *DNB* (John Kay). See also D. J. Jeremy, 'Damming the Flood: British Government Efforts to Check the Outflow of Technicians and Machinery 1780–1830', *Bus. Hist. Rev., 5* (1977).

61. For background see D. J. Jeremy, *Transatlantic Industrial Revolution: The Diffusion of Textile Technologies Between Britain and America 1790–1830* (Oxford, 1981), pp. 264–66.

62. See Chapters 1 and 2 especially of A. E. Musson and E. Robinson, *Science and Technology in the Industrial Revolution* (Manchester 1969); and Musson's introduction to A. E. Musson (ed.), *Science and Technology in the Eighteenth Century* (London, 1972).

63. On Keir and such scientific influence see Musson and Robinson, ibid., pp. 31, 58, 87–8, 96, 121–7, 142–3 *passim.*

64. A good account is in W. O. Henderson, *Britain and Industrial Europe, 1750–1870* (Leicester 1972), Chapter 2.

65. For features of Russian backwardness, see W. L. Blackwell, *The Beginnings of Russian Industrialisation 1800–60* (Princeton, 1968).

66. R. P. Bartlett, *Human Capital, The Settlement of Foreigners in Russia 1762–1804* (Cambridge, 1979), p. 164 forward.

67. Ibid., pp. 178–9. S. J. Tomkireff, 'The Empress Catherine and Mathew Boulton', *The Times Literary Supplement,* 22 December 1950.

68. For Sweden see K. G. Hildebrand, 'Foreign Markets for Swedish Iron in the 18thc', *Scandinavian Economic History Review, VI,* (1958); E. Söderlung, 'The Impact of the British Industrial Revolution on the Swedish Iron Industry' in L. S. Pressnell (ed.), *Studies in the Industrial Revolution* (London, 1960), and Flinn's introduction to *Swedenstierna's Tour of Great Britain 1802–03* (Newton Abbot, 1973).

3. SCIENCE AND TECHNOLOGY IN THE BRITISH INDUSTRIAL REVOLUTION

1. T. S. Reynolds, 'Medieval Roots of the Industrial Revolution', *Scientific American, 251*, no. 1, July 1984, pp. 108–16.

2. See essays in R. M. Hartwell (ed.), *The Causes of the Industrial Revolution in England* (London, 1967); M. Fores, 'The Myth of a British Industrial Revolution', *History, 66* (1981), more prone to semantics; W. W. Rostow, *How It All Began* (London, 1975).

3. For an introduction to the Industrial Revolution the following should be read in order of their presentation: H. L. Beales, *The Industrial Revolution 1750–1850*, 1958 edition (of 1928, London), with new introductory essay; T. S. Ashton, *The Industrial Revolution 1760–1830* (London, 1948); P. Deane, *The First Industrial Revolution* (Cambridge, 1965), Chapters 1–8; Peter Mathias, *The First Industrial Nation* (London, 1969), Chapters 5–7; E. J. Hobsbawm, *Industry and Empire* (London, 1968), Chapters 1–4; Sidney Pollard, *The Genesis of Modern Management* (London, 1965), Chapters 2–4; P. Deane and W. A. Cole, *British Economic Growth 1658–1959* (Cambridge, 1969), Chapters 2, 5 and 6; S. D. Chapman, *The Cotton Industry in the Industrial Revolution* (London, 1972); John Hicks, *A Theory of Economic History* (London, 1969), Chapters 7–10; R. M. Hartwell, *The Industrial Revolution and Economic Growth* (London, 1971), Chapters 5–9 and 12; F. Crouzet, *The Victorian Economy* (London, 1982), Chapters 3, 4, 7–10; E. Mandel, *Late Capitalism*, English edition (London, 1979), Chapters 1–5; and the superb collection of essays, D. McCloskey and R. Floud (eds), *The Economic History of Britain since 1700*, Vol. I, 1700–1860 (Cambridge, 1981), especially those by Crafts, Cole, Jones, McCloskey, Feinstein and Hawke; P. Mathias and M. M. Postan (eds), *The Cambridge Economic History of Europe*, Vol. VII, Part I (Cambridge, 1978), essays by Solow and Temin (excellent), Feinstein, and Pollard. For those with no background, start with R. M. Reeves, *The Industrial Revolution 1750–1850* (London, 1971), Chapters 1 and 4 and the statistics in 5; F. J. Wright, *The Evolution of Modern Industrial Organisation* (London, 3rd edition 1967), Chapters 5–10; M. W. Flinn, *Origins of the Industrial Revolution* (London, 1966), Chapters 1, 4 and 5.

4. David S. Landes, *The Unbound Prometheus* (Cambridge, 1969), quote p. 1.

5. The statistics are from McCloskey and Craft essays in McCloskey and Floud (1981), op. cit. (n. 3); E. L. Jones, *Agriculture and the Industrial Revolution* (New York, 1974).

6. All figures of output and productivity growth are estimates, and because the latter depend on the former plus estimates of factor shares, they should be taken with salt.

7. This does not mean that manufacturing and services were of little importance prior to 1780, but that at that time the efficiency with which they produced output was far lower than after 1780 or 1800.

8. Beales, op. cit. (n. 3), p. 30.

9. J. H. Clapham, *An Economic History of Modern Britain* (Cambridge,

1938); Sidney Pollard, *Peaceful Conquest, The Industrialization of Europe 1760–1970* (Oxford and London, 1981), Chapter 1.

10. For such constraints see Pollard, ibid. For a theory of the spread from nodal points see J. G. Williamson, 'Regional Inequity and the Process of national development: A Description of the Patterns', reprinted in C. Needleman (ed.), *Regional Analysis* (London, 1968).

11. J. Langton, 'The Industrial Revolution and the Regional Geography of England', *Trans. of the Institute of British Geographers, 9,* (1984); D. Gregory, 'The Production of Regions in England's Industrial Revolution', *Journal of Historical Geography, 14* (1988).

12. M. Berg, *The Age of Manufactures, 1700–1820* (London, 1985).

13. See the many examples in Raphael Samuel, 'The Workshop of the World: Steam Power and Hand Technology in Mid-Victorian Britain', *History Workshop*, 3, Spring 1977.

14. See Feinstein's essay in McCloskey and Floud (1981), op. cit. (n. 3), pp. 128–42.

15. Ibid., S. Pollard, 'Investment, Consumption and the Industrial Revolution', *Econ. Hist. Review, 11* (1958); F. Crouzet (ed.), *Capital Formation in the Industrial Revolution* (London, 1972).

16. A. J. Field, 'On the Unimportance of Machinery', *Explorations in Economic History, 22* (1985).

17. See Chapman and other essays in J. P. P. Higgins and S. Pollard (eds), *Aspects of Capital Investment in Great Britain 1750–1850* (London, 1971); Chapman, op. cit., (n. 3); and his 'The Cost of Power in the Industrial Revolution in Britain: The Case of the Textile Industry', *Midland History, 1* (1971).

18. K. Honeyman, *Origins of Enterprise* (Manchester, 1982).

19. S. D. Chapman, 'The Textile Factory Before Arkwright: A Typology of Factory Development', *Business History Review, 58* (1974).

20. McCloskey, op. cit. (n. 3), p. 117.

21. Crafts in ibid., pp. 1–17, see also his 'Industrial Revolution in England and France', *Econ. Hist. Review, 30* (1977).

22. Crafts, ibid. (1981), pp. 7–8.

23. McCloskey in ibid., pp. 110–11.

24. The best account of the range of technological change is Chapter 2 of Landes, op. cit., (n. 4); L. F. Haber, *The Chemical Industry in the Nineteenth Century* (Oxford, 1958), Chapter 2; D. C. North, 'Sources of Productivity Change in Ocean Shipping 1600–1880', *Journal of Political Econ., 76* (1968); M. W. Flinn, 'The Growth of the English Iron Industry', *Econ. Hist. Review, 11* (1958); T. C. Barker, R. Dickinson, D. W. F. Hardie, 'The Origins of the Synthetic Alkali Industry in Britain', *Economica, 23* (1956); L. Grittins, 'The Manufacture of Alkali in Britain 1779–1789', *Annals of Science, 37* (1966).

25. McCloskey, op. cit. (n. 3), p. 114, Table 6.2.

26. Landes, op. cit. (n. 4), p. 87. The classic example being the followership-invention associated with the improved Watt engine.

27. Beales, op. cit. (n. 3), pp. 49–50.

28. Samuel, op. cit. (n. 13), p. 10.

29. For instance, C. K. Hyde, 'The Adoption of Coke Smelting by the British Iron Industry 1709–1790', *Explorations in Econ. History*, *10* (1973), and generally Cole in Floud and McCloskey (n. 3); for criticisms of demand arguments see several comments by McCloskey in ibid. and J. Mokyr, 'Demand Versus Supply in the Industrial Revolution', *Journal of Economic History*, *37* (1977).

30. Landes, op. cit. (n. 4), pp. 45–6, 60–61.

31. Ibid., p. 63.

32. Ibid., pp. 62–8; A. E. Musson and E. Robinson, 'Science and Industry in the Later Eighteenth Century', *Econ. Hist. Review*, *12* (1960).

33. Landes, ibid., p. 77.

34. Crafts, op. cit. (in McCloskey, note 3, 1981), pp. 11–16.

35. F. Rapp, 'Structural Models in Historical Writing: The Determinants of Technological Development During the Industrial Revolution', *History and Theory*, *21* (1982), quote p. 337.

36. C. A. Russell, *Science and Social Change* (London, 1983), p. 98; Landes, op. cit. (n. 4), p. 104; J. D. Bernal, 'Science, Industry and Society in the Nineteenth Century', in S. Lilley (ed.), *Essays in the Social History of Science* (Copenhagan, 1953); D. S. L. Cardwell, 'Science and Technology in the Eighteenth Century', *History of Science*, *1* (1962); A. Rupert Hall, 'What Did the Industrial Revolution in Britain Owe to Science?', in N. McKendrick (ed.), *Historical Perspectives, Studies in English Thought and Society* (London, 1974).

37. K. Mendelssohn, *Science and Western Domination* (London, 1976), p. 127.

38. Haber, op. cit. (n. 24); Russell, op. cit. (n. 36), pp. 99–103.

39. Donald Fleming, 'Latent Heat and the Invention of the Watt Engine', *Isis*, *43* (1952); F. M. Scherer, 'Invention and Innovation in the Watt–Boulton Steam-Engine Venture', *Technology and Culture*, *6* (1965); G. N. von Tunzelmann, *Steam Power and British Industrialization to 1860* (Oxford, 1978); S. Takayama, 'Development of the Theory of the Steam Engine', *Japanese Studies in the History of Science*, *18* (1979).

40. Charles Gillispie, in *Isis*, *48* (1957), quote p. 399.

41. C. Trebilcock, *The Industrialisation of the Continental Powers 1780–1914* (London, 1981), phrase p. 104.

42. Arnold Thackray, 'Science and Technology in the Industrial Revolution', *History of Science*, *9* (1970), and 'The Industrial Revolution and the Image of Science', in E. Mendelsohn and Thackray (eds), *Science and Values* (New York, 1973).

43. Adam Smith, *The Wealth of Nations*, Book V, p. 267.

44. William Henry, *A General View of the Nature and Objects of Chemistry and its Applications to Arts, and Manufacturers* (Manchester, 1799).

45. Ibid., p. 7.

46. Ibid., p. 21.

47. R. M. Birse, *Engineering at Edinburgh University* (Edinburgh, 1983), Table, p. 16.

48. *Glasgow Mercury*, 8 April 1784, pp. 113–14.

49. N. T. Phillipson and R. Mitchison (eds), *Scotland in the Age of Improvement* (Edinburgh, 1970); H. R. Trevor Roper, 'The Scottish Enlightenment', *Studies in Voltaire and the Eighteenth Century*, *58* (1967).

50. R. N. Smart, 'Some Observations on the Provinces of the Scottish Universities 1560–1850', in G. W. S. Barrow (ed.), *The Scottish Tradition* (Edinburgh, 1974); B. and V. Bullough, 'Intellectual Achievers: A Study of Eighteenth Century Scotland', *American Journal of Sociology*, *76* (1971); Rab Houston, 'The Literacy Myth? Illiteracy in Scotland 1630–1760', *Past and Present*, *96* (1982).

51. J. Schneider, 'The Definition of Eminence and the Social Origins of Famous English Men of Genius', *American Sociological Review*, *3* (1938).

52. T. C. Smout, 'Scotland and England: Is Dependency a Symptom or a Cause of Underdevelopment?', *Review*, *3* (1980).

53. N. T. Phillipson, 'Culture and Society in the 18thC Province: The Case of Edinburgh and the Scottish Enlightenment', in L. Stone (ed.), *The University in Society* (Princeton, 1975).

54. W. Ferguson, *Scotland, 1689 to the Present* (Edinburgh, 1968).

55. For the flow, see J. H. McCulloch, *The Scots in England* (London, 1935); R. A. Cage (ed.), *The Scots Abroad* (London, 1984).

56. U. R. Q. Henriques, *Religious Toleration in England 1787–1833* (London, 1961); R. G. Cowherd, *The Politics of English Dissent from 1815 to 1848* (London, 1959).

57. *Abstracts of Some Statutes, Kings College, Aberdeen* (Aberdeen, 1753); *Plan of Education Marishal College* (Aberdeen, 1755).

58. A. H. T. Robb Smith, 'Medical Education at Oxford and Cambridge Prior to 1850', in F. N. L. Poynter (ed.), *The Evolution of Medical Education in Britain* (London, 1966), Table p. 44; and see also for examples of influence in industry, A. and N. L. Clow, *The Chemical Revolution* (London, 1952); A. E. Musson and E. Robinson, *Science and Technology in the Industrial Revolution* (Manchester, 1969); C. H. Lee, *A Cotton Enterprise* (Manchester, 1972).

59. Ashton, op. cit. (n. 3), pp. 16–17.

60. Musson and Robinson, op. cit. (n. 58), and A. E. Musson (ed.), *Science and Technology in the Eighteenth Century* (London, 1972), especially the Introductory Essay.

61. A. Thackray, 'Natural Knowledge in Cultural Context: The Manchester Model', *American Hist. Review*, *79* (1974), quote p. 678.

62. Ibid., p. 682.

63. I. Inkster, *Studies in the Social History of Science in England During the Industrial Revolution* [PhD thesis, University of Sheffield, 1977].

64. J. Alderson, *Address to Members of the Hull Subscription Library on Annual Lectures* (Hull, 1804), quote p. 15.

65. *Gore's Liverpool Advertiser*, 24 December 1835.

66. Estimates based on *1851 Population Census of Great Britain, Sessions 1852–54* (British Parliamentary Papers), Irish University Reprint Series, Population: Vol. 11.

67. For the basis of this see Ian Inkster, 'The Context of Steam Intellect in Britain to 1851', in Ian Inkster (ed.), *The Steam Intellect Societies* (Nottingham, 1985).

68. *Gore's Liverpool Advertiser*, 18 February 1780. For a general survey of the eighteenth-century trends in the city see Ian Inkster, 'Scientific Culture and Scientific Education in Liverpool prior to 1812', in M. D. Stephens and G. W. Roderick (eds), *Scientific and Technical Education in Early Industrial Britain* (Nottingham, 1981).

69. *London Observer*, 12 August 1804; *Liverpool Chronicle*, 18 April 1804.

70. Ian Inkster, 'Popularised Culture and Steam Intellect: A Case Study of Liverpool', in Inkster, op. cit. (n. 67).

71. *Liverpool Mercury*, 12 July 1839.

72. Shown well in much of the work of Musson and Robinson; see also Inkster, op. cit. (n. 63).

73. *Doncaster Gazette*, 20 April 1822, 2 January, 13 March 1835.

74. Source: Census 1851, op. cit. (n. 66).

75. W. H. Chaloner, *The Skilled Artizans During the Industrial Revolution, 1750–1850* (London, 1969), quote p. 13.

76. R. B. Prosser, *Birmingham Inventors and Inventions* (Birmingham, 1881).

77. *The London Journal of Arts and Sciences*, *9* (1825).

78. William Newton, *Letters and Suggestions Upon the Amendment of the Laws Relative to Patents for Invention* (London, 1835), quote p. 97.

79. Quoted in ibid., pp. 78–9.

80. Several such organisations emerged in the early 1830s and acted as pressure-groups for institutional reform. Commonly scientific, engineering and patentee interests merged.

81. *Report of Select Committee on the Law Relative to Patents for Invention, House of Commons, 1829* (London, 1829), p. 132.

82. All patent data is from the original abridgements of specifications, republished in the mid-nineteenth century. A complete analysis of the social, geographical and educational character of patentees for the period 1750–1850 is currently being researched, with particular emphasis on detailed construction of localised inventive environs. For a history of the system in these years see H. I. Dutton, *The Patent System and Inventive Activity During the Industrial Revolution* (Manchester, 1984).

83. Each group represents a range of sums which overlap. These are drawn from contemporary accounts in tradesmen's manuals, handbooks etc.

84. Steam intellect ran ahead of the steam engine. Preliminary work suggests that the dominance of London is not a mere function of the location of the patent system, but a genuine reflection of the skill-character of major London districts.

85. Farey, 1829, op. cit. (n. 81), p. 113.

86. McCloskey, op. cit. (n. 3), p. 120.

87. Douglas C. North, *Structure and Change in Economic History* (New York, 1981), quote p. 169. For an early statement see Fraudenberger and Redlich, *Kyklos*, 1964. Our argument is different from North's in that it was the pressure of continuous flow (machine juxtaposition) which defined technological progress in the factory, and hence true factories must be defined as more than central power places monitoring workers. We follow Chapman's definition of the 'true' factory.

88. The best synopsis of this is in Berg, op. cit. (n. 12). However, this author does not *demonstrate* that productivity increased significantly in such areas of the economy, and opts for organisational rather than technical change as a key efficiency improver.

89. For instance, in the 'manufacturies' in engineering, etc. so emphasised by Marx, or in landlord industrial organisations, or in proto-factories.

90. This is the emphasis given by Chapman in op. cit. (n. 19).

91. Paul explained the economic failure of his rollers in small mills on the poor quality and habits of the workers. This, admittedly, does point to the possible usefulness of the 'monitoring' argument. But there seems nothing especially sensible about the thesis that larger establishments will necessarily monitor workers better than small ones, especially in the conditions of the day.

4. THE SCIENTIFIC ENTERPRISE: INSTITUTIONS AND THE DIFFUSION OF KNOWLEDGE IN THE NINETEENTH CENTURY

1. Meeting of the Third Section of Glasgow and West of Scotland Branch of the Society of Chemical Industry, 3 November 1885, *Journal of the Society of Chemical Industry*, *4* (1885), p. 651.

2. Hilary and Steven Rose, *Science and Society* (Harmondsworth, 1970), pp. 2–3.

3. Everett Mendelsohn, 'The Emergence of Science as a Profession in Nineteenth Century Europe', in Karl Hill (ed.), *The Management of Scientists* (Boston, 1963), pp. 40–41.

4. F. K. Ringer, *The Decline of the German Mandarins* (Harvard UP, 1969), pp. 45 forward. See also R. Paul, 'German Academic Science and the Mandarin Ethos 1850–1880', *Brit. Jrnl Hist. Science*, *17* (1984), pp. 1–29.

5. The figures are drawn from a number of sources. For a comparable table on Prussia see Peter Lundgreen, 'Educational Expansion and Economic Growth in Nineteenth Century Germany: A Quantitative Study' in L. Stone (ed.), *School and Society* (Baltimore, 1976).

6. See material in refs 3–6 above and 8 and 21 below.

7. T. E. Thorpe, 'Chemical Instruction and Chemical Industries in Germany', *Nature*, 8 May 1902, pp. 32–4.

8. F. K. Ringer, 'A New Vitality: The History of Education', *Journal of Interdisciplinary History*, *XII* (1982), pp. 657–63; C. E. McClelland, *State, Society and University in Germany 1700–1914* (New York, 1980); H. Titze, 'The Cyclical Overproduction of Graduates in Germany in the Nineteenth and Twentieth Centuries', *International Sociology*, *2*, no. 4 (December 1987).

9. See the essay by Reingold and N. Reingold and M. Rothenberg (eds), *Scientific Colonialism* (Washington, 1987).

10. See Chapter 8 below.

11. Thomson's lab. was first established in 1850, but only officially recognised sixteen years later. This should not be confused with Thomas Thomson's Chemical Laboratory at Glasgow after 1829, the year in which

chemical laboratory work began also at University College, London. The second wave of university laboratories, from around 1878 to 1900, were associated with the academic rise of engineering and electrical engineering. See W. H. Armytage, *A Social History of Engineering* (New York, 1961); C. Domb (ed.), *Clerk Maxwell and Modern Science* (London, 1963); R. Sviedrys, 'The Rise of Physics Laboratories in Britain', *Historical Studies in the Physical Sciences, 7* (1976), pp. 405–35.

12. Prospectus of the *University of Durham, Founded 1831* (Durham, 1833).

13. *Tradesman's and Mechanics' Almanack* (London, 1833), pp. 69–71; for an outline of the Scottish tradition and institutional set-up see Chapter 3 above and G. E. Davie, *The Democratic Intellect: Scotland and her Universities in the Nineteenth Century* (Edinburgh, 1961).

14. For change and resistance to change at Oxbridge see Roy MacLeod and Russell Moseley, 'Breadth, Depth and Excellence: Sources and Problems in the History of University Science Education in England, 1850–1914', *Studies in Science Education, 5* (1978), pp. 85–106; F. S. Taylor, 'The Teaching of Science at Oxford in the Nineteenth Century', *Annals of Science, 8* (1952), pp. 82–112; Jack Morrell, 'Science and the Universities', *History of Science, 15* (1977), pp. 145–52; R. Sviedrys, 'The Rise of Physical Science at Victorian Cambridge', *Historical Studies in the Physical Sciences, 2* (1970), pp. 127–38; David B. Wilson, 'Experimentalists Among the Mathematicians: Physics in the Cambridge Natural Science Tripos 1851–1900', *H.S.P.S., 12* (1981), pp. 325–43.

15. By far the most thorough study of the new universities remains Michael Sanderson, *The Universities and British Industry 1880–1970* (London, 1972).

16. Joseph Ben-David, 'The Rise and Decline of France as a Scientific Centre', *Minerva 8* (1970), pp. 160–79; the commentary on the above by T. N. Clark, *Minerva, 8* (1970), pp. 599–601; and the overview from a contemporary focus in Robert Gilpin, *France in the Age of the Scientific State* (Princeton, 1968).

17. H. W. Paul, 'The Issue of Decline in Nineteenth-Century French Science', *French Historical Studies, 7* (1972), pp. 416–50, the relative absence of a clear 'industrial revolution' in France does not, of course, preclude the possibility of successful and gradual industrialisation and rising incomes per capita; see Chapter 3 of Clive Trebilcock, *The Industrialisation of the Continental Powers 1780–1914* (London, 1981). Rostow persists in finding a 'take-off' in 1830–70 based on coal, iron, railroads and engineering; W. W. Rostow, *The World Economy* (London, 1978), p. 400.

18. For overviews see H. I. Sharlin, *The Convergent Century* (London, 1967); David Knight, *The Nature of Science* (London, 1976).

19. Robert Fox, 'Learning, Politics and Polite Culture in Provincial France; the *Societés Savantes* in the 19thc', *Historical Reflections, 7* (1980); 'Science, Industry and the Social Order in Mulhouse 1798–1871', *British Jrnl Hist. Sci., 17* (1984).

20. The *Ecole Pratique* was founded from the State budget as well as the Municipal Council of Paris, to the tune of 366 000 francs in 1892.

21. Peter Lundgreen, 'Educational Expansion and Economic Growth in Nineteenth Century Germany: A Quantitative Study', in L. Stone (ed.), *Schooling and Society, Studies in the History of Education* (Baltimore, 1976); 'German Technical Associations Between Science, Industry and the State 1860–1914', *Historical Social Research, 13* (1980), pp. 30–15; 'Technicians and Labour Market in Prussia 1810–1850', *Annals Cisalpines d'Histoire Sociale, 2* (1971), pp. 9–29; Peter Lundgreen and W. Fischer, 'The Recruitment and Training of Administrative and Technical Personnel', in C. Tilly (ed.), *The Formation of National States in Western Europe* (Princeton, 1975).

22. The first *Technische Hochschule*, that of Berlin, was only formed in 1879 as a result of the merger between the Trade Institute (State-run) founded in 1821 and the Civil Engineering Academy. For vocational and trade education after 1820 see K. Harney, 'The Emergence of the Technical School System in Prussia', in Ian Inkster (ed.), *The Steam Intellect Societies* (Nottingham, 1985), pp. 131–41.

23. J. T. Merz, *A History of European Thought in the Nineteenth Century* (London, 1896–1914).

24. G. Haines, 'German Influence Upon English Education and Science 1800–1866', Monog. 6. Connecticut College (New London, 1957); R. H. Samuel and R. H. Thomas, *Education and Society in Modern Germany* (London, 1949); Arthur Shadwell, *Industrial Efficiency, A Comparative Study of Industrial Life in England, Germany and America*, 2 vols (London, 1906).

25. Shadwell, ibid., Ch. XVII, and the work of Lundgreen and other modern commentators.

26. Ibid., p. 426.

27. J. J. Lee, 'Labour in German Industrialisation', in P. Mathias and M. M. Postan, *The Industrial Economies, The Cambridge Economic History of Europe VII*, Part I, (Cambridge, 1978), pp. 442–91.

28. The provincial *Gewerbschulen* developed in the 1820s to serve the State-run trade institutes and acted as specialised schools for the regional employment markets.

29. Shadwell, op. cit. (n. 24), p. 428.

30. Calculated from material in Shadwell, C. H. Creasey, *Technical Education in Training Schools* (London, 1905) and F. Rose, *Chemical Instruction in Germany and the Growth and Present Conditions of the German Chemical Industry*, Diplomatic and Consular Reports (London, 1901); P. Magnus, *Industrial Education* (London, 1888); Michael Sadler, *Report on Secondary Education in Liverpool* (Liverpool, 1904).

31. Drawn from complete lists in *The London Gazette*, 1860, Vol. 2, April–June.

32. Table 4.5 is a re-categorisation of the many occupational distinctions in *1841 Census Great Britain, Occupations with Preface; Population 5*, British Parliamentary Papers, IURS reprints, Ireland.

33. Charles More, *Skill and the English Working Class 1870–1914* (London, 1980); P. Robertson, 'Technical Education in the British Shipbuilding and Marine Engineering Industries 1863–1914', *Econ. Hist. Review, 27* (1974) and his 'Employers and Engineering Education in Britain and the United

States 1890–1914', *Business History*, *23* (1981).

34. For some material on popular literature see 'Science in Grub Street', *Nature*, *224*, 1 November 1969, pp. 423–58; A. J. Meadows, *Communication in Science* (London, 1974).

35. Ian Inkster, 'Introduction: The Context of Steam Intellect in Britain', in Ian Inkster (ed.), *The Steam Intellect Societies* (Nottingham, 1985), pp. 3–20.

36. This embraces associations charging less than £1 per annum membership, providing classes, and some fixed facility in science and technology; equipment, laboratories, libraries, demonstration rooms, etc.

37. For the nature of the estimation see Inkster, op. cit. (n. 35), pp. 12–16, 19. For different estimations see D. S. L. Cardwell, *The Organisation of Science in England* (London, 1957), e.g. p. 57.

38. Cuthbertson, op. cit. (n. 1), p. 652.

39. Lewis T. Wright, 'Opening Address to Nottingham Branch', 23 November 1887; *Journal of the Society of Chemical Industry*, *7* (1888), p. 27.

40. Shadwell, op. cit. (n. 24), p. 429, 431. For the view of economists at this time see E. W. Evans and N. C. Wiseman, 'Education, Training and Economic Performance: British Economists' Views 1868–1939', *Journal of European Economic History*, *13* (1984), pp. 129–148.

41. Drawn from the annual *Official Handbook of Manchester and Salford*, Manchester for years 1843–1900, housed at Manchester City Reference Library, Local History Section, 352642MB. Column (a) relates only to voluntary associations promoting education/steam intellect outside of the 'schooling' system: (b) excludes public libraries, Owen's College and the enormous growth of 'clubs', of which there were nearly 60 by the end of the century; (c) excludes all 'trade' associations and all groupings with no intellectual/educational functions. In terms of total population, provisions of the 'cultural enterprise' seemingly peaked around the 1860s and 1870s. From then, Owen's College and government intervention reduced the constituency available to the less formalised institutions. In an estimate which included some 100 or so small private evening schools, J. H. Hinton argued that by 1850 Manchester was supporting something in the order of 6000–7000 students in evening schools. Denominational associations contributed to about 50 per cent of this total. See John H. Hinton, *A Review of the Evidence Taken Before the Committee of the House of Commons on Education in Manchester and Salford* (London, 1852). Even the Revd C. Richson's estimate of 4500 evening school attenders suggests a large potential audience for science subjects by mid-century.

42. *Report of the Commissioners on the Working of the Laws Relating to Letters Patent for Inventions*, *Both Houses*, London, H. of C. (1865), Appendix I.

43. Ibid., p. 152.

44. Thus the Manchester data shows an expansion of technical interest in sewing machines (nearly 5000 requests for patents during 1857–1860), telegraph and signalling equipment, methods of gas manufacture, dyeing and colouring processes.

45. Nearly 14 000 patent specifications had been reprinted in a five-year

period and copies distributed throughout Britain and the colonies. By 1865 the number of printed specifications available to libraries was over 43 000, forming 1711 volumes. The patent office was quite convinced that this availability of information had reduced the duplication of technical development work and consequent wastage of effort.

46. By the 1860s the patent office had also instituted the printing and advertising of *foreign* patent specifications, and in this the British were far ahead of any other nation.

47. The cost of patenting in Britain fell considerably between the 1840s and the 1860s, but the cost of obtaining copies of particular specifications fell even further, from around £7 or £8 to 8 or 10 shillings. Specifications without drawings could be obtained for as little as 1 or 2 shillings. Most periodicals for mechanics or artisans contained full technical details of major new inventions. The time taken and complexity involved in obtaining patents had also been greatly reduced.

48. Ian Inkster, 'The Ambivalent Role of Patents in Technology Development', *Bulletin of Science, Technology and Society*, *2* (1982).

49. Much of what follows arises from a detailed study of patent data in Britain and Australia in the years 1850–1914. The latter data allows an analysis of the patent 'invasion' of a small economic system by large British, European and American firms, and provides documentation of the technical history and marketing strategies of such firms. For a brief review of the Australian setting see Ian Inkster and Jan Todd, 'Support for the Scientific Enterprise 1850–1900', in R. Home (ed.), *Australian Science in the Making* (Cambridge, 1988), pp. 102–32, and Chapter 8 below.

50. J. M. Stopford, 'The Origins of British-based Multinational Manufacturing Enterprises', *Business History Review*, *48* (1974), pp. 303–34.

51. The best summary is A. D. Chandler, 'The United States: Evolution of Enterprise' in P. Mathias and M. M. Postan (ed.), *The Industrial Economies, Cambridge Economic History of Europe*, *VII*, no. 2, (Cambridge, 1978), pp. 70–133.

52. *Nature*, 14 November 1901, pp. 44–5. For excellent detail on the institutional structure in Germany see L. F. Haber, *The Chemical Industry during the Nineteenth Century* (Oxford, 1958).

53. See also Chapter 5 below and Chandler, op. cit. (n. 51), and his *The Visible Hand* (Cambridge, Mass., 1977), and (ed.) *Giant Enterprise* (New York, 1964); (ed.), *Managerial Innovation at General Motors* (New York, 1979); R. B. Davies, *Peacefully Working to Conquer the World: Singer Sewing Machines in Foreign Markets 1854–1920* (New York, 1976); G. R. Cooper, *The Sewing Machine: Its Invention and Improvement* (Washington, 1976); D. A. Hounshell, *From the American System to Mass Production 1800–1932* (Baltimore, 1984); L. S. Reich, *The Making of American Industrial Research, Science and Business at G. E. and Bell 1876–1926* (Cambridge, 1985); A. Heerding, *The History of N. V. Philips Gloeilamp enfabrieken*, Vol. I (Cambridge, 1986).

54. Reich, ibid.; G. D. Smith, *The Anatomy of a Business Strategy: Bell, Western Electric and the Origins of the American Telephone Industry* (Baltimore, 1985); David A. Noble, *America by Design: Science, Technology and the Rise of Corporate Capitalism* (New York, 1977).

55. For examples of the importance of patent regulations see Noble, ibid., Chapters 6 and 7. Smith, ibid., Chapter 1 and *passim* and N. R. Danieljan, *ATT & T: The Story of Industrial Conquest* (New York, 1939), Chapter 5. See also V. S. Clark, *History of Manufactures in the United States, 1860–1914* (Washington, 1928).

56. Constructed from material in Haber, op. cit. (n. 52) and *Chemist and Druggist*, 21 November 1896, 759; *Chemical Trade Journal*, 19 March 1910; *Board of Trade Journal*, 3 February 1910.

57. Haber, op. cit. (n. 52), p. 124; Rose, op. cit. (n. 30).

58. Haber, ibid., pp. 126f.

59. 'The Coal Tar Colour Industry in Germany and England', *Nature*, 12 December 1901, pp. 138–9.

60. The British firms were Read, Holliday & Co. (28 patents), Levinstein (19), Clayton Aniline Co. (21), Claus and Ree (9), Brooke, Simpson and Spiller (7), W. G. Thompson (2).

61. Haber, op. cit. (n. 52) gives excellent details of these firms, esp. p. 130 passim.

62. *Nature, 65* (1901), p. 138; see also *Nature, 34* (1886), a paper by Meldola warning of German chemical supremacy.

63. *Industries*, 29 October 1886, pp. 473–5.

64. W. H. G. Armytage, *The Rise of the Technocrats, A Social History* (London, 1965), pp. 177–8.

65. 'Work of the U.S. Bureau of Chemistry During 1910', *Oil, Paint and Drug Repository*, 13 November 1911.

66. These figures are from Rose, op. cit. (n. 30) and F. M. Perkin, 'Chemical Instruction and Chemical Industries in Germany', *Nature, 65* (1901), pp. 174–6.

67. The ratios in the polytechnics appear particularly good given the rate of expansion, around a 200 per cent increase in student numbers between 1887 and 1900.

68. *Nature*, 26 December 1901, p. 175. But contemporary German opinion included the view that the English method of 'heuristic' laboratory teaching was superior to that of Germany, especially against a background of classical learning, e.g. Karl T. Fischer, *Der Naturwissenschaftliche Unterricht in England, insbesondere in Physik und Chemie* (Leipzig, 1901).

69. P. Lundgreen, B. Horn, W. Krohn, G. Kuppers, R. Paslack, *Staatliche Forschung in Deutschland 1870–1980* (Frankfurt, 1986).

70. Philip Magnus (ed.), *Report on a Visit to Germany, With a View of Ascertaining the Recent Progress of Technical Education in the Country* (London, 1896), pp. 8–10.

71. *Nature, 86* (1911), 9 March, pp. 69–70.

72. Membership could thus be acquired by private enterprises as well as by individuals as long as this was approved by an elected Senate; the latter comprised subscribers and specialist *Gelehrte* (Savants).

73. See *Nature, 86* (1911), 23 February.

74. *Board of Trade Journal*, 17 August 1911; *British and Colonial Druggist*, January 1960; *U.S. Consular Reports*, July 1910.

75. *Chemical Industries, 34* (1911), pp. 588–9.

76. For a recent review of the role of the state from 1850 see Peter Alter, *The Reluctant Patron, Science and the State in Britain 1850–1920* (Oxford, 1987).

77. J. B. Poole and Kay Andrews (eds), *The Government of Science in Britain* (London, 1972); A. J. Taylor, *Laissez-faire and State Intervention in 19thc Britain* (London, 1972); Oliver MacDonagh, 'The Nineteenth Century Revolution in Government: A Reappraisal', *Historical Journal* I (1958), pp. 52–67.

78. See Alter, op. cit. (n. 76), Chapter 1; Cardwell, op. cit. (n. 37), Chapters 2 and 3.

79. These figures exclude *pensions* for services rendered to the State in the field of technology–industry. The calculations are from Colquhoun's *Treatise* of 1815 and *Report from the Select Committee on the Law Relating to Patents for Invention* (House of Commons, London, 1829) (Appendix).

80. 3 Geo. II c.29.

81. *Report 1865*, op. cit. (n. 42), Appendix I, p. 149.

82. These included the major provincial associations of Manchester and Birmingham, as well as the Society of Arts, Patent Law League and United Inventors Association.

83. This included reprinting of the old series (1623–1852) but not all costs of distribution or organisation, etc.

84. For a survey see G. W. Roderick and M. D. Stephens, 'Mechanics' Institutes and the State', in Ian Inkster (ed.), *Steam Intellect*, op. cit. (n. 35), pp. 60–72.

85. Shadwell, op. cit. (n. 24), p. 431.

86. Roy M. MacLeod, 'Statesmen Undisguised', *American Historical Review, 78* (1973), pp. 386–405.

87. Roy M. MacLeod, 'The Royal Society and the Government Grant: Notes on the Administration of Scientific Research 1849–1914', *Historical Journal, 14* (1971), pp. 323–58; Alter, op. cit. (n. 76), pp. 60–74.

88. R. A. Proctor, *The Wages and Wants of Science Workers* (London, 1876 reprinted 1970), pp. 45–6; Roy M. MacLeod, 'Science and the Treasury: Principles, Personalities and Policies 1870–1895', in G. L'E. Turner (ed.), *The Patronage of Science in the Nineteenth Century* (London, 1976).

89. *Nature, 65*, 27 March 1902, p. 487.

90. Alter, op. cit. (n. 76), pp. 138–49; Russell Moseley, 'The Origins and Early Years of the National Physical Laboratory: A Chapter in the Pre-History of British Science Policy', *Minerva, 16* (1978); Edward Pyatt, *The National Physical Laboratory: A History* (Bristol, 1983).

91. *The Japanese Industrial Laboratory*, Department of Agriculture and Commerce of the Japanese Imperial Government (Tokyo, 1909); *Journal of the Society of Chemical Industry, 29* (1910), p. 301, and various issues of ibid., 1906–9.

92. For this and its context, Chapters 11 and 12 of G. B. Sansom, *The Western World and Japan* (New York, 1949).

93. F. A. Bather, 'Natural Science in Japan', *Natural Science, IV*, January 1894, p. 190.

94. Reprinted in A. T. Simmons, 'Education and Progress in Japan', *Nature, 69*, 3 March 1904, pp. 416–18.

95. 'Why Japan is Victorious', *Nature*, *72*, 8 June 1905, pp. 128–9.

96. E. G. Holtham, *Eight Years in Japan* (London, 1883).

97. Henry Dyer, *Dai Nippon, The Britain of the East* (London, 1904), pp. 1–4.

98. *Dai Nippon*, ibid., pp. 4–5.

99. Holtham, op. cit. (n. 96) and Pat Barr, *The Deer Cry Pavillion* (New York, 1968), pp. 30–33.

100. *Japan Weekly Mail*, 5 February 1870, pp. 25–7.

101. E. J. Reed, *Japan: Its History and Traditions*, 2 vols (London, 1880), Vol. 2, pp. 143–5.

102. *Dai Nippon*, op. cit. (n. 97), p. 5.

103. John Perry, *Practical Mechanics* (London, 1883), esp. v–vi, pp. 214–15.

104. W. E. Griffis, *The Mikado's Empire* (London, 6th edn., 1980).

105. *Japan Weekly Mail*, 8 October 1870, pp. 476–78.

106. *Dai Nippon*, op. cit. (n. 97), pp. 6–7.

107. Griffis, op. cit. (n. 104), pp. 563–77, 358–9.

108. David Murray, *History of Education in New Jersey* (Washington, 1899), pp. 287–99, 227.

109. David Murray, *Outline History of Japanese Education* (New Jersey, 1876), David Murray, *Japan* (London, 6th edition, 1894), *Dictionary of American Biography*, pp. 358–9.

110. For a further treatment see Chapter 4 of Ian Inkster, *Science, Technology and the Late Development Effect*, Institute of Developing Economies (Tokyo, 1981).

111. *Japan Echo*, 1 November 1890, pp. 20–21; John Milne, *Seismology* (London, 1898), pp. 2–3, 305f.

112. *Japan Echo*, 1 November, pp. 20–21, 15 December, pp. 104–05, 1890, *The Engineer*, 6 May 1898, *Dai Nippon*, op. cit. (n. 97), p. 10.

113. *Dai Nippon*, ibid., pp. 10–11.

114. *Höchi Shimbun*, 27 November 1890, p. 3. For the importance of development projects (as against entire industries or individual enterprises) see Chapter 2 below, Chapters 8 and 9 above and Ian Inkster, 'Prometheus Bound: Technology and Industrialization in Japan, China and India Prior to 1914 – A Political Economy Approach', *Annals of Science*, *45* (1988).

115. *Japan Herald* 23 September 1871; *Japan Weekly Mail*, 30 September 1871.

116. *Japan Weekly Mail*, 30 September 1871, pp. 551–6.

117. Griffis, op. cit. (n. 104), pp. 601–5.

118. Barr, op. cit. (n. 99), pp. 26–7.

119. D. McCloskey (ed.), *Essays in a Mature Economy* (Cambridge, Mass., 1970); S. Nicholas, 'Total Factor Productivity Growth and the Revision of Post-1870 British Economic History', *Econ. Hist. Review*, *35* (1982), pp. 83–97; F. Crouzet, 'Western Europe and Great Britain: Catching Up in the First Half of the 19thc', in A. J. Youngson (ed.), *Economic Development in the Long Run* (London, 1972); P. O'Brien and C. Keyder, *Economic Growth in Britain and France 1780–1914* (Oxford, 1978).

120. D. Landes, *The Unbound Prometheus* (Cambridge, 1969), pp. 235–6.

121. Joseph Ben-David, *The Scientists' Role in Society: A Comparative Study* (New Jersey, 1971), pp. 125, 134.

5. TECHNOLOGY, ECONOMIC BACKWARDNESS AND INDUSTRIALISATION – GENERAL SCHEMA

1. Quoted in Eugene Tarlé, *Bonaparte* (New York, 1937), pp. 300–304.
2. For which see S. Griboyedov's play *The Misfortune of Being Clever*, published in 1823 and translated by Bernard Pares in John Cournos (ed.), *A Treasury of Russian Life and Humour* (New York, 1943); see also M. Blinoff (ed.), *Life and Thought in Old Russia* (University Park, Pennsylvania, 1961).
3. See the opening paragraph to the preface of Andrew Ure, *The Philosophy of Manufactures* (3rd edition, London, 1861).
4. H. Jomini, *Summary of the Art of War* (1st US edition, New York, 1868), pp. 48–9.
5. For good brief accounts of the relationship between military technology and tactics see J. E. C. Fuller, *The Conduct of War 1789–1961* (London, 1961).
6. John Yeats, *Recent and Existing Commerce*, 3rd edition (London, 1887), 'Tabulated List of Towns or Trade Centres Important in Business' (pp. 453–506). The list used 71 'local industry' categories.
7. In ibid. each centre's major productions are listed, and the first two listed in each case have been used to construct Table 5.3; production categories do not in fact = × 2 total centres, as some centres were single-product areas and others were difficult to judge with any accuracy.
8. P. Bairoch, 'International Industrialization Levels from 1750 to 1980', *Journal of European Economic History*, *11* (1982), see Tables pp. 281, 294, 296.
9. Calculated from summaries in H. K. Work, 'Metallurgy in the Nineteenth Century', *Journal of Chemical Education*, *28* (1950), pp. 364–8.
10. L. F. Haber, *The Chemical Industry During the Nineteenth Century*, (Oxford, 1958).
11. R. M. Hartwell, *The Industrial Revolution and Economic Growth*, (London, 1971), p. 12.
12. Marion J. Levy, Jr, *Modernisation: Latecomers and Survivors* (Basic Books, New York), 1972, pp. 12–13.
13. Only at one point does Gerschenkron refer to his approach as an 'illustrative model' and this may simply describe his typified version, as at *Continuity in History and Other Essays* (Cambridge, Mass., 1968), pp. 37–8. Henceforth *Continuity*.
14. Alexander Gerschenkron, *Economic Backwardness in Historical Perspective, A Book of Essays* (Cambridge, Mass., 1962), p. 1. Henceforth Gerschenkron, *EBHP*.
15. Most clearly expressed in Alexander Gerschenkron, *Europe in the Russian Mirror* (Cambridge, 1970), lecture 4, e.g. p. 99.
16. Gerschenkron, *EBHP*, pp. 21–3.

17. Ibid., p. 24.

18. Gerschenkron, 'Reflections on the Concept of "Prerequisites" of Modern Industrialisation' in *EBHP*, p. 33.

19. Ibid., pp. 34–6, 44, 46–7.

20. Ibid., pp. 41, 44.

21. Alexander Gerschenkron, 'Mercator Gloriosus', *Economic History Review* (1971), 643–66 (663).

22. Alexander Gerschenkron, 'The Discipline and I'; *Journal of Economic History, 27* (1967), pp. 443–59. As important as the 'smoothing' effects of aggregation is the simple point that such aggregates are unreliable as historical evidence. At another place, with reference to Rostow, Gerschenkron writes of the difficulty in using national income statistics: *Continuity*, op. cit. (n. 13), p. 35. See also his 'Problems in Measuring Long Term Growth in Income and Wealth', *Journal of the American Statistical Association, 52*, (1957), pp. 450–70.

23. *EBHP*, p. 8. On the notion of 'obstacles' and the criticism of them as absolute conditions facing universal solutions see A. O. Hirschman, 'Obstacles to Development: A Classification and a Quasi-Vanishing Act', *Economic Development and Cultural Change, XII*, (1965), and Paul M. Sweezy, 'Obstacles to Economic Development' in C. H. Feinstein, (ed.), *Socialism, Capitalism and Economic Growth* (Cambridge, 1967). Hirschman cites Gerschenkron's approach in his analysis of cases where alleged obstacles turn out to be 'not only substitutable, but outright dispensable' (p. 389).

24. *EBHP*, 'Prerequisites', pp. 46, 47.

25. John P. McKay, *Pioneers for Profit: Russian Industrialisation 1885–1913* (Chicago, 1970). A formal statement by Gerschenkron on the labour–technology link reads: 'The inadequate labour supply was substituted for by introduction of modern labour-saving technology . . . the importation of technology and qualified personnel from abroad being a substitution for the missing prerequisites of indigenous knowledge and deficiency in educational background', *Russian Mirror*, op. cit. (n. 15), pp. 99–104.

26. Gerschenkron, 'Comment on Schumpeter' in D. L. Spencer and A. Woroniak (eds), *The Transfer of Technology to Developing Countries* (New York, 1968), pp. 83–8, (86).

27. 'Mercator Gloriosus, op. cit. (n. 21), p. 664; J. Hicks, *A Theory of Economic History* (London, 1969), pp. 145–8.

28. R. Minami, *Power Revolution in the Industrialisation of Japan 1885–1940* (Hitotsubashi University, 1979), typescript, p. 308. I would like to thank Dr Minami for allowing me access to this typescript prior to the English-language publication of his text.

29. Ibid., p. 314.

30. For 'settlement' see Masaru Saito, 'Introduction of Foreign Technology in the Industrialisation Process', *The Developing Economies, 15* (1977).

31. W. W. Rostow, *The World Economy* (London, 1978), p. 65; E. J. Hobsbawm, *The Age of Capital, 1848–1875* (London, 1977), p. 49.

32. For the crucial importance of economies of specialisation for producers' goods (cf. economies of scale for consumer goods) see Nathan

Rosenberg, 'Capital Goods, Technology and Economic Growth', *Oxford Economic Papers, 15* (1963).

33. Such seeming quantities are only ever estimates; see Chapter 5 of James Foreman-Peck, *A History of the World Economy* (Brighton, 1983).

34. D. C. M. Platt, *Foreign Finance in Continental Europe and the United States, 1815–1870* (London, 1984). Platt shows that financial flows were frequently speculative and were smaller in their *net* flow than has often been argued. However, this does not impinge upon arguments which emphasise the sometimes strategic value of imported capital (e.g. on technologies), the impact on developers of investment which might be considered small in 'donor' terms, the impact of capital flow on interest rates in both donor and receiver nations, or the frequent importance of capital transfers occurring outside large companies or government securities.

35. Rondo E. Cameron, *France and the Economic Development of Europe, 1800–1914* (Princeton, 1961), p. 403. Cameron's exhaustive account illustrates his point.

36. For this and similar examples, ibid., pp. 375–97.

37. Rostow, op. cit. (n. 31), p. 144.

38. Ibid., Table, p. 4.

39. But see also Foreman-Peck, op. cit. (n. 33), p. 145.

40. There is much conjecture in this field. For older treatments of North America see C. Johnson, *A History of Emigration from Britain to North America, 1763–1912* (London, 1913) and R. T. Berthoff, *British Immigrants in Industrial America 1790–1950* (Cambridge, Mass., 1953).

41. For recent studies see H. Keil and J. B. Jentz (eds), *German Workers in Industrial Chicago, 1850–1910: A Comparative Perspective* (Dekalb, Illinois, 1983); D. Fitzpatrick, *Irish Emigration* (Dublin, 1984); K. A. Miller, *Emigrants and Exiles* (Oxford, 1985); and the series of papers on different immigrations into Britain in *History Today*, Vol. 35, 1985.

42. Although Brunel's *Great Western* and *Great Britain* are well-known early examples, the steamship only slowly surpassed the improved sailing ship, and this required innovations in fuel-economy, e.g. Alfred Holt's compound engines, and those which allowed the replacement of iron by steel, e.g. the Cunard liners. Steam navigation also awaited the erection of a system of worldwide coaling stations.

43. Insulation of cables followed from scientific work, as did improvements in electrical telegraphy. By 1862 the world telegraph system covered some 150 000 miles, 50 per cent of which was on the continent. For a very brief but clear account of the scientific work involved see T. K. Derry and T. I. Williams, *A Short History of Technology* (Oxford, 1960), pp. 627–9.

44. Sidney Pollard, *Peaceful Conquest* (Oxford, 1981), pp. 164–84 (quote p. 176).

45. Hobsbawm, op. cit. (n. 31), p. 83.

46. For a range of evidence see W. O. Henderson, *Britain and Industrial Europe 1750–1870* (Leicester, 1972).

47. On Liebig's importance in Britain see A. and N. L. Clow, *The Chemical Revolution* (London, 1952); J. B. Morrell, 'The Chemist Breeders:

The Research Schools of Liebig and Thomas Thompson', *Ambix*, *19* (1972); G. K. Roberts, 'The Establishment of the Royal College of Chemistry: An Investigation of the Social Context of Early-Victorian Chemistry', *Hist. Studs, Phys. Sci.*, *7* (1976); A. W. Hofmann, *Introduction to Modern Chemistry* (London, 1866); B. Lepsius, *August Wilhelm von Hofmann* (Leipzig, 1905).

48. For subsequent developments see W. G. Rimmer, *Marshall's of Leeds, Flax-Spinners, 1788–1886* (London, 1960).

49. J. D. Scott, *Siemens Brothers 1858–1958, An Essay in the History of Industry* (London, 1958).

50. D. C. Hague, *The Economics of Man-Made Fibres* (London, 1957).

51. W. O. Henderson, *The Industrial Revolution on the Continent 1800–1914* (London, 1967).

52. For general criticism see C. Trebilcock, *The Industrialisation of the Continental Powers, 1870–1914* (London, 1981), throughout and Chapters 1 and 6; for specific instances see A. Kahan, 'Government Policies and the Industrialisation of Russia', *Journal of Economic History*, *27* (1967); J. G. Williamson and L. J. de Bever, 'Savings, Accumulation and Modern Economic Growth: The Contemporary Relevance of Japanese History', *Journal of Japanese Studies*, *4* (1978); F. B. Tipton, 'Government Policy and Economic Development in Germany and Japan: A Skeptical Revaluation, *Journal of Economic History*, *XLI* (1981). For a more general approach from a development/structural perspective, R. H. Green, 'The Role of the State as an Agent of Economic and Social Development in the Less Developed Countries', *Journal of Development Planning*, *6* (1974). See also the brief but close assessment of Sidney Pollard, op. cit. (n. 44), pp. 159–63.

53. William de la Rive, *Reminiscences of the Life and Character of Count Cavour* (London, 1862), esp. pp. 55, 58–9.

54. Alexander Vucinich, 'Politics, Universities and Science', in T. G. Stavron (ed.), *Russia Under the Last Tsar* (Minneapolis, 1969), pp. 154–78.

55. I. T. Berend and G. Ranki, *Hungary, A Century of Economic Development* (New York, 1974).

56. J. J. Lee, 'Labor in German Industrialisation' in P. Mathias and M. M. Postan (eds), 'The Industrial Economies', *The Cambridge Economic History of Europe*, *VII*, Part 1 (Cambridge, 1978), pp. 442–97; T. K. Derry and T. I. Williams, *A Short History of Technology* (London, 1960), pp. 704–5. Apprenticeship yields greater control to the employer over what is taught, when and where, and usually includes some provision for containing the human capital thus created within the bounds of the firm. Premiums and lower wages only allow recoupment of expenses and cannot themselves take sufficient account of the effect of workers leaving the firm after training. In Germany, such movement of skilled workers was to the advantage of the larger firms. On the other hand, a high level of enterprise involvement means that the State looses control over the training process.

57. Vucinich, op. cit. (n. 54); C. E. Black, 'The Nature of Imperial Russian Society', in P. W. Treadgold (ed.), *The Development of the USSR* (London, 1973), pp. 186–7, and his (ed.), *The Transformation of Russian Society* (Cambridge, Mass., 1960); Olga Crisp, 'Labour and Industrialisation

in Russia' in Mathias and Postan, op. cit. (n. 56), pt. ii, pp. 308–415.

58. Ibid., p. 392; Vucinich, op. cit. (n. 54), pp. 158–70.

59. Yoichi Yano, 'The Development of Technical Education in Meiji Japan: An Interpretation', in Ian Inkster (ed.), *The Steam Intellect Societies, Essays on Culture, Education and Industry circa 1820–1914* (Nottingham, 1985). See also Chapter 6 below.

60. A point made by Inkster and Nicholas in *ibid.*, essays 7 and 13.

61. M. Tapley and M. Simmonds, 'International Diversification in the Nineteenth Century', *Colombia Journal of World Business*, Summer 1982, esp. pp. 67–8.

62. There is much debate on the determinants of transnational production prior to and after 1914. See M. Wilkins, *The Emergence of Multinational Enterprise: American Business Abroad from the Colonial Era to 1914* (Cambridge, Mass., 1970); J. M. Stopford, 'The Origins of British-Based Multinational Manufacturing Enterprises', *Business History Review*, Autumn 1974 and in the same issue the paper by L. G. Franko.

63. A. G. Kenwood and A. G. Loughheed, *Technological Diffusion and Industrialisation Before 1914* (London, 1982), esp. Chapter 11.

64. *Ibid.*, p. 179.

65. The literature on patents and patent history is large. For summaries see G. H. Fox, *Monopolies and Patents* (Toronto, 1947); F. Machlup and E. Penrose, 'The Patent Controversy in the Nineteenth Century', *Journal Economic History*, *X* (1950); Chapter 4 of W. W. Rostow, *How It All Began* (London, 1975); Ian Inkster, 'The Ambivalent Role of Patents in Technology Development, *Bulletin of Science, Technology and Society*, *2* (1982).

66. This was nowhere better illustrated than in the differences between patent regulations in Germany and Britain in the post-1852 years. See Haber, op. cit. (n. 10), pp. 198–204.

67. Ibid., p. 199.

68. Henderson, op. cit. (n. 46), p. 31; on Platt machinery see also Chapter 6 below.

69. E. Schiff, *Industrialisation Without National Patents, The Netherlands and Switzerland to 1912* (Princeton, 1971).

70. *Report of the Commissioners Relating to Letters Patent for Invention*, House of Commons and House of Lords (London, 1865), p. 153 (Table).

71. Though the high percentage of French patenting might have reflected the large role of French capital in European industrialisation, it was almost certainly some indication of differing legislation, e.g. *Report from Select Committee on Laws Relative to Patents for Inventions*, House of Commons (London 1829), pp. 67–8, 11, 222; ibid., p. 169.

72. *Twelth Report of the Controller-General of Patents, Designs and Trade Marks, with Appendices for the Year 1894* (London, 1895); *Chemist and Druggist*, 22 February 1887, p. 336; *Journal of the Society of Chemical Industry*, *XX* (1901), p. 515; *Journal of the Franklin Institute*, Vols 113 (p. 61), 109 (p. 309). See also Section 8.3 below.

73. On the inadequacies of the US system see *Journal of the Franklin Institute*, *72* (1862), p. 271f, *22* (p. 158) and *23* (p.197) (1837); *94* (1873), p. 229.

74. *Ding. Polytechnic Journal*, *295* (1894) pp. 160–64. The 10 groups are my own but coincide with those used by contemporaries: see *Journal of Society of Chemical Industry*, *XIV* (1895), pp. 406–9.

75. For example, instrument-making, watchmaking and a variety of local 'public-works' technologies.

76. *Historical Statistics of the United States, Colonial Times to 1970*, Vol. 2, pt 2, pp. 954–9.

77. To 1836 US patents were granted 'on demand' without real examination. Consequently, statistics of patents *issued* during this period are more comparable to subsequent statistics of *applications* than to subsequent statistics of 'issues'. To 1836 only aliens who had resided in US for two years could apply; from 1836 to 1861 aliens paid higher fees.

78. *Report of the Select Committee on Letters Patent, Proceedings of Committee, Appendices*, House of Commons (London, 1871), Append. 4, p. 203.

79. *The Official Australian Journal of Patents, Statistical Tables*, *29* (Melbourne, 1919), p. xv; *21* (1914), p. xv; *13* (1910), p. xiv, I (Supplement) (1904), xiv.

80. For some substantiation of these generalisations see the separate treatment of these economies in part 5 of Rostow op. cit. (n. 31). For a brave attempt at national rankings in terms of GNP per capita and literacy levels see L. G. Sandberg, 'Ignorance, Poverty and Economic Backwardness in the Early Stages of European Industrialisation', *Journal of European Economic History*, *11* (1982), p. 687 (Table).

6. INDUSTRIALIZATION: WINNERS AND LOSERS

1. I. T. Berend and G. Ranki, *Hungary, A Century of Economic Development* (London, 1974).

2. Gerschenkron, *EBHP* (*op. cit.*), various parts; T. H. von Laue, 'The Chances for Liberal Constitutionalism' *Slavic Review*, *XXIV* (1965); idem, *Why Lenin? Why Stalin?* (Princeton, 1964); and idem, in C. E. Black (ed.), *The Transformation of Russian Society* (Cambridge, Mass., 1960).

3. A. Podkolzin, *A Short Economic History of the USSR* (Moscow, 1968), p. 45.

4. Trebilcock, op. cit. (n. 52 of Chapter 5), p. 208.

5. The best account of Witte's policy remains T. H. von Laue, *Sergei Witte and the Industrialisation of Russia* (New York, 1963).

6. For further details of which see Trebilcock and von Laue above as well as J. P. MacKay, op. cit. (n. 25 of Chapter 5).

7. For general yet thorough accounts of the group and their peripheralisation see Pollard, op. cit. (n. 44 of Chapter 5), Chapters 5 and 6, and Trebilcock, op. cit. (n. 4), Chapter 5.

8. For spread effects see J. G. Williamson, 'Regional Inequality and the Process of National Development, a Description of the Pattern' in L. Needleman (ed.), *Regional Analysis* (Harmondsworth, 1968).

9. A. Gerschenkron, 'The Rate of Industrial Growth in Russia since 1885', *Journal of Economic History*, Supplement 7 (1947); R. A. Roosa,

'Russian Industrialists and State Socialism, 1906–17', *Soviet Studies*, *23* (1972).

10. The term is von Laue's and relates to the 'gamble' that foreign indebtedness would be removed by the success of industrial policy before such capital inflow was halted.

11. I. L. Bell, 'On the Separation of Carbon, Silicon, Sulphur and Phosphorus in the Refining and Puddling Furnace and in the Bessemer Converter', *Journal of Iron and Steel Institute*, *11* (1877), parts I and II, and *12* (1878), part I; George Snelus, 'On the Removal of Phosphorus and Sulphur During the Bessemer and Siemens Martin Processes of Steel Manufacture', ibid., *13* (1879).

12. D. L. Burn, *The Economic History of Steelmaking 1867–1939* (Cambridge, 1940). 'Basic' because Thomas and Percy Gilchrist introduced limestone in the firebricks of the converter and in the 'charge', which combined with phosphorus to form basic slag, which when pulverised produced fertiliser.

13. As suggested in Table 5.5 of Chapter 5. Of course, the production of *pig* iron was complementary, and improved due to innovations in furnaces, 'rapid driving'. On the finishing side there were continued improvements in rolling (reversing-mills) and continuous milling.

14. For diffusion and adaptation of the J. S. Macarthur and R. W. and W. Forrest patent (using potassium cyanide solutions to selectively dissolve and therefore extract gold from complex physical and chemical combinations in natural ores) see James Park, *The Cyanide Process of Gold Extraction* (London, 1900).

15. Mirks Lamor, *The World Fertiliser Economy* (Stanford, 1957). The major nutrients in commercial fertilisers are nitrogen, phosphorus and potassium. For phosphorus see n. 12. Nitrogen fixation (from the air) may be performed naturally by micro-organisms. Haber and Bosch produced synthetic ammonia (NH_3) using coke as a source of hydrogen.

16. David Landes, *The Unbound Prometheus* (Cambridge, 1969), p. 202.

17. Pollard, op. cit. (n. 44 of Chapter 5 above), presents the most sophisticated case for regional analysis.

18. A. O. Hirschman, *Development Projects Observed* (Washington, 1967), p. 1.

19. Ibid., p. 1.

20. Incremental decision-making in a combination of related projects may serve to create a larger 'decision-making' system in the industrial economy as a whole. For this and its relationship to the Gerschenkron schema (which seems to impose large burdens on the decision-making system during the 'industrial drive') see B. Klein and W. Meckling, 'Applications of Operations Research to Development Decisions', *Operations Research*, *6* (1958); C. E. Lindblom, 'The Science of "Muddling Through"', *Public Admin. Review*, *19* (1959); A. O. Hirschman and C. E. Lindblom, 'Economic Development, Research and Development, Policy Making: Some Converging Views', *Behaviourial Science*, *7/8* (1962–63); and the discussion in Chapter 1 of Ian Inkster, *Science, Technology and the Late Development Effect* (Tokyo, 1981).

21. G. Ahlström, *Engineers and Industrial Growth* (London, 1982).

22. E. Gazza, *Ansaldo 1853–1953* (Genova, 1953), esp. pp. 63, 86, 100–109, 224.

23. Peter Temin, *Iron and Steel in Nineteenth Century America* (Cambridge, Mass. 1964); Chapter 10 of N. Rosenberg, *Perspectives on Technology* (Cambridge, 1976).

24. On which see ibid. and Chapters 1 above and 7 below.

25. Estimates vary: see R. Fremdling, 'Railroads and German Economic Growth: A Leading Sector Analysis with a Comparison to the United States and Great Britain', *Journal of Economic History, 37* (1977); R. H. Tilly, 'Capital Formation in Germany in the Nineteenth Century', in P. Mathias and M. M. Postan, *The Cambridge Economic History of Europe*, Vol. 7, Part 1 (Cambridge, 1978), esp. pp. 414–41. Jenks had calculated that some 50 French companies had raised £80 million in the London capital market by 1845. For a sceptical view of this see Platt, *op. cit.* (n. 34 of Chapter 5), pp. 18–27.

26. J. Schmookler, *Invention and Economic Growth* (Cambridge, Mass., 1966).

27. P. T. Ellsworth, 'The Terms of Trade Between Primary Producing and Industrial Countries', *Inter-American Economic Affairs, 10*, (1956). Of course, much depends on the years chosen as appropriate terminals. See P. Bairoch, *The Economic Development of the Third World Since 1900* (Berkeley, 1977), Chapter 6.

28. For this and a judgement on the terms of trade effects see James Foreman-Peck, *A History of the World Economy* (Brighton, 1983), pp. 110–12.

7. TECHNOLOGY, ECONOMIC BACKWARDNESS AND INDUSTRIALISATION – THE CASE OF JAPAN

1. *Japan Weekly Mail*, 25 June 1870; *The Chrysanthemum* (Yokohama) II, No. 7, July 1882; F. A. Bather, 'Natural Science in Japan', *Natural Science IV*, January–June 1894. For sophisticated versions and qualifications along the same lines see the papers by Charles Moore, Miyamoto Shōson, Yukawa Hideki in Charles A. Moore (ed.), *The Japanese Mind, Essentials of Japanese Philosophy and Culture*, (Tokyo, 1973).

2. Reprinted in A. T. Simmons, 'Education and Progress in Japan', *Nature, 69*, 3 March 1904.

3. 'Why Japan is Victorious', *Nature, 72*, 8 June 1905. For several years war was seen as a sign of a healthy Japan, e.g. F. Coleman, *Japan Moves North* (London, 1918).

4. The best English-language surveys of the pattern of Japanese industrial and economic growth include Sake Tsunoyama, *Concise Economic History of Modern Japan* (Bombay 1965); J. W. Dower (ed.), *Origins of the Modern Japanese State, Selected Writings of E. H. Norman* (New York, 1975); R. P. Sinha, 'Unresolved Issues in Japan's Early Economic Development', *Scottish Journal of Political Economy, 15–16* (June 1969); K. Takahashi, *The Rise and Development of Japan's Modern Economy* (Tokyo, 1969); Yoshihara Kunio, *Japanese Economic Development, A Short Introduction* (Tokyo and Oxford, 1979);

E. S. Crawcour, 'Aspects of Economic Transition in Japan, 1840–1906', *Papers in Far Eastern History, 24*, September 1981 (Australian National University); and for a good analytical summary, Chapter 30 of Benjamin Higgins, *Economic Development, Problems, Principles and Policies* (New York, revised edition 1968). The best general textbook remains W. W. Lockwood, *The Economic Development of Japan, Growth and Structural Change*, (Princeton, 1968, expanded edition). The most recent treatment of economic development in a wider context is Jean-Pierre Lehmann, *The Roots of Modern Japan* (London, 1982).

5. T. C. Smith, *Political Change and Industrial Development in Japan: Government Enterprise 1868–80* (Stanford, 1955); K. Ohkawa and H. Rosovsky, *Japanese Economic Growth – Trend Acceleration in the 20th Century* (Stanford, 1973), quote pp. 39–40.

6. Ibid., pp. 218–19.

7. For examples see T. C. Smith, op. cit. (n. 5), Ian Inkster, 'Meiji Economic Development in Perspective: Revisionist Comments Upon the Industrial Revolution in Japan', *The Developing Economies*, XVII (1979); W. W. Lockwood op. cit. (n. 4); Mitsutomo Yuasa, 'The Role of Science and Technology in the Economic Development of Modern Japan', *12th Congress International d'Histoire des Sciences Paris 1968, XI*, Paris, 1971. For a sample of contemporary laudatory Western accounts of introduced 'best-technique' see *The Engineer, 82* (1896), *84* (1897); *Chemical News, 40* (1899); *Engineering*, October 1897, February 1898; *Indian and Eastern Engineer*, 27 June 1896; *Natural Science, IV*, January–June 1894; *Harpers Weekly*, January 1898. For official reports see *Board of Trade Journal*, December 1896 and *Bulletin of the (US) Department of Labour* January 1896.

8. For estimates of the industrial and economic growth rates see Ohkawa and Rosovsky, op. cit. (n. 5), K. Ohkawa, M. Shinohara, and L. Meissner, *Patterns of Japanese Economic Development: A Quantitative Appraisal* (New Haven and London, 1979), Chapter 9 of Lloyd G. Reynolds, *Image and Reality in Economic Development* (New Haven and London, 1977). The best summary promises to be Professor Ryoshin Minami's *Economic Development of Japan: A Quantitative Study*, in preparation for Macmillan, London.

9. The best of the early work in English is Thomas C. Smith, 'Farm Family By-employments in Preindustrial Japan', *Journal of Economic History, 29* (1969) and 'Pre-Modern Economic Growth: Japan and the West', *Past and Present, 60* (1973). See also, for general descriptions, D. Lloyd Spencer, 'Japan's Pre-Perry Preparation for Economic Growth', *American Journal of Economics and Sociology, 17* (1957–58).

10. The best account of the link between social control, urbanism and communications remains John Whitney Hall, 'The Castle Town and Japan's Modern Urbanisation', *Far Eastern Quarterly, 15* (1955–56). See also Takeo Yazaki, *Social Change and the City in Japan, From Earliest Times Through the Industrial Revolution* (Tokyo, 1968); David Kornhauser, *Urban Japan* (London, 1976); Matsuyo Takizawa, *The Penetration of Money Economy in Japan* (New York 1927, 1968).

11. The relative deprivation of groups amongst the samurai class has given rise to a literature on the origins of Japanese entrepreneurship: see

T. Hirschmeier, *The Origins of Entrepreneurship in Meiji Japan* (Harvard, 1964); Hirschmeier and Tsunehiko Yui, *The Development of Japanese Business 1600–1973* (London, 1975); Kozo Yamamura, *A Study of Samurai Income and Entrepreneurship* (Cambridge, Mass., 1974).

12. For Tokugawa culture and politics; J. W. Hall and M. B. Jansen (eds), *Studies in the Institutional History of Early Modern Japan* (Princeton, 1968); D. Magerey Earl, *Emperor and Nation in Japan* (Seattle, 1964); H. Webb, *The Japanese Imperial Institution in the Tokugawa Period* (New York, 1968). For a most accessible, brief account see Chapter 2 of Tetsuo Najita, *Japan, The Intellectual Foundations of Modern Japanese Politics* (Chicago and London, 1974). There is much of use in Lehmann, op. cit. (n. 4), pp. 130–62.

13. Several economic historians would either fail to recognise this period or mark it as merely 'transitional', between neo-feudalism and growth acceleration. We would argue that this interpretation arises from the insufficiency of statistical data for the years prior to 1881–5, and that the institutional formation of this phase was essential (i.e. necessary but not sufficient) to the production of industry in the 1880s and beyond. For a discussion of such matters see Chapter V of Ian Inkster, *Science, Technology and the Late Development Effect: Transfer Mechanisms in Japan's Industrialisation circa 1850–1912*, Institute of Developing Economies (Tokyo, 1981).

14. For such elements in European nations see Chapter 5 above, section III. For Japan see Tetsuo Najita, op. cit. (n. 12), Chapter 3, S. Tōbata, *The Modernisation of Japan*, Institute of Developing Economies (Tokyo, 1966); K. Nakamura, *The Formation of Modern Japan as Viewed from Legal History*, Centre for East Asian Cultural Studies (Tokyo, 1962); G. O. Totten III, 'Adoption of the Prussian Model for Municipal Government in Meiji Japan: Principles and Compromises', *The Developing Economies, XV* (1977). Between the foundation of the Ministry of Justice in 1871 and the systematisation of a Commercial Code in 1881 (eventually adopted in 1890) there was much confusion due to the early enthusiasm for French law. German codes were victorious in the 1890s.

15. For example of which see the articles in the special issue 'Adaptation and Transformation of Western Institutions in Meiji Japan', in *The Developing Economies, XV*, No. 4, December 1977.

16. Inflation was a *political* problem; Hiroshi Shinjo, *History of the Yen* (Tokyo, 1962).

17. See the papers by Saxonhouse and others in Hugh Patrick (ed.), *Japanese Industrialisation and its Social Consequences* (Berkeley, 1976); A. Gordon, *The Evolution of Labour Relations in Japan: Heavy Industry 1853–1955* (Harvard, 1985).

18. See n. 13 on doubts as to the genesis of this phase. As to its completion, there was little of a watershed character in either the political transition to the Taisho period (1912–26) or the outset of war in Europe in 1914. It could with some justice be argued that characteristics of industrial revolution remained into the heavy industrialisation and imperial years of the 'Shōwa Restoration' in the 1930s.

19. For agriculture and cotton see Section 7.4 below. For a summary of

foreign technology and improvements in railways see Harada Katsumasa, 'Japan's Discovery, Imports and Technical Mastery of Railways', *Project on Technology Transfer, Transformation and Development: The Japanese Experience*, United Nations University (Tokyo, 1979), paper HSDRJE-12/UNUP-51. For shipbuilding see S. A. Broadbridge, 'Technological Progress and State Support in the Japanese Shipbuilding Industry', *Journal of Development Studies, 1* (1964–65), Section II and T. C. Smith, op. cit. (n. 5).

20. For the patent system see Section II of Ian Inkster, 'On "Modelling Japan" for the Third World. (Part One)', *East Asia, 1* (1983); *Bureau des Brevet d'Invention, Lois concernant la protection de la proprieté Industrielle dans L'Empire du Japan* (Exposition Universelle de 1900 Paris), Paris 1900–1901; *Board of Trade Journal*, June 1896, February 1897.

21. W. W. Rostow, *The World Economy* (London, 1978), p. 422.

22. Official foreign loans in the early years were fairly negligible, but exclude borrowing arrangements made between Western traders and Japanese private enterprise, e.g. in the mining sector. Between 1896 and 1913 there was an acceleration in Japan's borrowing, perhaps amounting to a total of $800 million, or representing one-third of capital formation in Japan in those years. Most investment was by bond issue. To this might be added the considerable inflows as indemnities from China between 1895 and 1902.

23. J. G. Williamson and Leo J. de Bever, 'Savings, Accumulation and Modern Economic Growth: The Contemporary Relevance of Japanese History', *Journal of Japanese Studies, 4* (1978).

24. R. Huber, 'Effect on Prices of Japan's Entry into World Commerce After 1888', *Journal of Political Economy, 78* (1971).

25. For Japanese trading relations see Section 7.2, the references in n. 49 and Toshihiko Kato, 'Development of Foreign Trade' in Shibusawa Keizō (ed.), *Japanese Society in the Meiji Era* (Tokyo, 1958), and, for the political–diplomatic background see Marinosuke Kajima, *A Brief Diplomatic History of Modern Japan* (Tokyo, 1965). From 1871 there was pressure within Japan to revise the unequal treaties (which opened Japan to foreign goods), but only from 1894 did Japan begin to achieve customs autonomy, and only in April 1911 was the asymetric relationship finally ended.

26. For which see Hiroshi Shinjo, op. cit. (n. 16) and the excellent survey Koichi Emi, *Government Fiscal Activity and Economic Growth in Japan 1868–1960* (Tokyo, 1963). During the 1870s land tax revenue was of crucial importance.

27. For a good contemporary account of which see F. O. Adams, *A History of Japan* (London, 1874).

28. For a laudatory account see H. Tennant, 'The Commercial Expansion of Japan', *The Contemporary Review*, LXXXI (1897); for modern statistical exercises see Ippei Yamazawa and Akira Hirata, 'Industrialisation and External Relations: Comparative Analysis of Japan's Historical Experience and Contemporary Developing Countries' Performance', *Hitotsubashi Journal of Economics, 18* (1978).

29. The Portugese period is undergoing revaluation. See for a recent account C. R. Boxer, *Portugese Merchants and Missionaries in Feudal Japan* (London, 1986).

30. This 4-volume Japanese-edited compilation was the work of Christovao Ferreira (1580–1652) a *korobi bateren* (fallen padre) who had converted to Buddhism after inquisition by the Tokugawa shogunate. The book is important in combining introduced Western knowledge alongside Confucian commentaries.

31. G. B. Sansom, *The Western World and Japan* (New York, 1949, 1973); idem, *A History of Japan III 1615–1867* (London, 1964).

32. G. K. Goodman, *Japan: The Dutch Experience* (London, 1986), a rewriting of his classic *The Dutch Impact on Japan* (1967); J. Maclean, 'The Introduction of Books and Scientific Instruments into Japan 1712–1854', *Japanese Studies in the History of Science, 13* (1974).

33. Baron K. Suyematsu, *The Risen Sun* (London, 1905).

34. Thus the *Fūsetsu-sho* of the Dutch factory 'captains' became detailed reports on such subjects as railroadisation, French development in the Panama isthmus and military technologies.

35. Encapsulations were probably of most importance, as they injected elements of Japanese culture or values into the presentation of Western knowledge, and thereby hastened acceptance, e.g. the 5-volume *Taisei Shichikin Yakusetsu* on Western metallurgy, chemistry and pharmacy.

36. On the importance of *daimyo* pre-Meiji activity see T. C. Smith, op. cit. (n. 5); and especially pp. 317–89 of Dower, op. cit. (n. 4).

37. Detailed examples are provided in Sake Tsunoyama, op. cit. (n. 4). It was Satsuma that sent the technological mission to England in 1867 which resulted in the introduction of British cotton-spinning technology into Kagoshima.

38. See references in n. 10.

39. For such categories see R. R. Nelson and S. G. Winter, 'In Search of a Useful Theory of Innovation', *Research Policy, 6* (1977); F. R. Bradbury, V. T. P. Jervis, R. Johnson and A. W. Pearson, *Transfer Processes in Technical Change* (Rijn, Netherlands, 1978); B. J. Loasby, *Choice, Complexity and Ignorance* (London, 1976).

40. See essays in A. W. Burks (ed.), *The Modernisers: Overseas Students, Foreign Employees and Meiji Japan* (London, 1985). For earlier work, D. E. Smith and Y. Mikami, *A History of Japanese Mathematics* (Chicago, 1914); Masao Watanabe 'Science Across the Pacific', *Japanese Studies in the History of Science, 9* (1970); Marlene Mayo, 'The Western Education of Kume Kunitake', *Monumenta Nipponica, 28* (1973), P. W. van der Pas, 'Japanese Students of Mathematics at the University of Leiden During the Sakoku Period', *Japanese Studies History of Science, 14* (1975); Yoshio Hara, 'From Westernisation to Japanisation: The Replacement of Foreign Teachers by Japanese Who Studied Abroad', *The Developing Economies, 15* (1977).

41. At the 1867 Paris exhibition a party of 66 from Japan attended and produced a 95-volume report on the economic, institutional and social advances displayed there. Japanese internal industrial exhibitions began in 1877 and by the 1890s they were attracting over a million visitors. Between 1867 and 1910 Japanese participated in at least 38 major international exhibitions.

42. Judging such activity in financial rather than environmental terms

tends to lead to a minimisation of the government's contribution; see F. B. Tipton, 'Government Policy and Economic Development in Germany and Japan', *Journal of Economic History*, *XLI* (1981) and the discussion in Inkster, op. cit., (n. 20), Section III. For early enthusiasm see T. Ono, 'The Industrial Transition in Japan', *American Economic Association* (Transactions), *5* (1890), pp. 1–12.

43. For other emphases see the essays in Part II of W. W. Lockwood (ed.), *The State and Economic Enterprise in Japan* (Princeton, 1969).

44. UNCTAD, *Case Studies in the Transfer of Technology: Policies for Transfer and Development of Technology in Pre-War Japan*, UNCTAD Secretariat, 25 April 1978, TO/B/C.6/26 (Typescript), 47pp.

45. See Burks, op. cit. (n. 40), H. J. Jones, *Live Machines: Hired Foreigners and Meiji Japan* (Vancouver, 1980); N. Umetani, *The Role of Foreign Employees in the Meiji Era in Japan* Institute of Developing Economies (Tokyo, 1971); for contemporary accounts, E. G. Holtham, *Eight Years in Japan* (Japan, 1883); W. W. Griffis, *The Mikado's Empire* (London, 6th ed., 1890); Henry Dyer, *Dai Nippon* (London, 1904); Erwin Baelz, *Awakening Japan* (New York, 1932). For a summary of the function of employees see Ian Inkster, *Japan as a Development Model?* (Bochum, 1980), pp. 44–52.

46. Inkster, *ibid.*, pp. 52–9; Pat Barr, *The Deer Cry Pavilion* (New York, 1968); Inkster, op. cit. (n. 13), Chapter IV.

47. W. D. Wray, *Mitsubishi and the N.Y.K. 1870–1914* (Harvard, 1984).

48. The Diet itself was representative of perhaps only 1 per cent to 2 per cent of the population of Japan.

49. See Chapter 5. Excluding loans to the ex-samurai for occupational settlement in Hokkaido and elsewhere, something over a third of total government expenditure in the years 1868–75 was devoted to redemption of debts, civil control and samurai/*daimyo* compensation. A major control policy in these years was the *shizoku* (ex-samurai)-*jusan* system, designed to settle 13 000 families on new land at an estimated cost of 6 million yen.

50. Lehmann, op. cit. (n. 4), pp. 199–203, 259–75.

51. The most accessible accounts of educational investment during these years include Koichi Emi, 'Economic Development and Educational Investment in the Meiji Era', in M. J. Bowman, M. Debeauvais, V. E. Komarov and J. Vaizey, *Readings in the Economics of Education*, UNESCO (Paris, 1968); Koji Taira, 'Education and Literacy in Meiji Japan: An Interpretation', *Explorations in Economic History*, *8* (1970–71); R. P. Dore, 'The Importance of Educational Traditions: Japan and Elsewhere', *Pacific Affairs*, *45* (1972–73); G. C. Allen, 'Education, Science and Economic Development of Japan', *Oxford Review of Education*, *4* (1978).

52. *Gijutsu Kyōikushi* (History of Technical Education) in S. Umene (compiled and edited), *Sekai Kyoiku Shi Taikei*, *32*, Tokyo, 1981; N. Miyoshi, *Meiji no Engineer Kyōiku* (Tokyo, 1983); and the forthcoming 2 volumes edited by Toshio Toyoda, *Wagakuni ririkuki no jitsugyō kyoiku* (Vocational Education in the Meiji Era) (University of Tokyo Press, Tokyo). For an English-language treatment see Yoichi Yano, 'The Development of Technical Education in Meiji Japan – An Interpretation' in Ian Inkster (ed.), *The*

Steam Intellect Societies, Essays on Culture, Education and Industry 1820–1914 (Nottingham, 1985).

53. Ichirō Nakayama, *Industrialisation of Japan* (Tokyo, 1962).

54. A formulation dependent upon the assumption that innovative activity is stimulated by proximate economic circumstance, rather than by historically established value systems. See the treatment of France in Section 5.3 above.

55. This is a basic tenet of the large United Nations University Project on 'Technological Transfer, Transformation and Development: The Japanese Experience', coordinated by Takeshi Hayashi of the Institute of Developing Economies, Tokyo. English-language reports on this Project appear regularly in *Japanese Experience*, the newsletter published by the team and obtainable at IDE, 42 Ichigaya-Hommura Cho, Shinjuku-ku, Tokyo 162 Japan.

56. T. C. Smith, op. cit. (n. 9). More than 50 per cent of all output of the region was sold *outside* it, and 63 per cent of exports were salt and cotton.

57. The population of England and Wales during the British industrial revolution was around 6 million (1750) to 9 million (1801). Population sizes in representative *contemporary underdeveloped* nations are comparable to the British rather than the Japanese figure (5–10 million). See P. Deane and W. A. Cole, *British Economic Growth 1868–1959* (Cambridge, 1969), pp. 6–8, juxtaposed with M. Merhav, *Technological Dependence, Monopoly and Growth* (Oxford, 1969), pp. 18–19. For Japan see N. Sekiyama, *Population Trends of Japan in Tokugawa-Era*, Pamphlet 2 in Series A of Demographic Researches (Tokyo, Ministry of Welfare, 1948); I. B. Taueber, 'Demographic Research in Japan', *Pacific Affairs*, *22*, December 1949, pp. 292–7. Such early estimates and interpretations therefrom have been recently assessed and criticised. See especially S. B. Hanley and K. Yamamura, *Economic and Demographic Change in Pre-Industrial Japan 1600–1868* (Princeton, 1977). However, apart from its anti-Marxian stance, the major thrust of this work seems to be that Tokugawa birth control was designed (or evolved in order) to *retain* a relatively high standard of living. It does not in any way show that population trends in the century prior to Meiji were similar to those in eighteenth-century Europe (especially Britain), nor that population pressure acted as a prerequisite to industrialisation.

58. T. Furushima, *Nihon hōken nōgyōshi*, Tokyo, 1946.

59. J. I. Nakamura, 'Human Capital Accumulation in Pre-Modern Japan'. *Journal of Economic History*, *XLI*, (1981).

60. Torsten Hagerstrand, 'The Diffusion of Innovations', *International Encyclopedia of Social Science*, *4*, (1968), pp. 174–8.

61. See references under notes 10 and 11 above as well as Toshio G. Tsukahira, *Feudal Control in Tokugawa Japan: Sankin Kotai System* (Cambridge, Mass., 1956).

62. For statements of the crucial importance of retained sovereignty in the Japanese case see P. Baran, *The Political Economy of Growth* (New York, 1957), Chapter 2 and pp. 150–51; A. Gunder Frank, 'The Development of Underdevelopment' in J. D. Cockcroft et al., *Dependence and Underdevelopment* (New York, 1972), p. 11.

63. Ian Inkster, 'On "Modelling Japan" for the Third World (Part Two)', *East Asia, 2*, (1984).

64. F. Moulder, *Japan, China and the Modern World Economy* (Cambridge, 1977).

65. For a summary of a complex theme see Ian Inkster, 'The Response to Relative Backwardness – Bureaucrats, Intellectuals and the Transfer of Western Knowledge and Technique to Meiji Japan', *Project on Technology Transfer, Transformation and Development*, United Nation's University (Tokyo, 1981), p. 51.

66. C. Kerr, J. T. Dunlop, F. H. Harbison, C. A. Myers, *Industrialism and Industrial Man* (London, 1962), quote p. 87.

67. M. Morishima, *Why Has Japan Succeeded?* (Cambridge, 1982).

68. B. S. Silberman, 'Elite Transformation in the Meiji Restoration: The Upper Civil Service 1868–73', in Silberman and H. D. Harootunian (eds), *Modern Japanese Leadership* (Tucson, 1966); idem, *Ministers of Modernisation* (Tucson, 1964).

69. Ibid. (1966), pp. 238–48.

70. *Nihon teikoku*, 1st edition (Tokyo, 1882).

71. The famous Satsuma rebellion of 1877 was not in fact *typical* of political dissent during the 1870s, but did represent a high cost on the system of social control.

72. For the notion that the past was created as part and parcel of the 'institutional substitution' programme see Sidney Crawcour, 'The Japanese Employment System' and essays by Cole and Fruin in the same issue of *Journal of Japanese Studies, 4* (1978); K. Taira, *Economic Development and the Labour Market in Japan* (New York, 1970); W. E. Griffis, *The Religions of Japan* (London, 1895), (esp. pp. 153–286 on Buddhism), M. B. Jansen, 'Tokugawa and Modern Japan', *Japan Quarterly, XII*, (1965); Conrad Totman, *Japan Before Perry* (Berkeley, 1981); R. E. Cole, 'The Theory of Institutionalisation', *Economic Development and Cultural Change, XX* (1971); the most recent summary is in Chapter 1 of Robert J. Smith, *Japanese Society, Tradition, Self and Social Order* (Cambridge, 1983). For theoretical approaches to selection of cultures and revival of the past see V. G. Childe, *Social Evolution* (London, 1951); J. H. Steward, *Theory of Cultural Change* (Urbana, 1955); idem, (ed.), *Irrigation Civilisations* (Washington DC, 1955); L. A. White, *The Evolution of Culture* (New York, 1959).

73. Ian Inkster, 'The Other Side of Meiji – Conflict, Conflict Management and the Industrial Programme circa 1868–1885', in G. McCormack and Yoshio Sugimoto (eds), *Japanese Society – Modernisation and Beyond*, (Oxford, 1987).

74. *Kinji Hiōron*, 3 January 1878, p. 2.

75. 'The New Regulations Respecting Political Meetings or Societies', *Japan Weekly Mail*, 10 April 1880, p. 471; J. L. Hoffman, 'The Meiji Roots and Contemporary Practices of the Japanese Press', *Japan Interpreter, XI* (1977), Inkster, op. cit. (n. 73).

76. A. M. Craig, 'Fukuzawa Yukichi: The Philosophical Foundations of Meiji Nationalism' in R. E. Ward (ed.), *Political Development in Modern Japan*

(Princeton, 1968); Tsuchiya Takao, 'Shibusawa Eiichi', *Japan Quarterly, XII* (1965); J. Hirschmeier, 'Shibusawa Eiichi: Industrial Pioneer', in W. W. Lockwood (ed.), *The State and Economic Development in Japan* (Princeton, 1965).

77. See notes 49, 82, and 95; pp. 389–434 of Dower, op. cit. (n. 4).

78. Whether the origins of *industrial dualism* in Japan lay in deep-seated social factors or was a feature of the pattern of technology transfers is a debated issue. On the general field, dealing mostly with post-Meiji processes, see S. Broadbridge, *Industrial Dualism in Japan* (London, 1966).

79. A. C. Kelley and J. G. Williamson, *Lessons from Japanese Development* (Chicago and London, 1974), quote p. 193.

80. Japanese Food and Agricultural Association, *A Century of Technical Development in Japanese Agriculture*, published by the Association (Tokyo, 1959); Y. Hayami and S. Yamada, 'Agricultural Research Organisation in Economic Development: A Review of the Japanese Experience' in Lloyd G. Reynolds (ed.), *Agriculture in Development Theory* (New Haven and London, 1975); Keizo Tsuchiya, *Productivity and Technological Progress in Japanese Agriculture* (Tokyo, 1976). For a clear and accessible summary in English see R. P. Dore, 'Agricultural Improvement in Japan 1870–1900', *Economic Development and Cultural Change, 9* (1960–61), which is based on prewar Japanese sources.

81. Dore, ibid., p. 72. On the failure of introduced technique see Jiro Iinuma, 'The Introduction of American and European Agricultural Science into Japan in the Meiji Era', in R. T. Shand (ed.), *Technical Change in Asian Agriculture* (Canberra, 1973). For the failure of acclimatisation in the woollen industry see Chapter 2 of Michael R. Johnson, 'Complementarity and Conflict: Australian and Japanese Relations in the Nineteenth Century' [unpublished B.Com.Hons Thesis, Department of Economic History, University of New South Wales, 1981].

82. M. Takahashi, 'The History and Future of Rice Cultivation in Hokkaido', *Project on Technology Transfer, Transformation and Development: The Japanese Experience*, United Nations University (Tokyo, n.d.), U.N.U. No. HSDRJE-22/UNUP-100.

83. Thomas C. Smith, *The Agrarian Origins of Modern Japan* (New York, 1966); I. Hatate, *Irrigation Agriculture and the Landlord in Early Modern Japan*, Institute of Developing Economies (Tokyo, 1978).

84. P. Francks, 'The Development of New Techniques in Agriculture: The Case of Mechanisation of Irrigation in the Saga Plain area of Japan', *World Development, 7* (1979); Akira Tamaki, 'Development of Local Culture and the Irrigation System of the Azusa', *Project on Technology Transfer, Transformation and Development: The Japanese Experience*, United Nations University (Tokyo, 1979), UNU No. HSDRJE-4/UNUP-50. Power technologies are neglected in our account. For good treatments see Ryoshin Minami, 'The Introduction of Electric Power and Its Impact on the Manufacturing Industries' in H. Patrick and L. Meissner (eds), *Japanese Industrialisation and its Social Consequences* (Berkeley, 1976); 'Mechanical Power in the Industrialisation of Japan', *Journal of Economic History, 37* (1977); 'Water

Wheels in the Preindustrial Economy of Japan', Institute of Economic Research, Hitotsubashi University, *Discussion Paper Series*, No. 7, 1978; 'Mechanical Power and Printing Technology in Pre-World War II Japan', *Technology and Culture*, *23* (1982).

85. Dore, op. cit. (n. 80), pp. 75–7.

86. T. Ogura (ed.), *Agricultural Development in Modern Japan* (Tokyo, 1963).

87. Iinuma, op. cit. (n. 81), p. 7.

88. Y. Hayami, 'Development and Diffusion of High Yielding Rice Varieties in Japan, Korea and Taiwan, 1890–1940', in Shand, op. cit. (n. 81), esp. pp. 12–14.

89. Y. Hayami and S. Yamada, 'Technological Progress in Agriculture' in L. Klein and K. Ohkawa (eds), *Economic Growth: The Japanese Experience Since the Meiji Era* (Homewood, 1968).

90. Y. Hayami and V. Ruttan, *Agricultural Development: An International Perspective* (Illinois, 1971).

91. Dore, op. cit. (n. 80), pp. 72–3.

92. See the graphs on p. 235 of Hayami and Yamada, op. cit. (n. 80).

93. On the factor bias of technological change see Y. Hayami and V. Ruttan, 'Factor Prices and Technical Change in Agricultural Development', *Journal of Political Economy*, *78* (1970); J. C. Fei and G. Ranis, *Development of the Labour Surplus Economy* (Irwin, 1964); R. Minami, 'The Turning Point in the Japanese Economy', *Quarterly Journal of Economics*, *82* (1968).

94. Kelley and Williamson, op. cit. (n. 79), pp. 193–6.

95. Thus the development by Japanese settled in Hokkaido of new varieties and fertilisers to permit reduction in the whole cycle of rice culture to the frost-free season. Japanese methods of variety-selection acted as a prerequisite to later use of appropriate Western techniques: Takahashi, op. cit. (n. 82), pp. 6–8. To which should be added '*ensuisen*', the development of field experiments for selection, checkrow planting.

96. Le Thanh Nghiep and Y. Hayami, 'Mobilising Slack Resources for Economic Development: The Summer–Fall Rearing Technology of Sericulture in Japan', *Explorations in Economic History*, *16* (1979).

97. Takeo Izumi, 'Transformation and Development of Technology in the Japanese Cotton Industry', *Symposium on Problems of Acclimatisation of Foreign Technology*, The United Nations University, Tokyo, February 25–29, 1980, printed by UNU Project/IDE Tokyo, 1980.

98. During which period the volume of exports in total grew at 15 per cent per annum, 80 per cent of which were raw and semi-manufactured silk and tea. But Japan faced an *effective* decline in the *terms* of trade because shipments were in the hands of foreign merchants and because Western traders took advantage of the dual gold–silver standard at a time of falling silver prices, i.e. Western traders paid in silver but demanded gold.

99. See Inkster, op. cit. (n. 45), pp. 57–9; *Chiugai Bukka Shimpo*, 10 January 1879, *Mainichi Shimbun*, 13 November 1879; *Japan Weekly Mail*, 27 December 1879.

100. Nghiep and Hayami, op. cit. (n. 96); Y. Kiyokawa, 'Entrepreneurship and Innovations in Japan, An Implication of the Experience of Technological Development in the Textile Industry', *The Developing Economies, XVIII* (1980).

101. Refs in note 100; *Japan Weekly Mail*, 21 December 1878, 21 June 1879; Katsuo Otsuka, 'The Transfer of Technology in Japan and Thailand: Sericulture and the Silk Industry', *Development and Change*, *13* (1982).

102. Otsuka, ibid., pp. 423–9.

103. Yoshi Tsurumi, *Japanese Business* (New York and London, 1978), pp. 21–5.

104. Ibid., pp. 25–32.

105. *Ibid.*, p. 26; Nakamura Takafusa, 'The Modern Industries and the Traditional Industries at the Early Stage of the Japanese Economy', *The Developing Economies*, *4*, (1966).

106. Izumi, op. cit. (n. 97), p. 1.

107. Ibid., pp. 2–5.

108. Gary R. Saxonhouse, 'A Tale of Japanese Technological Diffusion in the Meiji Period', *Journal of Economic History*, *34* (1974).

109. *Ibid.* The activities of the *Bören* appear similar to those of government and prefectural institutes in agriculture.

110. For the development of Japanese looms see Takeo Izumi, op. cit. (n. 97), pp. 16–22.

111. It follows that the pattern of imperial expansion in the later nineteenth century follows the pattern of industrialisation of increasingly powerful capitalist nations. A *single*-factor theory of imperialism is, therefore, of little historical use.

112. Using the term 'organic' in a way similar to that of Saint-Simon, where phases in development are seen as following an alternating pattern between the 'critical' and the 'organic'.

113. Mary Matossian, 'Ideologies of Delayed Industrialisation; Some Tensions and Ambiguities', *Economic Development and Cultural Change*, *VI* (1958), p. 228.

8. SCIENCE, TECHNOLOGY AND IMPERIALISM: (1) INDIA

1. A. K. Cairncross, *Home and Foreign Investment, 1870–1913* (Cambridge, 1953); A. J. P. Taylor, *The Struggle for Mastery in Europe*; J. A. Hobson, 'Capitalism and Imperialism in South Africa', *Contemporary Review* LXXVII (1900), idem, *Imperialism, A Study* (London, 1902); V. I. Lenin, 'Imperialism: The Highest Stage of Capitalism', reprinted in *Collected Works*, *XXII* (Moscow, 1964), pp. 185–305, idem, 'Notebooks on Imperialism', ibid., *XXXIX* (Moscow, 1968), pp. 405–36; R. Hilferding, *Das Finanzkapital* (Wien, 1927).

2. J. Gallagher and R. Robinson, 'The Imperialism of Free Trade, 1815–1914', *Economic History Review*, *IV*, (1953–54); D. K. Fieldhouse, 'Imperialism: An Historiographical Revision', *Economic History Review*, *XVI*,

(1961); R. Koebner and H. D. Schmidt, *Imperialism, The Story of a Political Word 1840–1960* (Cambridge, 1964). For adequate but uninspired summaries of the state of play, see P. J. Cain, *Economic Foundations of British Overseas Expansion 1815–1914* (London, 1980); D. K. Fieldhouse, *Economics and Empire* 1830–1914) (London, 1973). By far the best and briefest summary of the classical economic position is in Eric Stokes, 'Late Nineteenth Century Colonial Expansion and the Attack on the Theory of Economic Imperialism: A Case of Mistaken Identity?', *The Historical Journal*, *XXX* (1969), especially pp. 286–9. Readers should note that Stokes demonstrates that for Lenin imperialism was a post-colonial phenomenon, i.e. was essential to and part of the expansion of monopoly finance capital in the early twentieth century, and that the export of capital was not a dominant feature of late nineteenth-century colonialism, which was caused by a compound of strategic, trading, ideological and political factors. An excellent recent paper by Norman Etherington modifies and extends Stoke's reassessment to the theories of Luxemburg, Schumpeter and others; 'Reconsidering Theories of Imperialism', *History and Theory*, *XXI* (1982).

3. In 1870–72 India bought over 8 per cent of Britain's total exports and stood third among the purchasers of British goods. By the 1880s Britain supplied 80 per cent of Indian imports – A. K. Cairncross, op. cit. (n. 1), p. 189. In 1870 £160 million (of total £785 million) of British overseas investment was in India; in 1885, £270 million out of £1300 million. See B. R. Tomlinson, 'India and the British Empire, 1880–1935', *Indian Economic & Social History Review*, *XII* (1975). As to private returns on British investment, 'Great fortunes spring up like mushrooms in a day: primitive accumulation went on without the advance of a single shilling'; Karl Marx, *Capital*, *I*, reprinted (Moscow, 1961), p. 753.

4. D. R. Hendrick, 'The Tools of Imperialism: Technology and the Expansion of European Colonial Empires in the Nineteenth Century', *Journal of Modern History*, *51* (1970). Unfortunately this paper concentrates only upon such direct technological 'means' as transport and military advancements. But technological change affected both means and motives. New techniques in Europe and America allowed the substitution of new ores from Asia in the manufacture of steels, and new fibres from Asia in textile production.

5. A. P. Gerschenkron, *Continuity in History and Other Essays* (Cambridge, Mass., 1968); idem., lecture four of *Europe in the Russia Mirror* (Cambridge, 1970), wherein Gerschenkron lectures Carr; E. H. Carr, *What is History?* (London, 1961), especially Chapter 4, wherein Carr lectures all of us.

6. M. B. Brown, *The Economics of Imperialism* (London, 1974); cf. Lance E. Davis and Robert A. Huttenback, 'The Political Economy of British Imperialism: Measures of Benefits and Support', *Journal of Economic History*, *XLII* (1982). By measurements of social returns, which take account of the costs of empire maintenance, the latter authors suggest that by and large the benefits of empire were insignificant to the growth of the British economy. They conclude that after 1880 empire businessmen and investors succeeded in earning at best slightly more than competitive returns only by 'transfer-

ring a fraction of the true costs of their activities to friends and neighbours [i.e. taxpayers] in the United Kingdom' (p. 128). Therefore, in essence, 'Hobson was correct' (p. 130). For their refinement of this analysis see the impressive Davis and Huttenback, *Mammon and the Pursuit of Empire* (Cambridge, 1986). In 1880 Indian taxes supported 130 000 Indian sepoys and 66 000 British troops. The heavy costs of colonial acquisitions were remarked upon by Adam Smith in 1776: 'Under the Present System of Management Great Britain Derives Nothing But Loss from the Dominion Which She Assumes Over her Colonies', *The Wealth of Nations, II* (Everyman Library Edition, London, 1926), p. 112.

7. This situation is changing. See the bibliography (esp. pp. 935–6) in Gavin Williams, 'Imperialism and Development: A Critique', *World Development, 6* (1978).

8. M. Edelstein, 'Foreign Investment and Empire 1860–1914', in R. Floud and D. McCloskey (eds), *The Economic History of Britain Since 1700*; II: *1860 to the 1970s* (Cambridge, 1981), quote p. 89.

9. R. Hyam, *Britain's Imperial Century 1815–1914: A Study of Empire and Expansion* (London, 1976); K. N. Chaudhuri, *The Economic Development of India Under the East India Company, 1814–1858* (Cambridge, 1971).

10. Sepoy revolts against the EIC and more general risings against the British took place in 1806, 1824 and 1852. But see also Eric Stokes, *The Peasant and the Raj* (Cambridge, 1978), which argues that disturbances were not necessarily caused by institutional reforms.

11. E. N. Komarov, 'Colonial Exploitation and Economic Development', *2nd International Conference of Economic History* (Aix-En-Provence, 1962). For an account with the same tenor see the material on India in Jozsef-Nyilas (ed.), *The Changing Face of the Third World*, (Leyden, 1978), which draws mostly on earlier writing. For a more extended discussion along the same lines see Brian Davey, *The Economic Development of India* (Nottingham, 1975), esp. pp. 53–65 on agriculture and revenue policy, as well as A. I. Levkovsky, *Capitalism in India* (Bombay, 1966).

12. See P. Harnetty, *Imperialism and Free Trade: Lancashire and India in the Mid-Nineteenth Century* (Manchester, 1972) and Ian Inkster, 'The "Manchester School" in Yorkshire: Economic Relations Between India and Sheffield in the Mid-Nineteenth Century', *Indian Economic & Social History Review, XXIII* (July–September 1986).

13. For a fairly judicious summary of the recent literature on the Raj see N. Charlesworth, *British Rule and the Indian Economy 1800–1914* (London, 1982). For a clear analysis of issues see W. J. Macpherson, 'Economic development in India Under the British Crown' in A. J. Youngson (ed.), *Economic Development in the Long Run* (London, 1972).

14. Irfan Habib, 'Potentialities of Capitalist Development in the Economy of Mughal India', *Journal of Economic History, XXIX* (1969); 'The Technology and Economy of Mughal India', *Indian Economic and Social History Review, XVII* (1980).

15. Angus Maddison, 'The Historical Origins of Indian Poverty', *Banca Nazionale del Lavoro Quarterly Review, 23* (1970).

16. M. D. Morris, *The Emergence of an Industrial Labour Force in India: A Study of the Bombay Cotton Mills, 1854–1947* (Berkeley, 1965).

17. J. Hirschmeier, *The Origins of Entrepreneurship in Meiji Japan* (Harvard, 1964); J. Hirschmeier and T. Yui, *The Development of Japanese Business 1600–1973* (London, 1975).

18. Habib, op. cit. (n. 14) (1980), p. 32.

19. Helen Lamb, 'The Indian Merchant' in Milton Singer (ed.), *Traditional India: Structure and Change* (Philadelphia, 1959).

20. Romesh Dutt, *The Economic History of India in the Victorian Age*, 8th Impression (London, 1956); R. E. Frykenberg (ed.), *Land Control and Social Structure in Indian History* (Maddison, 1969).

21. K. N. Chaudhuri, 'India's International Economy in the Nineteenth Century: An Historical Survey', *Modern Asian Studies, 2* (1968); A. J. H. Latham, *The International Economy and the Underdeveloped World 1865–1914* (London, 1978).

22. Dutt, op. cit. (n. 20); D. Naoroji, *Poverty and UnBritish Rule in India* (London, 1871).

23. However, our analysis below does complicate this formulation. Much of the political, financial and industrial policy of British rule indirectly influenced the pattern, extent and profitability of foreign (especially British) investment in India, e.g. the railway guarantee system.

24. For definitions see T. Mukerjee, 'The Theory of Economic Drain: The Impact of British Rule on the Indian Economy 1840–1900' in K. E. Boulding and Mukerjee (eds), *Economic Imperialism* (Ann Arbor, 1972); and Chaudhuri, op. cit. (n. 21).

25. Angus Maddison, op. cit. (n. 15). Maddison included private remittances in his calculations. See n. 23 above.

26. British investment in India rose from perhaps £160 million in 1870 to £380 million in 1913, which actually represented a fall in the *proportion* of total British foreign investment which went to India. As Charlesworth points out (n. 13 above), this may still have represented a significant investment within India, especially in such modernised sectors as jute, mining and shipping.

27. Charlesworth, ibid., pp. 54–5; J. Strachey, *India, Its Administration and Progress* (London, 1903), p. 195.

28. G. Myrdal, *Asian Drama* (London, 1969), p. 455.

29. K. N. Chaudhuri, 'India's Foreign Trade and the Cessation of the East India Co's Trading Activities 1828–40', *Economic History Review, 19* (1966).

30. D. Thorner, 'Long-Term Trends in Output in India', in S. Kuznets, J. J. Spengler, W. E. Moore (eds), *Economic Growth: Brazil, India and Japan* (North Carolina, 1955); D. and A. Thorner, *Land and Labour in India* (London, 1962); J. Krishnamurty, 'The Distribution of the Indian Working Force, 1901–51', in K. N. Chaudhuri and C. J. Dewey (eds), *Economy and Society, Essays in Indian Economic and Social History* (New Delhi, 1979). The shift from 'decline' to 'constancy' resulted initially from D. Thorner's re-examination of the highly problematic census data-bases, especially for

1881 and 1891, which involved collapsing such census categories as 'agriculture and general labour', 'manufacture and trade' and so on. There is a particular problem with definitions of female labour.

31. We are making a clear distinction between employment 'performance' and employment 'impact', the latter determined by factors falling outside manufacturing employment as such, and reflecting our judgement that an alternative scenario (of sovereign economic interaction) might have resulted in *increased* employment in native industries.

32. K. Marx, *Critique of Political Economy* (London, 1859), pp. 134–5.

33. Defence expenditure represented 30 per cent of total Raj expenditure between 1870 and 1900, paid for out of land revenue. For various reasons the spinoffs from such military investment were likely to be less than those in an independent and multiply-victorious nation such as Meiji Japan: see K. Yamamura, 'Success Illgotten?: The Role of Meiji Militarism in Japan's Technological Progress', *Journal of Economic History, 37* (1977). As to conservative elements see Imran Ali, n. 43 below, pp. 108–177, 277–79, 313–58, 444–51.

34. S. Ambirijan, *Classical Political Economy and British Policy in India* (Cambridge, 1978); S. K. Sen, *Studies in Industrial Policy and the Development of India 1858–1914* (Calcutta, 1964); S. Bhattacharya, 'Laissez-faire in India', *Indian Economic & Social History Review, 2,* (1965); J. Tomlinson, *Problems of British Economic Policy, 1870–1945* (London, 1981).

35. T. Raychaudhuri, 'A Re-Interpretation of 19thc. Indian Economic History?' in M. D. Morris et al. (eds), *Indian Economy in the 19thc.: A Symposium* (Delhi, 1969).

36. M. J. T. Thavaraj, 'Rate of Public Investment in India 1898–1938', in T. Raychaudhuri (ed.), *Contributions to Indian Economic History II* (Calcutta, 1963).

37. D. R. Gadgil, *The Industrial Evolution of India in Modern Times* (Oxford, 1944); D. H. Buchanan, *The Development of Capitalist Enterprise in India* (London, 1966).

38. A. K. Bagchi, 'European and Indian Entrepreneurship in India, 1900–30' in E. Leach and S. N. Mukherjee (eds), *Elites in South Asia* (Cambridge, 1970).

39. B. B. Kling, 'The Origins of the Managing Agents in India', *Journal of Asian Studies, 26* (1966); for a negative analysis of the agencies see A. I. Levkovsky, *Capitalism in India, Basic Trends in its Development* (Bombay, 1966), esp. 111–56.

40. M. Tatemoto and M. Baba, 'Foreign Trade and Economic Growth in Japan, 1858–1937' in M. Klein and K. Ohkawa, *Economic Growth, The Japanese Experience* (Yale, 1968); I. Yamazawa, 'Industrial Growth and Trade Policy in Pre-War Japan', *The Developing Economies, 1* (1975). Between 1897 and 1899 Japan gained tariff autonomy. A change in the structure of tariffs was designed as a move away from revenue collection towards a protective infant-industry policy. For the model see K. Akamatsu, 'A Historical Pattern of Economic Growth in Developing Countries', *The Developing Economies* (1962).

41. Gadgil, op. cit. (n. 37), p. 137f. For examples of previous activity under EIC auspices see n. 79 below.

42. G. S. Chhabra, *Social and Economic History of the Panjab 1849–1901* (Jullindur City, 1962), esp. Chapter VIII.

43. I. Ali, 'The Punjab Canal Colonies, 1885–1940', [PhD thesis, Australian National University, Canberra, 1979], p. 434. I would like to thank Imran Ali for allowing me to benefit from his script long prior to its examination.

44. For a brief account see Chapter XII of E. E. Kellett, *A Short History of Religions* (London, 1954).

45. Invaders prior to Plassey (1757) included the Aryans, Huns, Arabs, Tartars, the later Portuguese, Dutch and French.

46. For summaries of such 'negative' elements see Eric L. Jones, *The European Miracle* (Cambridge, 1981), Chapter 10. R. Das Gupta, *Problems of Economic Transition, Indian Case Study* (Calcutta, 1970) pp. 19–23; Angus Maddison, op. cit. (n. 33).

47. Arnold T. Toynbee, *A Study of History*, Vol. 12 (Oxford, 1961), pp. 86–88, 364–5, 583–4.

48. V. Mishra, *Hinduism and Economic Growth* (Oxford, 1962); John Hurd, 'The Economic Consequences of Indirect Rule in India', *Indian Economic & Social History Review, XII* (1975).

49. Mishra, ibid.; see also E. Blunt, 'The Economic Aspect of the Caste System', in R. K. Mukerjee (ed.), *Economic Problems of Modern Indian, I* (Calcutta, 1938); Max Weber, *The Religions of India* (New York, 1958); M. D. Morris, 'Values as an Obstacle to Economic Growth in South Asia: An Historical Survey', *Journal of Economic History, 27* (1967).

50. G. Kaushal, *Economic History of India 1757–1966* (New Delhi, 1979), Chapter IV.

51. Ibid., p. 43.

52. Morris, op. cit. (n. 49) and op. cit. (n. 16); R. Das Gupta, 'Factory Labour in Eastern India: Sources of supply 1855–1946', *Indian Economic and Social History Review, 13* (1976); R. K. Newman, 'Social Factors in the Recruitment of the Bombay Millhands', in Chaudhuri and Dewey, op. cit. (n. 39).

53. M. Singer, 'Cultural Values in India's Economic Development', *Annals of the American Academy of Political and Social Science, 305* (1956), p. 86.

54. S. Sangwan, 'Science Policy in the East India Company in India', in A. Rahman (ed.), *Science and Technology in Indian Culture, A Historical Perspective* (New Delhi, 1984), quote, p. 173.

55. L. H. Brockway, *Science and Colonial Expansion, The Role of the British Royal Botanic Gardens* (New York, 1979).

56. D. Kumar, 'Economic Compulsions and the Geological Survey of India', *Indian Journal of History of Science, 17* (1972), idem, 'Science, Resources and the Raj: A Case study of the Geological Work in 19thc. India' [typescript, for forthcoming publication in *Indian Historical Journal*, (New Delhi)].

57. S. Ambirajan, 'Steam Intellect and the Raj: South India in the Nineteenth Century' in Ian Inkster (ed.), *The Steam Intellect Societies* (Not-

tingham, 1985); D. Kumar, 'Racial Discrimination and Science in Nineteenth Century India', *Indian Economic & Social History Review, XIX* (1982).

58. *Colonisation and Settlement (India): Report for the Select Committee with Proceedings, 1859, V* (Session I), House of Commons, 1859, questions 628–30.

59. W. W. Hunter, *The Indian Musalmans* (London, 1871); W. Adams, *Reports on Vernacular Education in Bengal and Bihar*, First Report (Calcutta, 1835).

60. W. J. Barber, *British Economic Thought and India, 1600–1858* (Oxford, 1975).

61. Notes 58 and 59 above, and as follows from the argument of the present chapter.

62. All contemporary authorities commented upon the lack of employment opportunities in engineering and general machine-making.

63. Sangwan, op. cit. (n. 54); the survey article by B. V. Subbaroyappa in D. M. Bose, S. N. Sen, B. V. Subbaroyappa (eds), *A Concise History of Science in India* (New Delhi, 1971).

64. J. S. Jha, 'Origin and Development of Cultural Institutions in Bihar Under the British Government, With a particular reference to the Bihar Scientific Society', *Journal of Historical Research, 8* (1965); H. Malik, *Sir Syed Ahmed Khan and Muslim Modernisation in India and Pakistan* (New York, 1980); V. A. Narain, 'The Role of Bihar Scientific Association in the Spread of Western Education in Bihar' in *Proceedings of the Indian History Congress* (Modern India Section) (Calcutta, 1969), pp. 421–4; the recent work of Irfan Habib and Deepak Kumar of the National Institute of Science, Technology and Development Studies (of CSIR), New Delhi has added further detail to the broad outline of Indian cultural 'response'; Irfan Habib, 'Institutional Efforts: Popularisation of Science in the Mid-19thC.', [typescript NISTADS], Deepak Kumar, 'Colonial Science and Indian Response 1820–90' [NISTADS, delivered at International Conference for the History of Science, Berkeley, 31 July 1985]. I would like to thank Irfan Habib and Deepak Kumar for allowing me sight of their early work.

65. Habib, ibid.

66. Kumar, op. cit., esp. pp. 12–17.

67. S. Nurullah and J. P. Naik, *A History of Education in India* (Bombay, revised, 1951); *Report of the Association for the Advancement of Scientific and Industrial Education of Indians* (Calcutta, 1912), esp. pp. 1–8.

68. Deepak Kumar, 'Science in Agriculture: A Study in Victorian India', in A. Rahman, op. cit. (n. 54 above).

69. S. N. Sen, 'The Character of the Introduction of Western Science in India During the 18th and 19th Centuries', *Indian Journal of History of Science, I* (1966); and in the same journal D. Kumar, 'Patterns of Colonial Science in India' (*15*, 1980) and 'Science in Higher Education: A Study of Victorian India' (*19*, 1984).

70. For illustrations see Edward Wilson, *Acclimatisation* (London, 1875), read before the Royal Colonial Institute; G. Bennett, *Acclimatisation, Its Eminent Adaptation to Australia* (Melbourne, 1862), pp. 8–18; Brockway, op.

cit. (n. 66); 'M.S. Report of A. C. C. Carleyle, Curator Riddell Museum, Agra on the Proposed Introduction of Indian Silk Worms and the Acclimatisation of Indian Timber Trees and Plants in Australia' National Library of Australia, M. S. Room; Nan Kivell 4427 (M.S. 4060) (1870?); F. Buckland, *The Acclimatisation of Animals* (London, 1861); T. Hutton, *Remarks on the Cultivation of Silk in India* (Calcutta, 1869) (also in *Journal of the Agricultural and Horticultural Society of India*, *I*, part 2, new series). In the 50 years to 1832 the *Journal of the Asiatic Society* published 500 papers in mathematics and physical sciences, 560 in zoology, 320 in geology and 80 in botany.

71. Quoted at greater length in Subbaroyappa, op. cit. (n. 63), p. 551.

72. A possible exception was Rourkee Civil Engineering College which emerged from the project activity of EIC/Public Works Department – see evidence of Peacock in *Fifth Report of the Select Committee on Indian Territories 1852–53*, House of Commons (London, 1853), questions 8090–94.

73. R. K. Gupta, 'Birbhum Silk Industry: A Study of its Growth to Decline', *Indian Economic & Social History Review*, *7* (1980).

74. M. J. Twomey, 'Employment in 19thC. Indian Textiles', *Explorations in Economic History*, *20* (1983), quote p. 38.

75. Kumar, op. cit. (n. 68). The dual effect of increased demand and the introduction of foreign varieties was to introduce disease into native silk production in a range of nations, including Japan. But, in this case, the independence of both government action and producer response was such as to permit its removal through improved, Western knowledge: see F. O. Adams (of the London-based Silk Supply Association), *A History of Japan* (London, 1874); *Japan Weekly Mail*, January–June 1870, April–August 1871, and the discussion in Ian Inkster, *Japan as a Development Model?* (Bochum, 1980), pp. 56–9.

76. On the decline of indigo production in the West Indies, the EIC brought planters from there to India and financed modernised production in Bengal and Bihar during the end of the eighteenth century.

77. Subbaroyappa, op. cit., (n. 63), pp. 563–4; Inkster (n. 12).

78. Kumar, op. cit., (n. 68).

79. *Select Committee* (1853), op. cit. (n. 72), questions 8067–8122. Considerable capital investments were involved, e.g. the Company sanctioned the Ganges Canal improvements in 1847 with an estimate of £1 million, and upon further calculation of costs added another £1.5 million. In the Punjab two amounts of £500 000 each were sanctioned for the Bari Doab alone. But (e.g. question 8073) returns were very high. The Madras Commissioners calculated that the total expenditure of twelve-and-a-half million rupees on the Godavery anicut (a cut across the river) had resulted in an *increased* revenue by 1851 of nineteen-and-a-half million rupees.

80. Evidence of the military engineer Colonel H. B. Turner, in op. cit., (n. 58), questions 475–658.

81. C. P. Simmons, 'Recruiting and Organising an Industrial Labour Force in Colonial India: The Case of the Coal Mining Industry C.1880–1939', *Indian Economic & Social History Review*, *13* (1976), and his earlier paper in the same journal (1976).

82. S. J. Koh, *Stages of Industrial Development in Asia* (Philadelphia, 1966); Harnetty, op. cit. (n. 12); Chaudhuri, op. cit. (n. 28); Latham, op. cit., (n. 21); C. P. Simmons, Helen Clay, Robert Kirk, 'Machine Manufacture in a Colonial Economy: The Pioneering Role of George Hattersley and Sons Ltd. in India 1919–43', *Indian Economic & Social History Review, 20* (1983). Our alternative scenario of free trading relationships would generate a period of capital-goods imports followed by the emergence of indigeneous (a) repair, (b) adaptive and (c) manufacturing enterprises within the 'receiver' nation, e.g. Japan after circa 1880. The first phase occurred very early in India and during c.1849–58 imports of machinery rose from £18 000 to £465 000 – but did not give way significantly to phases (a)-(c) until after 1914.

83. See the earlier estimates of N. B. Mehta, *Indian Railways: Rates and Regulations* (London, 1927); L. H. Jenks, *The Migration of British Capital* (New York, 1927); D. H. Buchanan, *The Development of Capitalist Enterprise in India* (New York, 1934); W. E. Weld, *India's Demand for Transportation* (New York, 1920); recent assessments include Macpherson in the *Economic History Review, 8* (1955) and Lehmann in *Indian Economic & Social History Review, 2* (1965); I. J. Kerr, 'Constructing Railways in India – An Estimate of the Numbers Employed 1850–80', *Indian Economic & Social History Review, 20* (1983); M. B. McAlpin, 'Railroads, Cultivation Patterns and Foodgrain Availability: India 1860–1900', *Indian Economic & Social History Review, 12* (1975); M. Mukherjee, 'Railways and Their Impact on Bengal's Economy 1870–1920', *Indian Economic & Social History Review, 17* (1980).

84. That is, in some large areas, the *social* returns of the railways may have been minimal; Mukherjee, op. cit. (n. 83), pp. 194–203.

85. Various references in 83 note above.

86. The metre gauge replaced the older 5'6" gauge, and caused high costs and confusion at the points where they intersected.

87. D. Thorner, *Investment in Empire* (Philadelphia, 1950); idem, 'Great Britain and the Development of India's Railways', *Journal of Economic History, 11* (1951); Mehta, op. cit. (n. 83); M. Kidron, *Foreign Investment in India* (London, 1965).

88. References in notes 83 and 87; Gadgil, op. cit. (n. 37); A. K. Sen, 'The Commodity Pattern of British Enterprise in Early Indian Industrialisation, 1854–1914', *Second International Conference of Economic History* (Paris, 1965).

89. Various references above. See also H. B. Lamb, 'India: A Colonial Setting' in H. F. Williamson and J. B. Buttrick (eds), *Economic Development: Principles and Politics* (New York, 1954).

90. For modern accounts which include estimates of the overall performance of the economy see Argus Maddison, *Class Structures and Economic Growth in India and Pakistan Since the Moghuls* (London, 1971); S. Cornish, 'Recent Writing in Indian Economic History', *Journal of Economic History, 37* (1977); M. D. Morris, 'Quantitative Resources for the Study of Indian History', in V. R. Lorwin and J. M. Price (eds), *The Dimensions of the Past* (New Haven, 1972); M. Mukherjee, *National Income of India: Trends and Structure* (Calcutta, 1969).

9. SCIENCE, TECHNOLOGY AND IMPERIALISM: (2) CHINA AND BEYOND

1. Irfan Habib, 'Potentialities of Capitalist Development in the Economy of Mughal India', *Journal of Economic History*, *XXIX* (1969); 'The Technology and Economy of Mughal India', *Indian Economic and Social History Review*, *XVII* (1980).

2. Angus Maddison, 'The Historical Origins of Indian Poverty', *Banca Nazionale del Lavoro Quarterly Review*, *23* (1970).

3. M. D. Morris, *The Emergence of an Industrial Labour Force in India: A Study of the Bombay Cotton Mills, 1854–1947* (Berkeley, 1965).

4. S. C. Thomas, *Foreign Intervention and China's Industrial Development, 1870–1911* (Boulder and London, 1984).

5. Andrew J. Nathan, 'Imperialism's Effects on China', *Bulletin of Concerned Asian Scholars*, *4* (1972).

6. Joseph Esherik, 'Harvard on China: The Apologetics of Imperialism', *Bulletin of Concerned Asian Scholars*, *4* (1972).

7. F. V. Moulder, *Japan, China and the Modern World Economy* (Cambridge, 1977).

8. For a more extended working out of this perspective see Ian Inkster, *Science, Technology and the Late Development Effect: Transfer Mechanisms in Japan's Industrialisation, circa 1850–1912* (Tokyo, 1981).

9. *Japan Weekly Mail*, 21 August 1880, p. 1075.

10. Toynbee, op. cit. (n. 47 of Chapter 8), p. 179.

11. See the bibliography in Thomas, op. cit. (n. 4); Albert Feherwerker, *China's Early Industrialisation* (Cambridge, Mass., 1958); M. J. Levy, 'Some Aspects of Industrialism and the Problem of Modernisation in China and Japan', *Economic Development and Cultural Change*, *10* (1962); W. W. Lockwood, 'Japan's Response to the West: The Contrast With China', *World Politics*, *9* (1956); S. N. Eisenstadt, 'Tradition, Change and Modernity: Reflections on the Chinese Experience' in Ping-ti-Ho and Tang Tsou (eds), *China in Crisis* (Chicago, 1968).

12. Alexander Eckstein, *China's Economic Development* (Ann Arbour, 1975), pp. 93–5.

13. Ibid., pp. 99–101; Ping-ti-Ho, *The Ladder of Success in Imperial China* (New York, 1962).

14. Eckstein, ibid., p. 101; M. I. Sladkovsky, *China and Japan in Past and Present* (London, 1975), Chapter 2.

15. Sladkovsky, ibid., pp. 20–22; Eric Jones, op. cit. (n. 46 of Chapter 8), pp. 171–219; T. Filesi, *China and Africa in the Middle Ages* (London, 1972).

16. Toynbee, op. cit. (n. 10), pp. 202–3; Eckstein, op. cit. (n. 12), p. 107f.

17. Eckstein, ibid., p. 99; Mark Elvin and G. W. Skinner (eds), *The Chinese City Between Two Worlds* (Stanford, 1974).

18. Rhoads Murphey, 'The Treaty Ports and China's Modernisation' in Elvin and Skinner, op. cit. (n. 121), p. 39; Mark Elvin, *The Pattern of the Chinese Past* (Stanford, 1969); in Elvin's detailed analysis China between the eighth and thirteenth centuries underwent an agrarian-based 'economic

revolution', which affected agricultural productivity, transportation, credit, trade, urbanism and technology (textiles), coincident with Needham's emphasis on progress in medicine and mathematics.

19. Compare Eckstein's analysis of 1960 with his comments in 1968 – op. cit., pp. 95, 123–26 and *passim*.

20. See the first chapter of Inkster, op. cit. (n. 8); on Europe see references to Chapters 2, 3, 4 and 5 above.

21. Murphey, op. cit. (n. 18), pp. 39–41, 23–24; Moulder, op. cit. (n. 7), pp. 45–70.

22. See ibid., and Chapters 5 and 6 above.

23. Eckstein, op. cit. (n. 12), pp. 127–30.

24. Dwight Perkins, *Agricultural Development in China 1368–1968* (Chicago, 1969); A. M. Tang, 'China's Agricultural Legacy', *Economic Development and Cultural Change, 28* (1979); E. Kerridge, *The Agricultural Revolution* (London, 1967). Part of the Gerschenkron hypothesis is that industrialisation under conditions of relative backwardness *may* occur in the absence of an agrarian revolution as a necessary *prelude* to success. See Chapter 5 above, and the debate on Tokugawa Japan, Chapter 7. See also n. 18 above.

25. New cereal and starchy root crops such as sweet potatoes increased *calorie* production per acre, and thus represent a new *product* line. The labour input into the cultivation of these crops is particularly high, and a more mixed product line should have allowed an increase of cropping per year, e.g. cereals or vegetables may be planted as a second crop in areas where a second paddy crop is impossible, e.g. the Yang-tsu valley. During the 1950s 85–90 per cent of the calories in the Chinese diet were derived from cereals, 5–10 per cent for starchy tubers. During the 1930s the per capita calorie intake for China was perhaps above that prevailing in Japan.

26. Eckstein, op. cit. (n. 12), p. 130.

27. Ibid., pp. 130f.

28. Jerome Ch'en, *State Economic Policies of the Ch'ing Government 1840–1895* (New York and London, 1980). The Taiping programme of 1851 included land reform, equalisation of land ownership, and a move to industrialisation and railway building. Western leaders were anxious about the 'democratic' tendencies of Taiping, a factor in the Second and Third Opium Wars (1856–58, 1859–60) during which the West supplied modern weapons against the rebels.

29. Most of which may be found in the literature, notes 4–7, 11, 18; see also, D. W. Perkins, 'Government as an Obstacle to Industrialisation: The Case of Nineteenth-Century China', *Journal of Economic History, 27* (1967).

30. Eckstein, op. cit., p. 110f.; Moulder, op. cit., pp. 98–127; Sladkovsky, op. cit., pp. 18–28.

31. Tan Chung, *China and the Brave New World*, Bombay, 1978; *Commercial Reports from H.M. Consuls in China 1878 and 1880–1881*, House of Commons, London, 1881 [No. 3:China], p. 133; Chi-ming Hou, 'External Trade, Foreign Investment, and Domestic Development: The Chinese Experience 1840–1937', *Economic Development & Cultural Change, X* (1961), p. 23.

32. *Commercial Reports* (1881), op. cit., pp. 133–5.

33. Calculated from information in ibid., p. 139 and G. C. Allan and A. G. Donnithorne, *Western Enterprise in Far Eastern Economic Development: China and Japan* (London, 1956), Table 4, p. 260.

34. Hou, op. cit., pp. 27–8; Allan and Donnithorne, ibid., p. 262.

35. This is, of course, true today also, even during the post-Mao era.

36. *Ibid.*, p. 34; K. Emi, *Government Fiscal Activity and Economic Growth in Japan 1868–1960* (Tokyo, 1968); A. C. Kelley and J. G. Williamson, *Lessons from Japanese Development, An Analytical Economic History* (Chicago, 1974); H. Rosovsky, *Capital Formation in Japan* (Berkeley, 1960).

37. Eckstein, op. cit., p. 118; Thomas, op. cit. (n. 4, pp. 112–20; Hou, op. cit., p. 26.

38. For a comparison with Japan see Chapter 7 above and Ian Inkster, *Japan as a Development Model?* (Bochum, 1980), pp. 52–6; Murphey, op. cit. (n. 18).

39. F. L. Hawks Pott, *A Short History of Shanghai* (Shanghai, 1928), pp. 132–6.

40. *Commercial Reports* (1881), op. cit., p. 143. Foreigners who gained 'transit certificates' from the authorities escaped the full burden of the internal transit tax (*likin*); the purchase price of the certificate was fixed at 50 per cent of the value of the relevant import or export duty.

41. Hawks Pott, op. cit., pp. 110–12; 'D.W.S.', *European Settlements in the Far East* (London, 1900), pp. 88–126.

42. Eckstein, op. cit. (n. 12), pp. 122, 113.

43. Murphey, op. cit. (n. 18), pp. 17–21.

44. The refusal of foreigners in China to attempt to learn the language, repeated in several sources, is in contrast to the seeming readiness of Europeans to learn Japanese; *Hochi Shimbun*, 27 November 1890; *Japan Echo*, 1 December 1890; W. E. Griffis, *Hepburn of Japan* (New York, 1913); Erwin Baelz, *Awakening Japan* (New York, 1932); Henry Dyer, *Dai Nippon, The Britain of the East* (London, 1904); N. Umetani, *The Role of Foreign Employees in the Meiji Era in Japan* (Tokyo, 1971).

45. *Commercial Reports* (1879), op. cit., p. 143.

46. Chung-li Chang, *The Income of the Chinese Gentry* (Seattle, 1962); Ping-ti Ho, op. cit. (n. 117).

47. T. G. Rawski, 'Chinese Dominance of Treaty Port Commerce and its Implications 1860–1875', *Explorations in Economic History*, 7 (1970).

48. Hawks Pott, op. cit. (n. 152), pp. 24–5, 50–53, 62, 93; W. F. Mayers, N. B. Dennys and C. King, *The Treaty Ports of China and Japan* (London and Hong Kong, 1867), pp. 353–63.

49. Hawks Pott, ibid., pp. 39–40, 65–68, 79–80, 125–7; Mayers, *et al.*, ibid., pp. 364–5, 384–6.

50. Hawks Pott, ibid., pp. 66–8.

51. Ibid., pp. 111–12.

52. Ibid., pp. 100–103.

53. The Shanghai–Woosung line was developed by the British and Americans in an effort to get around the stipulations set for the originally prepared Shanghai–Soochow line. The line was bought up and closed by the government in 1877. This case is often cited as a classic example of the

official myopia. In the years 1876–78 Western plans for telegraph development met the same resistance.

54. Hou, op. cit. (n. 31), pp. 39–40.

55. See generally in Ch'en, op. cit. (n. 28), ibid., pp. 31–7; Allan and Donnithorne, op. cit. (n. 33), pp. 149–60, 165–7; Eckstein, op. cit., pp. 117f.

56. Hou, op. cit., p. 36.

57. Chi-ming Hou, *Foreign Investment and Economic Development in China 1840–1937* (Cambridge, Mass., 1965); Chang Kia Ngau, *China's Struggle for Railroad Development* (New York, 1943).

58. Ch'en, op. cit. (n. 28), pp. 75–82.

59. *Ibid.*, p. 77.

60. *Ibid.*, p. 97; Thomas. op. cit. (n. 4), p. 94; Ngau, op. cit. (n. 57), pp. 25–6; H. Stringer, *China, A New Aspect* (London, 1929), pp. 14–20.

61. Hou, op. cit., pp. 25–6.

62. Moulder, op. cit., pp. 115–16; Thomas, op. cit., pp. 94f.; Ngau, op. cit., pp. 27–35.

63. Ngau, ibid., pp. 34f.

64. Stringer, op. cit. (n. 60), pp. 33–9, 13.

65. P. Reinsch, *Colonial Government* (New York, 1902).

66. Ibid.; Stringer, pp. 7–14; Esherick, op. cit. (n. 6), pp. 13–15.

67. Thomas, op. cit. (n. 4), pp. 131–47.

68. W. H. Moreland, *From Akbar to Aurangzeb* (New Delhi, 1923, reprinted 1972); C. McEvedy and R. Jones, *Atlas of World Population History* (Harmondsworth, 1978).

69. H. Stringer, *China, A New Aspect* (London, 1929), pp. 31–50.

70. Norton Ginsberg, *The Pattern of Asia* (London, 1958), pp. 125–6, 242–4, 537–8, 576–8.

71. Murphey, op. cit. (n. 18), p. 22.

72. A. Eckstein, K. Chao, and J. Craig, 'The Economic Development of Manchuria', *Journal of Economic History, 34* (1974); Garo Hirata, 'Manchukuo's Foreign Trade', *Contemporary Japan, 2* (June 1933); Allan and Donnithorne, op. cit. (n. 33), pp. 138 forward, 180–81; Murphy, ibid., pp. 50–51; Eckstein, op. cit. (n. 12), pp. 156–7.

73. W. W. Rostow, *The Stages of Economic Growth* (Cambridge, 1960), p. 56.

74. Thorner, op. cit. (n. 30 of Chapter 8 above), p. 218.

75. J. Foreman-Peck, *A History of the World Economy* (Brighton, 1983), pp. 135–8.

76. Measures included new civil and criminal codes, settlement of *shizoku* and *sotsu* (ex-samurai groups), household registration, the 1876 Sword Ban Order, Assemblies of local governors, and the 1880 Regulations for Transfer of Factories to Private Control.

77. Chapter 3 of Nisaburo Murakushi, 'Technology and Labour in Japanese Coal Mining', *Project on Tech. Transfer, Transformation and Development: The Japanese Experience*, United Nations University, limited distribution (Tokyo, 1980), no. HSDRJE-17/UNUP-82.

78. Fumio Yoshiki, 'Modernisation of Metal Mining in Japan', *Symposium*

on Problems of Acclimatisation of Foreign Technology, United Nations University (Tokyo, 1980); Yoshiro Hoshino, 'Technological and Managerial Development of Japanese Mining; The Case of the Ashio Copper Mine', *The Developing Economies*, *X* (1982).

79. Ken'ichi Iida, 'Origin and Development of Iron and Steel Technology in Japan', *Project* (n. 193 above) (Tokyo, 1980), no. HSDRJE-8/UNUP-89.

80. K. Yamamura, 'Success Illgotten?: The Role of Meiji Militarism in Japan's Technological Progress', *Journal of Economic History*, *37* (1977), (quote p. 113). For a comparison of the Japanese model of transfer via militarism with that of the Chinese attempt, see B. C. Hacker, 'The Weapons of the West: Military Technology and Modernisation in China and Japan', *Technology and Culture 18* (1977).

81. Yamamura, ibid., pp. 116–18.

82. Kurt Mendelssohn, *Science and Western Domination* (London,1976), pp. 141–2.

83. Joseph Needham, 'China's Philosophical and Scientific Traditions', *Cambridge Opinion*, *36* (1963); idem, for a stress on the agricultural–technical conditioning factors in Chinese science (as against cultural or philosophical forces) see various parts of his *Clerks and Craftsmen in China and the West* (Cambridge, 1970), e.g. pp. 71–82. Idem, with more emphasis on social relations, *The Grand Titration* (London, 1969), see pp. 14–54 with its emphasis on social relations and the earlier supremency of China in *applying* principles to purposes; and the vast amount of material in the volumes of idem, *Science and Civilisation in China* (Cambridge, from Vol. I, 1954). For summaries of Needham's work see T. E. Ennis, 'The Role of Chinese Science and Technology in Modern Civilisation', *Eastern World*, *20* (1966); J. Chesneaux, 'Le "Miracle Chinois"', *La Recherche*, *3* (1972); Colin A. Ronan, *The Shorter Science and Civilisation in China* (Cambridge, 1978).

84. Mendelssohn, op. cit. (n. 82), p. 142.

85. J. Bartholomew, 'Japanese Culture and the Problem of Modern Science' in A. Thackray and E. Mendelsohn (eds), *Science and Values* (New York, 1974); Chapter 8 of L. S. Feuer, *The Scientific Intellectual (New York, 1963); S. Nakayama, Academic and Scientific Traditions in China, Japan and the West* (Tokyo, 1984).

86. Ian Inkster, 'Meiji Economic Development in Perspective: Revisionist Comments Upon the Industrial Revolution in Japan', *The Developing Economies*, *XVII* (1979). Thus Japanese bacteriology was established by Shibasaburō Kitazato, who trained in Germany, and Masanori Ogata who had studied at Pettenkofier's Hygiene Institute in Munich and at the Pathology Institute in Berlin, and their success was determined by government activity rather than general culture receptivity – see Bartholomew, op. cit. (n. 202), pp. 109–55, T. Brock (ed), *Milestones in Microbiology* (New York, 1961). For a general review, Ian Inkster, 'Prometheus Bound: Technology and Industrialisation in Japan, China and India Prior to 1914 – A Political Economy Approach', *Annals of Science*, *45* (1988).

10. CENTRE AND PERIPHERY: SCIENCE AND TECHNOLOGY IN AMERICA AND AUSTRALIA

1. R. L. Andreano (ed.), *New Views on American Economic Development* (Cambridge Mass., 1965); L. M. Hacker, 'The First American Revolution' in G. D. Nash (ed.), *Issues in American Economic History* (Lexington, 1964); J. R. T. Hughes, *Industrialisation and Economic History* (New York, 1970); J. W. McCarty, 'Australia as a Region of Recent Settlement in the Nineteenth Century', *Australian Economic History Review, 13* (1973); Barrie Dyster, 'Argentine and Australian Development Compared', *Past and Present, 14* (1979).

2. Donald Fleming, 'Science in Australia, Canada and the United States; Some Comparative Remarks', *Procs of Tenth International Congress of the History of Science 1962*, Vol. II (1964); George Basalla, 'The Spread of Western Science', *Science, 156*, no. 3775 (5 May 1967); Edward A. Shils, 'Towards a Modern Intellectual Community' in J. S. Coleman (ed.), *Education and Political Development* (Princeton, 1965); idem., *The Intellectuals and the Powers and Other Essays* (Chicago, 1971), esp. Chapter 17.

3. Jared Eliot, *Essays Upon Field Husbandry in New England and Other Papers, 1748–1762* (New York, 1934).

4. Brooke Hindle, *The Pursuit of Science in Revolutionary America 1735–1789* (Chapel Hill, 1956); W. Goetzman, *Exploration and Empire: The Explorer and the Scientist in the Winning of the American West* (Knopf, 1966).

5. Allan Pred, *The Spatial Dynamics of U.S. Urban–Industrial Growth 1800–1914* (Cambridge Mass., 1966).

6. S. B. Warner, Jr, *The Private City* (Philadelphia, 1968); R. G. Miller, *Philadelphia, The Federalist City, 1789–1801* (New York, 1976).

7. Carl and Jessica Bridenbaugh, *Rebels and Gentlemen, Philadelphia in the Age of Franklin* (New York, 1965), quote p. 356; see Chapter IX, 'The Love of Science'. For much data on late eighteenth-century and early nineteenth-century culture and institutions, E. P. Oberhottzer, *Philadelphia, A History of the City and Its People*, 4 vols (Philadelphia 1920), esp. Vol. II, Chapters XX and XXI.

8. C. W. Peale, *Introduction to a Course of Lectures on Natural History* (Philadelphia, 1799), quote p. 12, L. Kerr, *Wonder of This World, Charles W. Peale* (New York, 1968); C. C. Sellers, *Charles Wilson Peale* (New York, 1969).

9. Expenditures, letters and museum details are found in 'The Peale Manuscripts', [Case 15, Historical Society of Pennsylvania] and 'Peale–Sellers MS Collection' [Archives of the American Philosophical Society, Philadelphia, BP 31]. See also C. W. Peale, *Address Delivered to the Corporation and Citizens of Philadelphia* (Philadelphia, 1816).

10. Ian Inkster, 'Robert Goodacre's Astronomy Lectures and the Structure of Scientific Culture in Philadelphia', *Annals of Science, 35* (1978).

11. For the institutional structure of Philadelphia science in this interim period see E. T. Freedley, *Philadelphia and its Manufactures* (Philadelphia, 1858). In 1825 (*Philadelphia Gazette*, 31 October) there had already been voiced a demand for a 'Polytechnic College'. For the development of similar

popular, general and applied science institutions in Boston see M. A. Calvert, *The Mechanical Engineer in America 1830–1910* (Baltimore, 1967).

12. For differing views on this see J. C. Greene, 'Science and the Public in the Age of Jefferson', *Isis, 39* (1958); A. Hunter Dupree, *Science in the Federal Government* (Cambridge Mass., 1957); G. H. Daniels (ed.), *Nineteenth Century American Science: A Reappraisal* (Evanston, 1972); Nathan Reingold (ed.), *Science in America Since 1820* (New York, 1976); G. H. Daniels, 'The Process of Professionalisation in American Science, 1820–1860', *Isis, 58* (1967).

13. Whitfield J. Bell, Jr, 'The American Philosophical Society as a National Academy of Sciences 1780–1846', *10th Congress*, op. cit. (n. 3).

14. Lance E. Davis and Robert E. Gallman, 'Capital Formation in the United States During the 19thc', in P. Mathias and M. M. Postan (eds), *The Cambridge Economic History of Europe, VII*, pt 2, (Cambridge, 1978), Table 20, pp. 42–3.

15. The context of which is forcefully outlined in Chapter 8 of Charles E. Rosenberg, *No Other Gods, On Science and American Social Thought* (Baltimore and London, 1976).

16. Fleming, op. cit. (n. 3), p. 188.

17. For modern introductions to American economic growth see Davis and Gallman, op. cit., (n. 14), Douglas C. North, *Growth and Welfare in the American Past*, (Englewood Cliffs, New Jersey, 1966 and 1974); Peter Temin, *Causal Factors in American Economic Growth in the 19th Century* (London, 1975).

18. North, ibid., pp. 68–117, Temin, ibid., Chapters 4 and 5.

19. For the standard definition see p. 369 of John E. Sawyer, 'The Social Basis of the American System of Manufacturing', *Journal of Economic History, 14* (1954), pp. 361–79.

20. Nathan Rosenberg, *Technology and American Economic Growth* (New York, 1972); Russell I. Fries, 'British Response to the American System: The Case of the Small Arms Industry after 1850', *Technology and Culture, 16* (1975), pp. 377–403. The 1855 Whitworth and Wallis report is reproduced in Rosenberg (ed.), *The American System of Manufactures* (Edinburgh, 1969).

21. Paul Uselding, 'Elisha K. Root, Forging and the American System', *Technology and Culture, 15* (1974), pp. 543–68; Merritt Roe Smith, *Harper's Ferry Armory and the New Technology: The Challenge of Change* (Ithica, NY, 1977).

22. See Chapters 1, 2, and 4 of David A. Hounshell, *From the American System to Mass Production 1800–1932* (Baltimore and London, 1984).

23. Smith, op. cit. (n. 21).

24. For a summary of such matters see Nathan Rosenberg, 'Technology in Glenn A. Porter (ed.), *Encyclopedia of American Economic History: Studies of the Principal Movements and Ideas, 3 vols*, (New York, 1980), I, pp. 294–308.

25. H. J. Habakkuk, *American and British Technology in the Nineteenth Century* (Cambridge, 1962).

26. Rosenberg, op. cit. (n. 24), p. 300.

27. Ibid., p. 301.

28. *Ibid.*, see also, for specific techniques, Rosenberg, 'Technological Change in the Machine Tool Industry 1840–1910', *Journal of Economic*

History, *23* (1963), pp. 414–43; R. S. Woodbury, 'The Legend of Eli Whitney and Interchangeable Parts', *Technology and Culture*, *1* (1960), pp. 225–53; and his *History of the Lathe to 1850* (Cleveland 1961).

29. The attempts to perform this exercice represent a thread running through the literature cited (notes 14–17 above). See also; S. P. Lee and P. Passell, *A New Economic View of American History*, New York 1979, pp. 21–6, 52–61; Chapter 6 of A. W. Niemi, *U. S. Economic History* (Chicago, 2nd edn, 1980).

30. North, op. cit., pp. 81–2.

31. *Ibid.*, p. 82. For criticisms of the Habakkuk thesis – which argued that America's comparative abundance of land and capital promoted substitutions of capital for labour, ie. technique shifts leading to technological progress; H. J. Habakkuk, *American and British Technology in the Nineteenth Century*, Cambridge, 1962 – see Chapter 1 of Paul A. David, *Technical Choice, Innovation and Economic Growth* (Cambridge, 1975) and A. J. Field, 'On the Unimportance of Machinery', *Explorations in Economic History*, *22* (1985).

32. The transformation from 'learning economy' to innovative 'producing economy' could be very fast in some sectors, e.g. the period 1829–38 represented a learning phase in the construction of locomotive engines, but by the later 1840s the US was selling locomotives to Britain and Russia, and American engineers were involved in transport systems in Germany, France and Austro-Hungary. At the 1851 exhibition in London, the McCormick reaper demonstrated its economic superiority, as did the Pitt thresher at Paris in 1855. From this time began the technology export invasion of Europe – Goodyear rubber, oscillating engines, Singer's sewing machines, steamboat engines, and the Morse telegraph. Then followed the American inventive surge in agriculture-reapers, harvesters, the Appleby binder, the chilled iron plough and the disk harrow, as well as chemical fertilisers. See also the patent data below.

33. J. W. Oliver, *History of American Technology* (New York, 1956); M. A. Calvert, *The Mechanical Engineer in America 1830–1910, Professional Cultures in Conflict* (Baltimore, 1967).

34. Hounshell, op. cit (n. 22), esp. Chapters 1 –4.

35. Alfred D. Chandler, *The Visible Hand: The Managerial Revolution in American Business* (Cambridge, Mass., 1977), quote p. 89; and his 'The United States – Evolution of Enterprise' in P. Mathias and M. M. Postan (eds), op. cit. (n. 14), esp. quote p. 101. See also the contextual first chapter of R. W. Garnet, *The Telephone Enterprise* (Baltimore and London, 1985).

36. Rosenberg, op. cit. (n. 24), p. 298.

37. The general emphasis of Chandler. See also R. W. Fogel, *Railroads and American Economic Growth* (Baltimore, 1964); P. D. McClelland, 'Railroads, American Growth and the New Economic History: A Critique', *Journal of Economic History*, *18* (1968).

38. Chandler, *Visible Hand*, op. cit. (n. 35), quote p. 87.

39. For a sophisticated approach see Oliver E. Williamson, 'The Modern Corporation: Origins, Evolution, Attributes', *Journal of Economic Literature*, *19* (1981), pp. 1537–68, esp. pp. 1553–57.

40. Jim Hightower, 'Hard Tomatoes, Hard Times: Failure of the Land Grant College Complex', *Society, 10* (1972), pp. 10–1, 16–22. L. S. Reich, 'Research, Patents, and the Struggle to Control Radio: A Study of Big Business and the Uses of Industrial Research', *Business History Review, 51* (1977), pp. 208–35; David F. Noble, *America by Design: Science, Technology and the Rise of Corporate Capitalism* (New York, 1977).

41. North, op. cit. (n. 17), p. 148.

42. J. Schmookler, *Invention and Economic Growth* (Cambridge Mass., 1966); US Department of Commerce, Bureau of the Census, *Historical Statistics of the U.S., Colonial Times to 1970, 2*, pt. 2 (Washington, 1975), pp. 954–9. In fact we have not used Schmooklers 'estimated number of successful patent applications procedure', but the alternative contemporary sources (see Chapters 3 and 5 above); see subject index of *Journal of the Franklin Institute* for 1826–85, the only basic source for complete information on specifications and claims of patents in the US for 1826–43. For summary index see the *Journal CXX* covering Vols 1 to 69.

43. W. G. Armytage, *A Social History of Engineering* (London, 1979).

44. Until 1836 US patent registration involved little more than grants awarded upon the oath of the applicant that his invention was original. There was no formal examination as to novelty and other requirements. The relatively low level of foreign patenting in America resulted for its effective prohibition until 1836, and to 1861 aliens paid higher fees than residents/citizens on a spurious 'theory of reciprocity'. The act of 1836–7 undoubtedly tightened up the US patent system.

45. 'Return relating to patents for invention taken out for the United Kingdom, showing the number of Applications made from each of the undermentioned Cities and Towns during the years 1867, 1868, 1869', in *Report of the Select Committee on Letters Patent, Proceedings of Committee*, Appendices, House of Commons, London, 1871: Append. 4.

46. Pred. op. cit. (n. 5 above) esp. Chapter 3; Pred, 'Industrialisation, Initial Advantage and American Metropolitan Growth', *Geographical Review, 55* (1965); Irwin Feller, 'The Urban Location of United States Invention, 1860–1914', *Explorations in Economic History, 8* (1970–71).

47. Bureau of the Census, op. cit. (n. 42), pp. 958–9; 'German Patent Office Statistics, for 1894', *Journal of Soc. of Chemical Industry, XIV* (1895), pp. 406–9.

48. For a good example of which see Bruce Sinclair, *Philadelphia's Philosopher Mechanics* (Baltimore, 1974).

49. The large generalisations of this section are derived from more detailed material in Ian Inkster, 'Scientific Enterprise and the Colonial Model': Observations on Australian Experience in Historical Context', *Social Studies of Science, 15* (1985) and Ian Inkster and Jan Todd, 'The Support Structure for the Australian Scientific Enterprise circa 1851–1916', in R. Home (ed.), *Australian Science in the Making* (Cambridge, 1988), pp. 102–33.

50. P. Burroughs, *Britain and Australia 1831–55* (Oxford, 1967); J. J. Eddy, *Britain and the Australian Colonies 1818–31* (Oxford, 1969); W. P. Morrell,

British Colonial Policy in the Age of Peel and Russell (London, 1966).

51. D. N. Jeans, 'The Impress of Central Authority upon the Landscape: South–Eastern Australia 1788–1850', in J. M. Powell and M. Williams (eds), *Australian Space, Australian Time* (Melbourne, 1975), quote p. 5.

52. Inkster and Todd, op. cit. (n. 49), Section IV.

53. *Proceedings of the Zoological and Acclimatisation Society of Victoria, II,* (1873) (Melbourne, 1873); Edward Wilson, *Acclimatisation*, Royal Colonial Institute (London, 1875).

54. W. H. D. Le Souef, 'Acclimatisation in Victoria', *Report of the Second Meeting of the Australasian Association for the Advancement of Science, Melbourne, 1890.*

55. Inkster and Todd, op. cit. (n. 49), Section IV.

56. Todd from Inkster and Todd, op. cit.

57. *The Bulletin*, 17 July 1897, p. 19.

58. *The Australian Mining Standard*, 14 January, 1897, pp. 1588–9.

59. T. H. Houghton, 'Annual Address to Engineering Section, Royal Society of N. S. W.,' 18 May, 1898, *Journal of the R.S. of N.S.W., 32*, (1898), xii–xv.

60. For case studies of which see Section V of Inkster and Todd, op. cit.

61. As reported in the introduction to J. C. F. Johnson, *Getting Gold: A Practical Treatise for Prospectors, Miners and Students* (London, 1905, 4th edn).

62. R. K. Wilson, *Australia's Resources and their Development* (Sydney, 1980); N. G. Butlin, *Investment in Australian Economic Development, 1861–1900* (Cambridge, 1964).

63. For the four years; *New South Wales Statistical Register*, 1889 and 1894 (Sydney) and *The Australian Official Journal of Patents, Indexes to Proceedings*, I (1904), Melbourne, 1905. For details of sources on patenting and a fuller treatment of the subject see Ian Inkster, 'Intellectual Dependency and the Sources of Invention: Britain and the Australian Technological System in the Nineteenth Century', forthcoming in *History of Technology*.

64. By 1918 39 per cent of Australian patents were lodged by residents of N.S.W., 34 per cent by Victorians. Prior to 1901 the patent figures capture patent lodgements between the individual colonies; they nevertheless do indicate differences between the supply of and demand for innovations between states, if not the origin of inventiveness.

65. Applications per 10 000 inhabitants in Australia in 1904 were 6.4, in UK 7.0 in USA 6.6.

66. *Official Journal*, op. cit. (n. 63), for 1918–19.

67. This is particularly true of patents issued for ploughing, cultivating, harvesting, refrigeration, steam power, mining, sanitation, harbours, marine propulsion and locomotives and rolling stock.

68. That social or cultural factors explain the nature of the market and hence the spread of the 'American system' remains a feature of the literature and may be traced clearly back to Sawyer (n. 19) and beyond.

69. Chandler's periods are westward expansion (1815–50), railroad expansion (1850–1870s), the growth of the urban-based national market (1880s–1900s), applications of electricity and internal combustion engine

aided by new institutionalisation of R&D (early twentieth century).

70. M. C. Ughart and K. A. H. Buckley (eds), *Historical Statistics of Canada* (Cambridge, 1965) (Patents section).

71. For a rare exception, G. Bindon and D. P. Miller, 'Sweetness and Light: Industrial Research in the Colonial Sugar Refining Company, 1855–1910', in Home (ed.), op. cit. (n. 49), pp. 170–96.

11. TWENTIETH-CENTURY AFTERMATHS: SCIENCE, TECHNOLOGY AND ECONOMIC DEVELOPMENT

1. For an introduction to the economics of development see, in order of appearance: Walter Elkan, *An Introduction to Development Economics* (London, 1973); R. B. Sutcliffe, *Industry and Underdevelopment* (London, 1971); Benjamin Higgins, *Economic Development: Principles, Problems and Policies* (New York, 1959 and later editions); Paul Bairoch, *The Economic Development of the Third World Since 1900* (Berkeley, 1975); P. Streeton, 'The Frontiers of Development Studies: Some issues of Development Policy', *Journal of Development Studies*, Oct. 1967; Chapters 1 and 2 of Ralph Pievis, *Studies in the Sociology of Development* (Rotterdam, 1969); Benjamin Higgins, 'Economic Development and Cultural Change: Seamless Web or Patchwork Quilt?', in Manning Nash (ed.), *Essays on Economic Development and Social Change in Honour of Bert F. Hoselitz*, (Chicago, 1977).

2. G. Myrdal, *Economic Theory and Underdeveloped Regions* (London, 1963), quote p. 51.

3. Sutcliffe, op. cit. (n. 1), p. 327.

4. *Ibid.*, see also Elkan, n. 1, Chapter 8; P. Newman, *Malaria Eradication and Population Growth* (University Michigan Press, 1965).

5. E. A. Brett, *The World Economy Since The War: The Politics of Uneven Development* (London, 1985), p. 178.

6. Elkan, op. cit., n. 1, p. 33.

7. J. Boeke, *Economics and Economic Policy of Dual Societies* (New York, 1953).

8. *Draft Sixth Five-Year Plan 1978–83*, Indian Planning Commission, Government of India (New Delhi, 1979).

9. For an introductory commentary see L. Gomes, *International Economic Problems* (London, 1978), Chapters 1 and 7, and for a good summary of debate on the economic role of protectionism see P. R. Krugman, 'Is Free Trade Passé?', *The Journal of Economic Perspectives*, Vol. 1 (1987).

10. Boeke, op. cit. (n. 7); R. E. Baldwin, *Economic Development and Export Growth: A Study of Northern Rhodesia 1920–1960* (Berkeley, 1966).

11. James H. Street, 'The Institutionalist Theory of Economic Development', *Journal of Economic Issues, 21* (1987), quote p. 1861; Ian Inkster, 'The Institutionalist Theory of Economic Development, Technological Progress and Social Change', *Journal of Economic Issues 23* (1989).

12. For a brief consideration of such matters, see Inkster, ibid., p. 1244–5.

13. Adam Smith, *The Wealth of Nations* (1776) (London, Dent and Sons, 1937), Vol. 1, book 1, Chapter XI, p. 173.

14. Gerald W. Scully, 'The Institutional Framework and Economic Development', *Journal of Political Economy, 96* (1988), quote p. 661.

15. The data on institutions are taken from R. D. Gastil, *Freedom in the World* (Westport, Conn., 1982).

16. Scully, op. cit. (n. 14), p. 654. The reference is to Myrdal's, *Asian Drama* (London, 1968), pp. 67, 115–16.

17. This is the major reason why the dynamics of the Gerschenkron schema may not be directly applied to the post-1914 world: see Chapter 5.

18. Both approaches are outlined in Chapter 5. See also note 1 above.

19. V. W. Ruttan, 'Induced Institutional Innovation' in G. E. Schuh (ed.), *Technology, Human Capital and the World Food Problem*, (Minnesota, 1986), quote p. 127.

20. *Japan Weekly Mail*, 10 September 1870, pp. 422–5.

21. L. P. Jones and Il Sakong, *Government, Business and Entrepreneurship in Economic Development: The Korean Case* (Cambridge, Mass., 1980); E. S. Mason, *The Economic and Social Modernization of the Republic of Korea* (Cambridge, Mass., 1980).

22. For an excellent, institutional treatment of Japan in the 1950s see G. C. Allen, *Japan's Economic Expansion* (Oxford, 1965).

23. Between 1945 and 1952 Japanese economic policy was directed by SCAP (Supreme Commander of Allied Powers), whose early agenda concentrated only on extraction of reparations and limitation of Japan's strategic industries. Policy became more progressive from 1947, and in 1948 a *Nine Point Stabilisation Programme* was introduced. The outbreak of war with Korea in June 1950 changed the complexion of things. The two treaties of 1952 restored Japanese sovereignty but 'special procurement' spending by the US in Japan continued as a recovery device, e.g. during 1952–56 such spending reached $US 3380 million, equivalent in 1955 to 27 per cent of Japan's exports.

24. V. P. Chitale, *Foreign Technology in India* (New Delhi, 1973).

25. Ibid., pp. 7, 63 forward. Royalty payments increased from $163 million in 1961 to $652 million in 1970, 50 per cent of which went to the USA.

26. T. Ozawa, *Japan's Technological Challenge to the West 1950–74, Motivations and Accomplishments* (Tokyo, 1974); K. Ohkawa and H. Rosovsky, 'Post-War Japanese Growth in Historical Perspective' in L. Klein and K. Ohkawa (eds), *Economic Growth, The Japanese Experience Since the Meiji Era* (Illinois, 1968).

27. Y. Yoshino, *Japan's Multinational Enterprises* (Cambridge, Mass., 1976), pp. 11–12; R. Vernon, 'International Investment and International Trade in the Product Cycle', *Quarterly Journal Economics, 53* (1966).

28. Yoshino, ibid., pp. 12–16; GATT, *Japan's Economic Expansion and Foreign Trade, 1955 to 1970* (Geneva, 1971).

29. Chitale, op. cit. (n. 24), p. 69 forward.

30. S. Sekiguchi and T. Horiuchi, 'Myth and Reality of Japan's Indus-

trial Policies', *The World Economy*, *8* (1985); Hugh Patrick, 'Japanese High Technology Industrial Policy in Comparative Context' in Patrick and L. Meissner (eds), *Japan's High Technology Industries, Lessons and Limitations of Industrial Policy* (London and Tokyo, 1986); C. Johnson (ed.), *The Industrial Policy Debate* (San Francisco, 1984).

31. Well summarised in Yoshino op. cit. (n. 27), Chapter 1; R. Caves and M. Uekusa, 'Industrial Organisation' in H. Patrick and H. Rosovsky, *Asia's New Giant*, (Washington, 1976); R. P. Dore and K. Taira, *Structural Adjustment in Japan 1970–1982* (Geneva, 1986).

32. K. Sato, 'Did Technical Progress Accelerate in Japan?' in Shigeto Tsuru (ed.), *Growth and Resources Problems Related to Japan* (Tokyo, 1978).

33. See pp. 158–60, 177–78 of Sato, ibid.; Oshima commentary in Tsuru, ibid., p. 184; Science and Technology Agency, *Indicators of Science and Technology* (Tokyo, 1977).

34. *Patent Office, Japan*, Annual Report, 1981 (Tokyo, 1981).

35. Keizai Koho Centre, *Japan 1983, An International Comparison* (Tokyo, 1983), p. 18.

36. *The Eastern Economist*, 23 July 1943, p. 340.

37. Bepin Behari, *Economic Growth and Technological Change in India* (Delhi, 1974), quote p. 132.

38. Ibid., pp. 133–7.

39. B. D. Nag Chaudhuri, *Technology and Society, An Indian View*, Indian Institute of Advanced Studies, Simla, 1979, pp. 57–61; Ministry of Information and Broadcasting, Government of India, *India 1986, A Reference Manual* (Delhi, 1987), pp. 106–46.

40. Chitale, op. cit. (n. 24), pp. 58 forward; of a total of 1696 million rupees of remittances abroad in the period 1971–73, royalties composed 132 million, payments for technical knowhow, 252 million.

41. Charan Singh, *India's Economy Policy, The Ghandian Blueprint* (Delhi, 1978), pp. 69–71.

42. For a clear summary of the radical position see Usha Menon, *Science for the Nation*, Delhi Science Forum (New Delhi, 1987).

43. Ibid., pp. 10–12.

44. This is not dependent upon equity control but on the TNC control over knowledge: of 1041 foreign collaborations in 1985, 24 per cent involved foreign capital.

45. P. M. Pillai, 'Technology Transfer, Adaptation and Assimilation', *Economic and Political Weekly* (India) *14* (1979), quote p. 47.

46. Usha Menon, 'World Bank and Transfer of Technology: Case of Indian Fertiliser Industry', *Economic and Political Weekly*, *15* (1980); D. N. Duravalla, 'Design and Development of Fertiliser Plants in Developing Countries with Special Reference to India' in UNIDO, *Fertiliser Production: Technology and Use* (New York, UN, 1968), and various reports in *Economic and Political Weekly*, 28 October 1972, 27 July 1977, 10 May 1980, 21 April 1979.

47. K. K. Subrahmanian, 'Collaborative Agreements and their Impact on Assimilation and Diffusion of Know-how and Outgo of Resources' in

M. Gibbons, P. Gummett, B. Udgaonkar (eds), *Science and Technology Policy in the 1980s and Beyond* (London, 1984), pp. 259–63.

48. Ian Inkster, 'Appropriate Technology, Alternative Technology and the Chinese Model: Terminology and Analysis', *Annals of Science*, *46* (1989), pp. 263–76.

49. The World Bank, *China, Long-Term Development Issues and Options* (Baltimore and London, 1984).

50. R. Baum (ed.), *China's Four Modernisations* (Boulder, 1980); K. Reiitsu, 'The Bearers of Science and Technology have Changed', *Modern China*, *5* (1979), pp. 187–230.

51. For an epitome of the old view, presented in order to clarify the new, see Ma Hong, 'On China's New Strategy for Economic Development', in *Almanac of China's Economy 1981, With Economic Statistics for 1949–1980*, Economic Research Centre (New York and Hong Kong, 1982), pp. 299–318.

52. It was the ethos of the 1978 Congress which was important. Science and technology were formally acknowledged as having played a role in China's modernisation and scientists, though intellectuals, were hailed as a productive segment of the working class. The Chinese were searching for expertise/professionalism without isolation/elitism.

53. Most of this new alignment took place at the province and country levels. Even the national symposia of forestry, ecology, agronomy and so on took strong regional characteristics.

54. Policy Research Office, Chinese Academy of Science, 'The Sciences in China' in *Almanac 1981*, (op. cit., n. 51), p. 684.

55. These were in addition to already-existing specialised research institutes in the provinces, e.g. in Liaoning Province there were 15 of these employing some 7000 personnel. Provincial reponsiveness to science may well have been hastened in cases where academic scientists were given major provincial political positions, e.g. Wang Jinling as Vice-governor of Heilongjiang, Zheng Shouyi, Vice-mayor of Qingdao.

56. Policy Research Office, op. cit. (n. 54), pp. 686–7; The World Bank, op. cit. (n. 49), p. 118.

57. The Park was established in 1980 as a centre for high technology. By 1985 the Park had approved 60 investment proposals, 50 already operating, specialising in computers, precision instruments, material science, biochemical engineering. Enterprises receive a five-year tax holiday, exemption from foreign trade duties, an option for joint venture with government, training programmes and low-interest financing.

58. Samuel P. S. Ho and R. W. Huenemann, *China's Open Door Policy: The Quest for Foreign Technology and Capital* (Vancouver, 1984); N. T. Wang, *China's Modernisation and Transnational Corporations* (Lexington, 1984).

59. W. C. Kirby, 'Joint Ventures, Technology Transfer and Technocratic Organisation in Nationalist China 1928–1949', *Republican China*, *12* (1987), pp. 3–21.

60. A. Rahman, *Science and Technology in India* (New Delhi, 1984), Table 10, p. 193, and Table 1017 p. 595 of *Statistical Abstract of the United States 1982–83*, 103rd edn, Bureau of the Census (Washington, 1982).

61. The basic sources for the statistics in this Section are UNESCO, *World Summary of Statistics on Science and Technology* (Paris, 1970); UNESCO, *Statistical Yearbook* (Paris, 1986); Business International Corporation, *Worldwide Economic Indicators, Annual Comparative Statistics for 131 Countries with World and Regional Reports*, (New York, 1984).

62. In 1972 66.4 per cent of all Japanese GERD was provided by the business sector.

63. From data in various tables of UNESCO, 1986 (n. 61).

64. The 'productive sector' must not be confused with 'manufacturing' for it includes also agriculture, extraction, utilities, construction, transport and a miscellany of other activities and embraces all industrial and trading establishments which produce and distribute goods and services for sale.

65. UNESCO, 1986 (n. 61), V-1.

66. US National Science Foundation, *National Patterns of Science & Technology Resources*, Washington DC, Annual.

67. J. J. Servan-Schrieber, *The American Challenge* (London, 1968); E. F. Denison, *Why Growth Rates Differ* (Washington, 1967).

68. It is impossible to detail the copious material under this heading. For one related sector, education, see B. Duke, *The Japanese School: Lessons for Industrial America* (New York, 1986); M. White, *The Japanese Educational Challenge: A Commitment to Children* (New York, 1987).

69. R. J. Barber, *The Politics of Research* (Washington, 1966); D. K. Price, *The Scientific Estate* (Boston, 1965).

70. J. S. Dupre and S. A. Lakoff, *Science and the Nation*, Englewood Cliff, 1962; C. H. Danhoff, *Government Contracting and Technological Change* (Washington, 1968).

71. The growth of the NASA space budget 1958–65 was spectacular, and declined thereafter to the early 1970s. See J. B. Weisher, *Where Science and Politics Meet* (New York, 1965) and Leslie Sklair, *Organised Knowledge*, (St Albans, Herts, 1973).

72. C. Norman, 'A New Push for a Federal Science Department', *Science*, *226* (December 1984), 1398–9; J. A. Remington, 'Beyond Big Science in America: The Binding of Inquiry', *Social Studies of Science*, *18* (1988), pp. 45–72.

73. George Keyworth, Ronald Reagan's former science adviser, seems to have had a strong belief in the efficiency of centralism.

74. During the mid to late 1980s 'Star Wars' research has absorbed most funds for state-of-the-art technologies; Remington, op. cit., (n. 72), p. 64.

75. D. Shapley and R. Roy, *Lost at the Frontier: US Science and Technology Policy Adrift* (Philadelphia, 1985); T. Schultz, 'An Unpersuasive Plea for Centralized Control of Agricultural Research', *Minerva*, *21* (1983); David Dickson, *The New Politics of Science* (New York, 1984); A. Weinberg, 'Unity as a Value in Administration of Pure Science', *Minerva*, *22* (1984).

76. *Statistical Handbook of Japan, 1987*, Statistics Bureau Management and Coordination Agency (Tokyo, 1987), p. 139.

77. *Nature*, *313* (1985), p. 253.

78. Issued by MITI in 1979, the document marked the explicit acknowledgement of the transition from 'using foreign technology' to 'original developments on our own'. Of course, this new thrust is not only a phenomenon of catch-up, but of new American alertness and limitations over technology-exporting in the Reagan administration.

79. Since 1985 the STA has taken over from the MESC as a principal supplier of funds for smaller projects in basic research at the same time that it continues its support of big science. Other major target areas for MITI are aircraft, optoelectronics, new alloys and mechatronics (electronic machinery). MITI's institutional innovation is the setting up of a new centre to provide loans for basic research by private industry and new joint research projects.

80. Sheridan Tatsuno, *The Technopolis Strategy: Japan, High-Technology and the Control of the Twenty-First Century* (New York, 1986), pp. 28–31.

81. Tatsuno, ibid., esp. Chapters 3–6.

82. This is of particular relevance to our findings on nineteenth-century institutions in Chapter 4. The continued success of the British industrial economy from 1840 to 1880 or even later was a function of a more diffused 'culture' (in a limited, institutional sense) of ST, found in workshops, mechanics' institutes, polytechnics, professional clubs and enterprise, rather than in government or centralised institutions.

83. Tatsuno, op. cit. (n. 80), Chapters 3, 5, 9 and 11.

84. At several points Gary Saxonhouse has argued (quite convincingly) that *government* industrial targeting is not as important to high technology development as is at times claimed and that much of government economic intervention merely substitutes for the demonstrable *lack* of competitiveness and market-induced information dispersal found in Japan. Gary R. Saxonhouse, 'What is All this About "Industrial Targeting" in Japan?', *The World Economy*, *6*, no. 3 (1983), pp. 253–73.

85. See the theoretical perspectives in Andre Gunder Frank, *Latin America: Underdevelopment or Revolution* (New York, 1969), esp. Chapter 1; I. Wallerstein, 'The State and Social Transformation', *Politics and Society*, *I* (1971), pp. 359–64; Johan Galtung, *Concept and Theories of Development* (Oslo, 1983); G. Palma, 'Underdevelopment and Marxism', *Thames Papers in Political Economy*, IDS, Sussex, Summer 1978.

86. J. Saravanamuttu, 'The Political Economy of Japan's Involvement in ASEAN: Some Theoretical and Policy Implications', *East Asia 3* (1985), pp. 169–93.

87. Kwan-Chi Oh, 'Korea as a Market for Technology', *East Asia 3* (1985), pp. 25–56.

88. *Ibid.*, pp. 34–5; S. P. S. Ho, 'South Korea and Taiwan: Development Prospects and Problems in the 1980s', *Asian Survey*, December 1981.

89. Oh, op. cit. (n. 87), p. 31.

90. *Ibid.*, pp. 31–2.

91. Andras Hernadi, 'Export-Oriented Industrialisation and its Successes in the Asia-Pacific Region', *East Asia 3* (1985), p. 3.

92. G. Ranis, 'Industrial Sector Labor Absorption', *Econ. Devpt and Cultural Change XII* (1973), Hernadi, ibid., p. 4 (quote).

93. Hernadi, ibid., p. 4.

94. S. P. S. Ho 'South Korea and Taiwan: Development Prospects and Problems in the 1980's, *Asian Survey*, December 1981.

95. Kwan-Chi Oh, 'Korea as a Market for Technology', *East Asia 3* (1985), pp. 25–56.

96. Ibid., p. 46.

97. *Ibid.*, pp. 46–55; Korea Industrial Research Institute, *Industrial R&D Survey* (Korea, 1984).

98. Ian Inkster, 'Partly Solid and Partly Sophistical – Adam Smith as an Historian of Economic Thought', *Quarterly Review of Economics and Business*, forthcoming 1990.

99. Ian Inkster, 'On Modelling Japan for the Third World (Part One)', *East Asia 1* (1983), pp. 155–87, esp. pp. 171–80.

100. Ian Inkster, 'The Other Side of Meiji – Conflict and Conflict Management', in G. McCormack and Yoshio Sugimoto (eds), *The Japanese Trajectory Modernisation and Beyond*, (Cambridge, 1988), pp. 107–80.

101. For recent prognoses on Japanese R&D see J. Bell, B. Johnstone and S. Nakaki, 'The New Face of Japanese Science', *New Scientist*, 21, March 1985; 'R&D by Japanese Private Sector', *Focus Japan*, March 1985 (a survey by STA of 763 firms); M. Moritani, 'Japanese Technology: Potential and Pit Falls', *Japan Echo*, 13 (1986); M. Moritani, *Japan's Technological Innovation and Trade Structure* (Tokyo 1985); A. M. Anderson, *Science and Technology in Japan* (Harlow, Essex, 1984); N. Makino, 'High Technology in Japan: Its Present and Future', *Nippon Steel Forum*, April 1984.

102. D. J. de Solla Price, *Little Science, Big Science* (New York, 1965), quote p. 14.

Name Index

Abramovitz, Moses 4
Adelman, Irma 30
Adams, Henry 89
Adelung, Christian 47
Ahmad, Syed 220
Alcock, Michael 55
Ali, Imran 215, 354
Ambirajan, S. 219
Ansted, David T. 105
Armstrong, Robert 106
Arnott, Neil 75
Arrow, K. 28
Ashton, T. S. 76
Auer, H. 116
Aurangzeb 208

Babbage, Charles 12, 83
Babington, William 42
Baines, Edward 242
Baird, Charles 53–4
Bairoch, Paul 137–8
Barber, E. G. 33
Basalla, George 249
Bather, E. A. 123
Beales, H. L. 62, 68, 80
Beckmann, Johann 38
Belinsky, V. 132
Bell, Whitfield 252
Bernoilli, Daniel 48
Berthollet, C. L. 71
Bessemer, Henry 161
Binns, T. W. 105
Binswanger, H. B. 23
Birkbeck, George 75, 83
Birr, Kendall 10
Black, Joseph 76
Bodmer, J. G. 51
Boerhaave 47
Borchers, Wilhelm 113
Boulton, Matthew 9, 42
Boyle, Robert 50
Bradbury, Frank 25
Brassey, Thomas 178
Brown, L. A. 16
Brunel, M. I. 51, 334
Bullough, B. & V. 74
Bunge, N. K. 171
Burnet, Thomas 34

Cairncross, A. K. 205
Calvert, Frederick C. 106
Cameron, Rondo E. 148
Capron, Horace 197
Cardwell, D. S. L. 38
Carnot, Sadi 12, 44
Caro, Heinrich 103, 166
Carter, C. F. 10
Chaloner, W. H. 80, 82
Chamberlayne, John 48
Chambers, Ephraim 38, 48
Chandler, Alfred D. 260, 367
Chapman, S. D. 65

Chardonnet, Count de 154
Clapham, J. H. 63
Clarke, John 34
Clay, Henry 82
Clegg, Samuel 105
Cockerill, James 153
Coiquet, F. 245
Collier, John 161
Condorcet, Marquis de 44
Cooper, Thomas 42
Corneille, Thomas 37
Crafts, Nick 66–8, 86
Crisp, Olga 158
Curry, James 75
Cuthbertson, John 46
Cuthbertson, J. Neilson 89–90, 94, 107
Cyert, R. M. 26

Dale, Thomas 47
Dalhousie, Lord 208
Dallowe, Timothy 47
Dalton, John 42, 76
Dana, S. L. 111
Darnton, Robert 37
Das Gupta, R 217
David, Paul 15, 25, 26
Davidou, Denis 131
Davis, Lance E. 350–1, 364
Davy, Humphrey 76
Deacon, Henry 310
de Bever, Leo J. 187
de Cavour, Camille 156
de Gerard, Philippe 153
de Magellan, J. H. 42
de Solla Price, Derek 303
de Thiersant, Mons 230
Denison, E. F. 6–8
Derham, William 34
Desaguliers, Theophilus 34, 47, 314
des Campomanes, Conde 54
des Penaflorida, Conde 46
Divers, Edward 124
Donnithorne, A. G. 237
Dove, D. J. 48
Dyer, Henry 124–25

Eckstein, Alexander 232–3, 237
Edelstein, M. 207
Eliot, Jared 250
Eliot, R. J. 50
Elkan, Walter 273
Enos, J. L. 13
Euler, Leonard 48

Farey, John 82, 84
Ferreira, Christovao 343
Fischer, J. C. 154
Fishlow, A. 5
Fleming, Donald 249, 253
Fouguier, Pierre 52
Fourcroy, A. F. 44
Frankland, Edward 105
Franklin, Benjamin 47–8, 76, 250–1
Frederick the Great (Prussia) 54
Freeman, C. 9
Furushima, Toshio 194

Gadgil, D. R. 215
Galbraith, J. K. 24–5
Gasgoyne, Charles 53
George, K. D. 26
Gerschenkron, Alexander 140–4, 168, 206, 277, 332–3, 338, 359, 369
Gilfillan, S. C. 5
Gillispie, C. C. 44
Gordon, George 34
Gould, J. D. 6–7
Gramme, Z. T. 157
Graumann, Arthur 113
Green, A. G. 116
Griffis, W. E. 125–6
Gupta, R. K. 222

Habakkuk, H. J. 256, 259, 365
Haber, L. F. 70
Habib, Irfan 210–11, 355
Hagerstrand, T. 16–18, 67
Hahn, R. 38
Haigh, J. P. 103
Haldane, R. B. 91
Hall, Danuel 52

Hamilton, David 27
Hanchwitz, Ambrose G. 50
Harding, William 314
Harris, John 37
Harrison, John 49, 304
Hartwell, R. M. 139–40
Haswell, William 266
Hawks Pott, F. L. 236–37
Hawksbee, Francis 34
Heertje, Arnold 15
Henry, Thomas 42
Henry, William 71
Hernàdi, Andràs 296
Herzen, Alexander 132
Hicks, John 143
Higgins, Benjamin 23, 29
Higgins, Jean 29
Higgins, William 42–43
Hilferding, R. 205
Hirschman, Albert O. 277
Hobsbawm, Eric 80, 152
Hobson, J. A. 205
Holker, John 50–2, 59
Holtham, E. G. 124
Hope, James 75
Hou Chi-ming 239–40
Houghton, T. H. 266
Howitt, H. H. 266
Hownshell, David 259–60
Huber, Richard 187–8
Hughes, John 153
Humboldt, Alexander 156
Hu Yaobang 286
Hunter, William 75–6
Huntsman, Benjamin 51, 153, 304
Hurter, Ferdinand 114, 116
Huttenback, Robert A. 350–1
Huxley, T. H. 25, 266

Indad Ali Syed 220

Jackson, Humphrey 20
Jaeger, C. H. 112
Jeans, D. N. 265
Jomini Henri, Baron 133

Kankrin, E. F. 243
Kant, I. 33
Kaushal, G. 217
Kay, John 49–50, 304
Keir, James 51, 76
Kennedy, John 65, 76
Kikuchi, O. 126
Kipling, Rudyard 123
Kirwan, Richard 42, 315
Klein, Felix 117
Knoop, Ludwig 154
Komarov, E. N. 209–11
Krause, E. M. 266
Kronick, D. A. 41
Kumar, Deepak 355
Kuznets, Simon 3, 72
Kyd, Robert 218

Lagrange, J. 35
Lambert, Charles 153
Landes, David 28, 61, 68, 128, 176
Lavoisier, A. L. 44, 316
Lay, H. N. 124
Lebelye, Charles 50
Leblanc, Nicholas 45, 310, 316
Leclerc, Thomas 52
Leibenstein, H. 26
Leibnitz, G. W. 32–3
Le Neve Foster, P. 264
Lenin, V. I. 172, 204, 350
Leonhardt, August 103
Leontief Wassily 8
Levy Marion, J. 140
Lie Kun-yi 240
Liebig, Justus 93, 152, 156
List, Friederick 133
Lockyer, Norman 185
Lombe, John and Thomas 2, 11
Lomonosov, M. V. 317
Louis XIV (France) 46
Lowell, Francis Cabot 258
Lundgreen, Peter 101

MacLeod, Roy 122
Macquer, P. J. 51
Magnus, Heinrich 156
Manby, Aaron 153

Mansfield, E. 15–16
Marcet, A. J. C. 75
Marggraf, A. S. 35, 39
Marshall, John 153
Martin, Benjamin 38
Martin, Pierre 175
Marshall, Alfred 66, 278
Marx, Karl 205, 213, 224, 275, 350
Masson, David 266
Mather, Cotton 251
Mathews, R. C. O. 25, 27
Matossian, Mary 204
Maxwell, James Clerk 99
Mazzini, Giuseppe 133
McCloskey, Donald 60, 66–67, 78, 86
McConnel, James 65
McKay, John P. 143
Mendelsohn, Everett 94
Mendelsohn, Kurt 69, 246–47
Menon, Usha 284
Merz, J. T. 102
Middleton, Conyers 34, 313
Mill, James 212
Mill, John Stuart 78, 276
Minami, Ryoshin 143, 333
Mohammed Ali (Egypt) 155–6
Mokyr, Joel 8
Mond, Ludwig 110
Money, John 42
Moore, E. G. 16
More, Charles 105
Morishima, Michio 195
Morris, Cynthia T. 30
Morris, James 52
Morris, Morris D. 211, 217, 222
Moulder, Frances 229
Mueller, Ferdinand 265
Müller, John 51
Müller, William 152
Murphey, Rhoads 237, 244
Murray, David 126
Murray, Reginald A. F. 266
Musson, A. E. 68–9, 77
Myrdal, G. 213, 271–2, 276–7

Nairne, Edward 42
Needham, Joseph 28, 362
Nelson, R. R. 15
Newton, William 82–3
North, Douglass 19, 87–8, 259, 261–2, 323
Nurkse, R. 21

Ohkawa, K. 185
Oldknow, Samuel 65

Park, Chung Hee 277
Pattison, William 178
Paul, Lewis 50, 88, 304, 324
Peacock, Thomas Love 223
Peale, Charles W. 251–2
Pearce, Zachray 34, 313
Pearson, George 42, 315
Perkins, Jacob 51
Perry, John 125
Peter the Great (Russia) 46
Phillipson, Nicholas 74
Pillai, P. M. 283
Platt, D. C. M. 334
Playfair, Lyon 128
Podkolzin, A. 170
Pol, A. N. 153
Pollard, Sidney 60, 63, 67, 131, 151–2
Postlethwaite, Richard 79
Pred, Allan 251, 262
Prévost d'Exiles, A. F. 318
Priestley, Joseph 42–3, 46, 76, 254, 316
Proctor, Richard 122

Rabout-Pomier, J. A. 314
Raphael, Samuel 68
Rapp, Friedrich 69
Rawski, T. G. 237–8
Ray, John 34
Raychaudhuri, Tapan 214
Rees, Abraham 38
Reiter, H. 103
Reutern, M. K. 171
Rittenhouse, David 251
Robb-Smith, A. H. T. 75

Roberts, Richard 153, 306
Robinson, E. 68–9
Rockefeller, John 253
Roebuck, John 76, 304
Rose, Hilary and Steven 91
Rosenberg, Charles 248
Rosenberg, Nathan 6, 15, 20, 256–8
Rosovsky, H. 185
Rostow, W. W. 187, 203, 228, 244
Roux, Jean 55
Rowlinson, H. G. 209
Russell, Colin 69
Rutherford, Ernest 250
Ruttan, W. V. 23

Saint Simon, Henry 89, 349
Sanderson, Michael 325
Sawyer, John E. 22
Saxonhouse, Garry 202, 373
Scheele, C. W. 71
Schmookler, Jacob 9–10, 180, 366
Schneider, Joseph 74
Schultz, Thomas 28
Schumpeter, Joseph 8, 293
Shadwell, Arthur 101–2, 107
Shaw, Peter 34
Shibasaburö, K. 362
Shils, Edward 35, 249
Siemens, Werner 119
Siemens, William 154
Silberman, B. S. 195
Silliman, Benjamin 111
Singer, Milton 218
Skully, Gerald 276
Slater, Samuel 50
Smith, Adam 70, 276–7, 298, 351
Smith, Juniu 82
Smith, Thomas C. 185, 193
Solo, R. 22
Solow, Robert 5, 26
Solvay, Ernest 310
Sowell, Thomas 27
Spencer, Adam 47
Spencer, Herbert 25
Stancliffe, John 79
Stewart, Dugald 71
Stiglitz, J. E. 27
Stokes, Eric 350
Stringer, H. 241–2
Sturve, L. P. 106
Sutcliffe, R. B. 272
Switzer, Stephen 314

Tatsuno, Sheridan 292
Temin, Peter 5
Thackray, Arnold 70, 77
Thavaraj, John 214
Thomas, Stephen 229
Thomason, Edward 82
Thompson, Edward 43
Thomson, Thomas 324
Thomson, William 99
Thorner, D. 244
Tindal, Matthew 313
Todd, Jan 265
Toynbee, Arnold 216, 230
Trebilcock, Clive 167
Trelfall, Richard 266
Tucker, Jeremiah 66
Twomey, M. J. 222–3

Ure, Andrew 133
Uvarov, S. S. 132
Van Crell, Lorenz 47
Van der Burg, P. 190
Van Marum, Martinus 46
Verdiev, A. 55
Vernon, R. 279
Voelker, J. A. 220
Voltaire, F. M. A. 37
Von Hoffman, A. W. 93
Von Moltke, Count Helmuth 133
Von Neumayer, G. B. 265
Vyschnegradskii, I. A. 171

Wakefield, Edward George 264
Walker, Adam 42
Walkinghame, Frances 38
Wang, Jinling 371
Watt, James 9, 76

Wedgewood, Josiah 42, 76
Wehnelt, Artur 114
Westinghouse, George 262
Wheatstone, C. 174
Whewell, William 128
Whiston, William 34
White, Lynn T. 202
Whitehead, A. N. 11
Whitworth, Joseph 255
Williams, B. R. 10
Williams, G. A. 43
Williamson, J. G. 187, 197, 199
Winthrop, John 250
Witte, Sergei 171–2, 174
Wright, Lewis T. 107

Yamamura, Kozo 246
Yano Yoichi 158

Zheng Shouyi 371

Subject Index

acclimatisation 250–1, 253, 264–5, 268
adaptive adoption/diffusion 14–17, 24, 56–9, 623, 66, 167, 192, 201–2, 246, 256–9, 261, 263, 268–9, 276, 278, 283–4, 296
AEG (Germany) 161–2
Africa 205–6
agricultural schools 197, 199, 253
agricultural technology 3–4, 62, 64, 151–2, 156, 170, 186, 197–200, 211, 231, 233, 250, 253, 258, 265, 267, 338, 348, 358–9, 365
Akabane Engineering Office 191
Aktievolaget Seperator (Sweden) 112
Aligarh Scientific Society (1864) 220
All Japan Cotton Spinning Association 202
American Academy of Arts and Sciences 251
American Association for the Advancement of Science 252
American agriculture 4, 250–2
American Philosophical Society 250–1
American system of manufactures 254–64
American technology 22, 50–1, 255–64
American Telephone and Telegraph 113
American thought 33
Amoskeag Manufacturing Co. 257
Amsterdam Coffee House 34
Apollo Project 291
Appleby Binder 365
Apprentices and Artizans 79–86, 102–5, 157–8, 211, 259, 269, 335
appropriate technology 30–1, 56, 167–8, 284–8, 300–1
appropriation of technology 197–202, 247
areas of recent settlement 182, 248–9, 268–70
Arkwright Mill 50, 52, 65, 88
artificial silk 154
assembly line 11
Association of South East Asian Nations 295
Astrakhan silk mills 55
Australasian Association for the Advancement of Science (1888) 265–6
Australasian Institute of Mining

Engineers 267
Australia 147–9, 165–6, 179, 248–9, 264–70, 275, 328, 367
Austro–Hungarian Compromise (1867) 169
Austro–Prussian War 133

Badische aniline works 115–56
Baird iron works (Russia) 53–4
Baku oil (Nobels) 156
Ballarat school of mines 266
Belgium 162, 166, 176, 178, 180
Bell Telephone 114
Bengal Association 220
Berlin Academy of Science 32–3, 35
Berlin Technical Institute 98, 326
Bessemer and Martin steelmaking 170, 174–5, 178–9, 245, 306
Bharat heavy plants and vessels 283
Bihar Scientific Society (1868) 220
Birmingham 42–3, 82–3, 100, 111, 118
Board of Agriculture (US) 253
Board of Longitude 121
Bombay cotton industry 217–18
Boston Manufacturing Co. 258
boundary conditions 158, 167–8, 182–3, 247
Boyle Lectures 34
Brahmins (Priests) 216
British Association for the Advancement of Science 94, 116, 123, 185
British Board of Ordinance 255
British corporations 112–13, 115, 162–5, 329
British economy (general) 21–2, 28, 60–88, 128, 139, 144–5, 151–1, 162, 204, 206–7, 235, 297–8, 373
British patenting 41, 82–6, 104, 106, 122, 163, 262–3, 327–8
British Society of Chemical Industry 89–90
British Science 39, 42–3, 47, 99–100, 120–3
British technology 51, 82–6, 90, 103–5, 155
British Westinghouse 112
Bureau de Commerce 44
Bureau of Animal Husbandry (US) 253
Burroughs Wellcome 113
Bussangaku 189–90

Cambridge 36–9, 73, 99–100
Canada 270
Canton system 234
capital 6, 64–6, 141, 146–9, 155, 159, 169, 172–3, 179, 187, 210, 217, 225, 231, 235, 237, 264, 301, 334, 339, 350, 356
capital goods sector 18, 28, 146, 246, 256–7, 263, 284
Carnegie Institution (Washington) 111
caste 209, 211, 216–17
catalytic oxidation 310
catching up 137–9, 280–2, 301–2, 373
Cavendish Laboratory (Cambridge) 110, 266
challenge and response 140–1
Chapter coffee house (London) 42
chemical developments 39–40, 69–71, 98, 114–16, 118, 138–9
China (general) 181, 194–5, 202, 227–47, 276–7, 284–8
China Association for Science and Technology 286
Chinese Communist Party 285–6
Chinese economy (general) 28, 30, 284–5, 358–9
Chinese technology 14, 371
CIBA (Swiss Chemische Industrie) 162
Colonial Land and Emigration Commission (1840) 265

colonialism, imperialism 22–3, 74–5, 119, 121, 133–4, 147, 181–3, 205–8, 227–30, 233–6, 248–9

Doabs (wastelands) 215
doctrine of lapse 208
Doncaster science 80
Dublin 42
Dunlop 113
Du Pont (Corporation) 113
Dusseldorf 107

East India Company 99, 208–9, 212, 218–19, 222, 356
Eastman Kodak 113
Ecole des Ponts et Chaussées 100
Ecole Polytechnique 44, 93, 100, 153, 156
Ecole Pratique 100, 325
economic development 1–3, 29–31, 215
economic dualism 172–3
economic growth 1, 29, 215
economic sovereignty 183, 194–6, 207, 209–10, 227–30, 234, 241–2, 244–7, 283–4, 298–9, 303
economic underdevelopment 29–30, 172–3, 181–3, 209–10, 212–15, 221–6, 246–7, 270–4, 283, 288–90, 294, 299–300
Edict of Nantes 317
Edinburgh 36, 73, 75
Edison patents 161–2
education 28, 44, 46, 73–4, 79–81, 101–9, 122, 126–7, 142, 156–8, 192, 199, 218–21, 252–4, 290, 313, 327
effective distance 18
Egypt 155–6, 177, 206
electric dynamo 156–57
electrical telegraphy 334
electrolytic refining 245–46
Elkington and Co. (Birmingham) 154
enclave development 57–9, 63, 172, 214, 222, 225–6, 244–7, 275, 301
Encyclopédie 37–8
endowments 98, 110–11
Energy Research Institute (Korea) 297
Engel's law effects 182
engineering efficiency 12
ensuisen (rice technology) 198, 348
enterprise innovation/R&D 16, 92, 110–16, 129, 260–4, 293–4
entrepreneurship 298–9
expeditions 252
export-promotion strategy 294

Fachschulen 102
factory production 64–5, 87–8, 254, 260, 296, 323
Farbenfabriken Bayer Co. 115
financial institutions 140–2, 143, 147, 159–60, 239–40, 280
Foreign Capital Inducement Act (Korea) 295
foreign communities 191, 228, 236–9, 265, 360
foreign investment 146–9, 212–13, 235, 236–7, 241–2, 244–5, 283, 334, 342, 352
foreign trade and technology 21–2, 54, 57, 144–6, 167, 182–3, 187–8, 212–23, 234–5, 254–5, 266–7, 279–80
France 33, 37, 51–3, 134–8, 145, 153, 176–7, 325
Franco-Prussian War 133
Franklin Institute 263–4
French Academie 35, 37, 44–5, 48, 157, 316
French science 43–46, 100–1
Fusetsugaki 189, 343

Ganges Canal 356
Ganz Electrical Works (Hungary) 157

Gara-bö (rattling spindle) 200
General Electric 111, 113, 161, 192
Genova–Torino Railway 178
German chemical industry 114–16, 118, 120, 138–9, 146, 164
German science and technology 96–8, 101–3, 117–20, 148–9, 163–4, 166
Germany (states) 38, 47, 54, 96, 134–9, 148–9, 171–3, 174–80, 188, 204, 263
Gewerbefreiheit 145
Gilchrist–Thomas Process 4, 306, 338
Godavery Anicut 356
Goodyear Rubber 365
Göttingen Society for the Promotion of Applied Physics 117
government employment 93–4, 117, 157, 253, 265–7
government intervention 19, 49, 53–5, 92, 116–23, 130, 154–8, 168–83, 190–6, 200–2, 232, 236, 239–42, 245–7, 253, 265–7, 269, 273, 276–7, 280, 282–8, 299
Great Exhibition (1851) 61, 103, 128, 181, 255, 365
Guadalajara Woollen Mill 56

Haber–Bosch process 4, 175, 338
Hanseatic League 51
Hanyang iron complex 239
Harper's Ferry (US) 255–6
Harvard 98
Hayange furnaces 50
Hinduism 211, 216–18, 230
Hindustan lever 283
Hsikechaung–Lutai Canal 241
Huguenots 47–8, 50, 317–18
human capital 55, 73, 101–9, 259, 261–3, 294 (*see also* skills)
Hungary 157, 168–70, 173

Ilbert and Co. (China) 23
Imperial Chemical Institute (Germany) 120
Imperial Fibres Syndicate Ltd 112
Incandescent Heat Co. (UK) 112
incremental decision-making 68–9, 298–301, 338
incremental technological change 13, 20, 66–7, 70–2, 82–6, 308, 310
India (general) 175, 177, 181, 205–6, 234, 242–7, 273, 282–4, 288–90, 302, 350
Indian Association for the Advancement of Science (1869) 220
Indian First Five Year Plan 282
Indian Mutiny (1857) 208
Indian Sixth Five Year Plan (1978–83) 274
indigenous entrepreneurs 151, 192–3, 209–10, 213–14, 221, 225, 234, 239–40, 245, 259, 262, 272, 278–9, 299, 340
indigenous peoples 249
indirect transfer mechanisms (defined) 144, 187–8
Indonesia 216, 295
inducements to innovation 138–9, 178–9
induction watt-meter 157
industrial drive (concept) 61, 141, 157, 167–1, 182, 186–7
industrialisation (general) 59, 61, 131–42, 165–6, 168–76, 298–9
Industrial Revolution 51, 60–88, 126, 168, 173–6, 182, 187, 284, 295, 298, 319, 325
industrial targetting 280, 293
industrial technology Research Institute (China) 287
information flow 16–18, 27, 46–9, 51, 54, 56–8, 72–80, 83–4, 87, 105–7, 127–8, 189–93, 198–9, 202, 236–8,

250, 253, 259–60, 268–70, 277, 281–2, 286, 297
Inland Revenue Department (UK) 121
innovation (concept) 5, 8–13, 27, 258, 345
innovation environment 191–2, 258–64
Institut Chimique (Nancy) 116
institutional substitutions(concept) 140, 144, 156–9, 168, 185, 190, 196, 204, 278, 285, 333, 346
institutionalisation 92–3, 95–6, 100, 124, 128–30, 140–2, 144, 156–60
institutions and economics 26–9, 67, 86–8, 95–6, 128–9, 230–2, 260–1, 274–7, 278, 285–6
institutions and science 89–130, 285–8
institutions and technology 23, 28, 30, 62, 87–8, 95, 144, 260–2, 274–7, 285–8, 290–7, 300, 302
intellectual property rights 112–14, 261
interchangeable parts 255–7, 259
International Telegraph Union 151
international trade/economy 271–2, 274–5, 278, 295
invention (concept) 5, 8–13, 69, 82–6
inventions and innovations (1700–1880) 304–6
Inventors' Aid Association (London) 106
investment banks 160
isoquant (concept) 14–16, 26
Italy 133, 143, 149, 153, 159, 168, 172–3, 178

Japan 19, 113, 123–8, 138, 143, 158, 159, 168, 181, 184–204, 215, 229, 242–7, 275, 278–82, 284, 288–91, 299, 301–2, 341–3, 369
Japan–NIC relationship 295–6
Japanese Industrial Laboratory 123
Jardine, Matheson and Co. 236, 240
Java 275
Jitsugaku 189
joint stock companies 169
joint ventures 283, 287–8
Journal de Commerce 156
Journal Polytechnique 94

Kagoshima Spinning Mill 185, 343
Kaifang Zhengce 287
Kaiping Railway Company 241
Kaiser Wilhelm Gesellschaft 118–19
Kaminoseki (Choshu) 193
Keiretsu 278–80
Kenkon Bensetsu (Heaven and Earth) 189
Kew Gardens 218
Khadakwasala Dam 215
Kharif (Autumn Harvest) 211
Kharkhanas 211
Kobusho (Public Works Department, Japan) 124–7
Korea (South) 277, 295–7
Korean War 279, 282
Korobi Bateren (fallen padre) 343
Kshatriyas (warriors) 216
Kuge (court aristocracy) 195
Kyoshin-kai (exhibitions) 191

laboratories 93, 98–9, 102, 110–11, 114–23, 129, 324–5, 329
labour-absorbing technical change 143, 199–202, 217–18, 278, 296, 348, 359
labour-saving technical change 256–7, 333, 365
Laffitte and Blount (Bankers) 160

La Sociedad Espanola de la Construccion Naval 155
Latin America 145–6, 149, 166, 212, 252, 288, 294–5
Lawrence Science School (Harvard) 98
leading sector complex 64
learning and learning by doing 52–5, 256–9
Leblanc process (inorganic chemical) 20, 45, 138–9, 310
Leopoldine Academy 32
Lever Bros 113
Liaoning Scientific and Technological Commission 286, 371
liberal parliamentarianism 170, 176, 252, 276–7, 298
licence agreements 279, 282, 295
Liebig Laboratory 93, 156
Linnaean Society 35, 41
Lister and Co. 215
Liverpool 2, 42–3, 79–80, 100, 111, 122
Liverpool School for the Encouragement of Arts, Science, Trade and Commerce 79
Liversey, Hargreaves and Co. 305
Lombe Silk Mill 12, 88
London Philosophical Society (1780s) 315
Lowell Mills (Massachusetts) 18, 257
Lunar Society (Birmingham) 50

Macaulay's Minute (1835) 219
machine tools 18, 257–8
machinofacture (concept) 64, 257
Malaya 175
managing agency system 214
Manchester 100, 108–9, 111, 118, 327
Manchuria 232, 236, 244
Maoism 285
Marconi Wireless Telegraph Co. 112
Marine Coffee House 37
market forces and technology 58–9, 68–9, 76–7, 87, 170, 176, 256, 260–2, 266, 270, 276–7
Matsumoto 277
McCormick Agricultural Machinery 256, 261, 365
mean information field 17–18, 76–80
mechanics' institutes 79–80, 82
medieval technology 13–14, 60
Meiji Japan (miscellany) 24, 169, 184–204, 218, 233, 236–7, 245–7, 277, 299, 301, 353
Meister, Lucius and Bruning Co. (Hochst) 115–16
Melbourne 265, 267
mental capital 50–5, 75, 301
mergers and technology 261
Merrimack Manufacturing Company 111
Mexican War (1848) 256
migration of skills 20, 48, 51, 149–50, 178, 256, 269, 317, 333
military science and technology 44–5, 53–4, 125, 131–4, 154–6, 169–70, 185, 190–1, 206, 212, 245–7, 255–6, 291–2, 353, 362
mining technologies 146, 150, 172, 175, 245–6, 265–7
Ministry of International Trade and Industry (MITI, Japan) 279–80, 292–4, 373
Ministry of the Economic Planning Board (Korea) 295
Mita Botanical Experiment Yard (1874) 197
Mitsubishi 192, 245, 292
Mitsui Bussan 201–2
Morozov Company (Russia) 154

Morrill Act (1862) 253
Morse Telegraph 365
most favoured nation 226–27
Mughal India 210–11
Munich Hygiene Institute 362
Murray William Co. (Engineers) 106
Museum of National Manufactures 83
mutual funds 159

Nanbangaku (1543–1639) 189
NASA 372
National Bureau of Standards (US) 117
National Physical Laboratory (UK) 123
National Science Congress (China) 286, 371
National Science Foundation (UK) 291
natural theology 33–4
Navigation Acts 248
Neighbourhood effects 67, 194
New College of Arts and Sciences (Manchester) 42
newly industrialising countries (NICs) 272–4, 278–9, 289, 294–7
New Russia Company 153
Newtonianism 33–4
Nirayama Ransho Honyuko Kata 190
Nishigahara experimental farm 198
Nodankai (agricultural discussion societies) 198
Northrop Automatic Loom 202
nuclear power 31, 312

occupation forces (Japan) 278
Oldknow Cotton Mill 65
opium 234–5, 238, 359
Oratory, The (London) 34
organisational technology (concept) 10–11, 255–62, 281, 323–4
Osaka arsenal 124, 246
Osaka–Köbe Railroad 186
Osaka Spinning Co. 185, 201–2
Oxford 36, 73

Pakistan 287
Pandits 219
Paris Exposition (1885) 181, 343, 365
Parliamentary Committees/Commissions 83,100, 110, 122, 356, 366
Parsons Turbine 192
Patent Journal and Inventor's Magazine 106
patents (miscellany) 10, 12–13, 20, 41, 44, 49, 82–6, 104, 106, 109, 111–15, 121–2, 153, 159–66, 192, 262–3, 267–8, 279, 281, 295, 304–7, 310, 315, 323, 327–9, 336–7, 366–7
patronage 33, 45, 48–9, 51–2, 92, 119–20, 129–30, 253
peripheral intellectuals 249–54, 264–7
Perkin Analine Dye 161, 306
Philadelphia 47–8, 52, 251–2
Philadelphia Centennial Exhibition (1876) 181
Philadelphia Museum (1784) 252
Philippines 289–90, 295
Philadelphian Polytechnic College 363
Physikalich Technische Reichenstalt 119
phytochemistry 266
Pitt Thresher 365
Plassey (1757) 354
Platts Bros. Co. (Oldham) 146, 161, 202
political economy of intellectuals 300–1
polytechnics 90, 93, 118–20, 126, 329
population 149, 173, 186, 194, 243, 250–1, 258, 265, 269, 345

Practical Mechanics' Journal 106
Presidential Committee on Industrial Competitiveness (1985) 292
product cycle 279, 292–93, 297
production centres, location (1887) 135–7, 332
production function 1, 4–6, 10
production responsibility system (China) 285
productivity 2, 4–6, 64, 67, 185, 254
professional technical associations 108–9
property rights 27
proto-industrial development 63, 67, 134–8, 170–1, 181, 186, 296
Prussia 133, 171, 176, 181
public goods 170, 180, 254
Public Undertakings Committee (India) 283
Punjab economy 215–16, 356
Putney College of Civil Engineering 99, 105

R&D (corporate) 9–10, 27, 289–94
R&D (general) 274–7, 279–80, 282–3, 285–6, 288–94
R&D world distribution 288–90
radical culture 42–43, 63
railway guarantee system 352
railways 124, 127, 133, 147, 150–2, 160, 169, 176–83, 186–7, 224–5, 235, 240–3, 260, 268, 365
Rainbow coffee house 48
Rangaku (Dutch learning 1720–1853) 189
Reagan Administration 372–3
regulations for transfer of factories to private control (Japan) 361
relative economic backwardness 54, 129–30, 137–44, 161, 165–6, 170–6, 190, 203, 206, 232, 244, 298–9
Rensselaer Polytechnic Institute 98, 252
research institutes/Programmes 91, 93, 97, 100, 111, 118–20, 174–5, 252–4, 266, 286–8
resources 3–4, 175, 249, 259
Rheinish-Nassonische Bergweks 112
Rhodesia 275
risk environment 299
Roberts, Dale and Co. (UK) 152
Robi (Spring Harvest) 211
Rono system (1880–95) 198
Rothamsted Agricultural Research Station 110, 306
Rouen textile factory (1750s) 51–53
Rourkee Civil Engineering College 356
Royal Botanical Gardens (Sidpur) 218
Royal College of Chemistry 93
Royal Dutch Shell 113
Royal School of Mines 99
Royal School of Naval Architecture 99
Royal Society of London 33, 37, 50–1, 122, 251
Ruhr ironworks 147–8
Russia 20, 48–9, 53–5, 131–2, 136, 140–1, 143, 153–4, 156–7, 162–3, 166, 168, 170–2, 174, 178–81, 228, 243, 317
Rutgers Scientific Schools 126
Ryotwari 210

St Petersburg Academy 48
Saga Plain 197–8
Sakoku (1637–1720) 189
Satsuma 343, 346
Savory and Newcomen engines 64
science and invention/technology 69–72, 90, 94, 114, 263–4, 269–70, 286–7

Science and Technology Agency (STA, Japan) 292, 373
science as information 70–2
science industrial park (China) 287, 371
science policy 274, 282, 286–97, 371–3
science–technology culture 38–45, 50–1, 58–9, 69–86, 91–4, 105–9, 116, 250–4, 259–60, 268–70, 327
scientific academies 35, 47–8, 252
scientific lectures 43, 47, 79–80, 105–6, 314
scientific publishing 36–8, 41, 47, 106, 314
scientific societies 35, 79–80, 99–100, 126–7, 220, 250–4, 265–6, 313
scientists and engineers 72–3, 102–3, 108, 152–4, 253, 259, 313–14
Scinde railway 223
Scotland 42, 73–6, 250
Siesmological Journal of Japan 126–7
selection environment 190–91
Sepoy rebellion 351
serf emancipation 169
service environment 192–3
Shanghai 236–9
Shanghai–Soochow railway 239, 360–1
Shansi remittance banks 231
Sheffield Science School (Yale) 98
Shibpur Engineering College 221
Shingawa glass factoy 245
Shinjuku Agricultural Experiment Station 197–8
Shizoku (ex-Samurai) 196, 344, 361
Siemens regenerative furnace 154, 175
Silk Supply Association 356
Silk Technology 188, 198–201, 222, 277, 356
Silkworm Research Institute (Japan) 200
Singapore 290, 295, 297
Singer–Prebisch thesis 182
Singer sewing machines 256, 261, 365
Sino–Japanese War (1894–95) 228
skills 52–6, 70–1, 104–5, 107–8, 127, 142, 148, 153–4, 158
Social distance 16–18, 42, 45, 68, 72, 74–80, 105–6, 231
social marginality 77
social system and industrialisation 63, 68, 72, 74–8, 184–6, 192, 211, 216–18, 221–3, 229–32, 239, 272–3, 275–7, 312, 367, 373
Société Academique des Arts 38
Société d'Encouragement pour l'Industrie Nationale 109
Société de la Vieille-Montagne 148–9
societies of arts 38–9
Society for the Promotion of Natural History 35
Society of German Naturalists and Natural Philosophers 94
Sogo Shosha 278
Solvay process (inorganic Chemical) 20, 90, 139, 170, 306, 310
Sorocold waterwheel 88
Spain 46, 49, 54, 56, 145, 155, 159, 172, 180, 189, 317
spatial diffusion 16–18, 134–7, 194, 224–5, 243, 269
specialisation 94
Springfield ordinance factory 255
staple exploitation 249, 254, 264–5, 268
Star Wars 372
State Scientific and Technological Commission (China) 296
steam engines 314
steel production 138, 168, 174–5, 179–80, 338

strategic elites 74–5, 78, 176, 190, 195–96, 204, 247
Stuttgart Polytechnic 102, 118–19
Sudras (farmers–peasants) 216
Suez Canal 150–1, 188, 228, 306
Surveyors and Engineer's College (London) 106
Sweden 53–4, 112, 147, 168, 178
Swiss engineers 153–4, 162
Switzerland 161–2, 166, 176

Taiping Civil War 133, 228, 240, 359
Taiping Tien Kuo (1851) 233
Taylor and Prandi Machine Co. (Italy) 178
technique (definition) 12, 14–15, 88, 282
Technisches Hochschulen 93, 96–7, 102–3, 326
technological adaptation 16–18, 56–9, 200–2, 255–63
technological change (definition) 2–4, 67, 88, 197–202, 256–61, 320
technological diffusion (definition) 13–20, 53, 56–9, 246, 256–7, 259–61, 282–3, 303
technological filtration 282
technological imperatives 23–24, 56–7, 167, 183, 301, 311–12
technological progress (definitions) 5–6, 12, 14–15, 61, 67, 150–2, 160, 174, 197, 256–8
technology transfer (general) 20–5, 30, 49–59, 90, 142–66, 171–6, 178–81, 197–202, 204, 215, 221–5, 245–7, 253–4, 256–60, 269–70, 277–85, 295–6, 310, 343
Technopolis 293
terms of trade 144–46
Test Acts 43
Thomas–Snelus basic steel 175
Tokyo Engineering College 124–6
Tokugawa economy (1603–1868) 186, 189–90, 193–4, 197, 211, 345
Tokugawa Office of Barbarian Writings 123
Tomioka Silk Mill 186, 200–01
transaction costs (definition) 19, 26–7, 61, 161
transfer mechanisms 21–2, 127–8, 142–6, 182, 187–93, 208, 276–84, 287–8, 294–7, 301
Transit of Venus (1769) 250–1
transnational corporations 24–5, 113, 159, 275, 283–4, 295, 297, 302, 370
transport improvements 134, 146, 147, 150–1, 186, 257, 334
Trans-Siberian railroad 228

unequal treaties 342
UNESCO 289
Union Carbide Co. 283
unitarians 42–3
United Alkili Co. (UK) 113–14, 116
United Shoe Machine Co. (US) 112
United States Bureau of Chemistry 117
United States corporations 111–12, 159, 260–4, 294
United States economic growth 4–5, 8, 28, 135, 146–50, 168, 178–9, 248–64, 268–70, 365, 367–8
United States patents 262–3, 337, 366
United States science and technology 98, 111, 162–4, 179, 254–64, 290–92
US Steel Corporation 111
universities 91, 93, 96–101, 111, 117–19, 252–4, 265–66
University of Durham 99
University of Heidelberg 114, 118

University of London 99–100, 118
University of Moscow 317
University of Strasburg 89, 118
USSR 12–13, 21

Vaishas (traders) 216
Vickers Company 159
Victorian Acclimatisation Society 265
Victorian Institute for the Advancement of Science 265
Victorian Science Board 265
Vienna Exposition (1873) 190
Viscose Spinning Syndicate Ltd 162

wages and technology 151–2
water-jacket blast furnace 245
water turbines 259
Watt steam engine 9, 18, 56, 70, 320
Western Syndicate Ltd. (UK) 112
Westinghouse Electric Co. 113, 161, 261–2
Westman Process Co. (US) 112
Woolf high-pressure engine 153, 305
Woolwich Academy 51
Woosung bar dredging 239
workshop production 63–5, 82–8, 102, 104–7, 171, 173, 193–4, 210–11, 213, 222–3, 244, 256–60, 282
World Bank 284
world trade 145–6
Wuertenberg 102–3

Yale 252
Yamburg textile complex 57
Yawata steel works 246
Yokosuka iron foundry/arsenal 185
Yokosuka naval arsenal 125, 246

Zaibatsu 278
Zamindar 209–10, 224
Zollverein (customs union) 171
Zurich Polytechnic 114, 116